3rd International Conference on Science, Technology, and Interdisciplinary Research (IC-STAR 2017)

IOP Conference Series: Materials Science and Engineering
Volume 344

Bandar Lampung, Indonesia
18 - 20 September 2017

Editors:

Ardian Ulvan
Irza Sukmana
Denni Kurniawan

ISBN: 978-1-5108-6160-2
ISSN: 1757-8981

Printed from e-media with permission by:

Curran Associates, Inc.
57 Morehouse Lane
Red Hook, NY 12571

Some format issues inherent in the e-media version may also appear in this print version.

Printed by Curran Associates, Inc. (2018)

For permission requests, please contact the Institute of Physics
at the address below.

Institute of Physics
Dirac House, Temple Back
Bristol BS1 6BE UK

Phone: 44 1 17 929 7481
Fax: 44 1 17 920 0979

techtracking@iop.org

Additional copies of this publication are available from:

Curran Associates, Inc.
57 Morehouse Lane
Red Hook, NY 12571 USA
Phone: 845-758-0400
Fax: 845-758-2633
Email: curran@proceedings.com
Web: www.proceedings.com

Table of contents

Volume 344

The 3rd International Conference on Science, Technology, and Interdisciplinary Research (IC-STAR)

18–20 September 2017, University of Lampung, Indonesia

Accepted papers received: 22 March 2018
Published online: 16 April 2018

Preface

Papers

IOP Conf. Series: Materials Science and Engineering **344** (2018) 011001 doi:10.1088/1757-899X/344/1/011001

Preface

The 3rd International Conference on Science, Technology and Interdisciplinary Research (IC-STAR 2017) was organized by University of Lampung, in Bandar Lampung, Indonesia, on September, 18 – 20, 2017. The IC-STAR is a foremost annual forum for sharing and highlight original and novel ideas, significant advances, the state-of-the-art, current status, and future challenges on all aspects related to the research and development in the field of science, engineering and technology, and their inter-discipline subjects. IC-STAR is also intended to be the platform for information dissemination, gathering, and networking for academia, researchers, engineers, and professionals within the areas. The conference was also planned to provide advanced orientation and understanding for related industries and governments to looking across industrial partnerships, business strategic, and policy and regulations, with expected for a wider range of participants and readers.

This year the scientific committee received 128 submitted papers, by which 84 papers was selected and presented in the conference. The peer-review processes have been done based on the originality, impact, novelty, and the close relation to the area of the conference. The presented papers were selected to be published in several journal partners, including the IOP Conference Series: Materials Sciences and Engineering. In this volume, 29 papers were selected from IC-STAR 2017 and its conference partner.

We are pleased and thankful for all distinguish authors and reviewers for their contribution that have made this publication possible. We do hope that you will enjoy this publication and find it as a useful guide and reference.

Editorial board:

Ardian Ulvan (*Dept. of Electrical Engineering, University of Lampung, Indonesia*)
Irza Sukmana (*Dept. of Mechanical Engineering, University of Lampung, Indonesia*)
Denni Kurniawan (*Dept. of Mechanical Engineering, Curtin University, Sarawak Malaysia*)

IOP Conf. Series: Materials Science and Engineering **344** (2018) 011001 doi:10.1088/1757-899X/344/1/011001

IC-STAR 2017 Committee

Advisory Board

Prof. Dr. Ir. Hasriadi Mat Akin – Rector of University of Lampung
Prof. Syafsir Akhlus – Rector of Universitas Maritim Raja Ali Haji, Chairman of Rector Forum BKS-PTN Barat
Prof. Akio Tomiyama – Kobe University, Japan
Prof. Hadi Nur – Universiti Teknologi Malaysia, Malaysia
Prof. Pavel Ripka – Czech Technical University in Prague
Prof. Toshiro Ohashi – Hokkaido University, Japan

Steering Committee

Prof. Bujang Rahman –Vice Rector, University of Lampung, Indonesia
Prof. Bustami Syam – Vice Rector, University of Sumatera Utara, Indonesia
Prof. Dr. Suharno – Dean of Faculty of Engineering, University of Lampung, Indonesia
Dr. Warsono – Director of Research Institute of University of Lampung
Prof. Warsito – Dean of Mathematics and Natural Science, University of Lampung, Indonesia.
Dr. Eng. Helmy Fitriawan – University of Lampung, Indonesia
Dr. Muh. Sarkowi – University of Lampung, Indonesia
Prof. Ing. Boris Simak – Czech Technical University in Prague, Czech Republic
Dr. Yusron Saadi – Dean of Faculty of Engineering, University of Mataram, Indonesia
Mr. M. Ikhsan – PT. Telkom, Indonesia
Dr. Jiri Krejci – Vodafone, Czech Republic
Mr. Terklin Sinulingga – Alstom Grid, United Kingdom

Organizing Committee

Ardian Ulvan – Conference Chair
Dewi A. Iryani – Secretary, Public Relation, and Sponsorship
Melvi - Treasurer
Gigih Forda Nama – Information and Conference Management System
Hery Dian Septama – Publication and Circulation
A. Yudi Eka Risano– Sponsorship
Afri Yudamson – Social event and Excursion
Novri Tanti – Logistics and Appurtenance

Scientific Committee

Irza Sukmana – University of Lampung, Indonesia – Chair
Denni Kurniawan – Curtin University, Sarawak, Malaysia – co-Chair

Scientific Committee Member

Basem Abu Izneid – Effate University, Saudi Arabia
Joni Agustian – University of Lampung, Indonesia
Anna Antonyova – University of Presov, Slovakia

IOP Conf. Series: Materials Science and Engineering **344** (2018) 011001 doi:10.1088/1757-899X/344/1/011001

Naseem Ahmed – Indian Institute of Technology Roorkee, India
Fitri Arnia – Syiah Kuala University, Indonesia
Edwin Azwar - University of Lampung, Indonesia
Yanuar Burhanuddin – University of Lampung, Indonesia
R.Y. Ferry Burhan – 10 November Institute of Technology, Indonesia
Homero Toral-Cruz – University of Quintana Roo, Quintana Roo, Mexico
Joy Rizky Pangestu Djuansjah – Universiti Teknologi Malaysia, Malaysia
Taha Abd. El-Rahman – University of Sherbrooke, QC, Canada
Ainul Ghurri – Udayana University, Indonesia
Cipta Ginting – University of Lampung, Indonesia
Lukmanul Hakim – University of Lampung, Indonesia
Agus Haryanto – University of Lampung, Indonesia
Khomaini Hasan – Biotechnology Research Centre, Bandung Institute of Technology, Indonesia
Udin Hasanudin – University of Lampung, Indonesia
Ismail Tasib Karas – Karabuk University, Turkey
Naoya Kasai – Yokohama National University
Mamoru Kikumoto – Yokohama National University
Irdika Mansur – Bogor Institute of Agriculture, Indonesia
Antonin Mazalek – University of Defence, Brno, Czech Republic
Agung Mataram – Sriwijaya University, Indonesia
Christia Meidiana – Brawijaya University, Indonesia
Lee Meyer – University of Kentucky, USA
Khairul Munadi – Syiah Kuala University, Indonesia
Noor Danish Mundari – LPU, India
Kaneko Nabuhiro -Yokohama National University, Japan
Maksum Pinem – University of Sumatera Utara, Indonesia
RR. Poppy Puspitasari – Universitas Negeri Malang, Indonesia
Chithra Reghuvaran – National Institute of Technology, Rourkela, Odisha, India
Uceu Fuad Hassan R.S. – GKSS, Germany
Emerson Pascawira Sinulingga – University of Sumatera Utara, Indonesia
Suherman – University of Sumatera Utara, Indonesia
Ahmad Saudi Samosir – University of Lampung, Indonesia
Shirley Savetlana – University of Lampung, Indonesia
Diding Suhandy – University of Lampung, Indonesia
Wikan Danar Sunindyo – Bandung Institute of Technology, Indonesia
Gatot Eko Susilo – University of Lampung, Indonesia
Takahasi Susume – Kyoei University, Japan
Irfan Syamsuddin – Makassar State Polytechnic, Indonesia
Josaphat Tetuko – Chiba University, Japan
Agung Trisetyarso – Telkom University, Indonesia
Mokhamad Fakhrul Ulum – Bogor Institute of Agriculture, Indonesia
Mustafa Usman – University of Lampung, Indonesia
Zuzana Vranova – University of Defense, Brno, Czech Republic
Dedy H.B. Wicaksono – Swiss German University, Indonesia

Documentation

IOP Conf. Series: Materials Science and Engineering **344** (2018) 011002 doi:10.1088/1757-899X/344/1/011002

Peer review statement

All papers published in this volume of *IOP Conference Series: Materials Science and Engineering* have been peer reviewed through processes administered by the proceedings Editors. Reviews were conducted by expert referees to the professional and scientific standards expected of a proceedings journal published by IOP Publishing.

IOP Conf. Series: Materials Science and Engineering **344** (2018) 012001 doi:10.1088/1757-899X/344/1/012001

Microstructures and mechanical properties of duplex low carbon steel

Alfirano*[1], U S Eben[1] and M Hidayat[2]
[1]Department of Metallurgical Engineering, Faculty of Engineering, University of Sultan Ageng Tirtayasa, Cilegon, 42435, Indonesia
[2]Iron Making Division, PT. Krakatau Posco, Cilegon, 42435, Indonesia

Email: alfirano@untirta.ac.id

Abstract. The microstructures behavior of duplex cold-rolled low carbon steel for automotive applications has been investigated. Intercritical annealing treatment is commonly used to develop a duplex low carbon steel containing ferrite and martensite. To get a duplex phase ferrite and martensite, the specimens were heated at inter-critical annealing temperature of 775°C - 825°C, for heating time up to 20 minutes, followed by water-quenched. The hardness of specimens was studied. The optical microscopy was used to analyze the microstructures. The optimal annealing conditions (martensite volume fraction approaching 20%) at 775°C with a heating time of 10 minutes was achieved. The highest hardness value was obtained in cold-rolled specimens of 41% in size reduction for intercritical annealing temperature of 825°C. In this condition, the hardness value was 373 HVN. The correlation between intercritical annealing temperature and time can be expressed in the transformation kinetics as $f\gamma/fe = 1-\exp(-Kt^n)$ wherein K and n are grain growth rate constant and Avrami's exponent, respectively. From experiment, the value of K = 0.15 and n = 0.461. Using the relationship between temperatures and heating time, activation energy (Q) can be calculated that is 267 kJ/mol.

1. Introduction

The high demand for materials of automotive industry with enhanced mechanical properties has brought to the development of wide variety of duplex phase steels. The structure of duplex phase steel consists of ferritic matrix and clusters of martensite. Martensite contribute to strength of base material and ferritic provide good ductility [1]. Dual phase steel is appropriate material for automotive industry because this material has good mixture of high strength with elongation [2]. The request for more light and fuel efficient vehicles has caught attention beside the fact that the duplex phase steel has a good tensile strength which is similar to plain carbon steels, and excellent ductility [3]. The ferrite phase offers the ductility while martensite provides strength. The properties of duplex phase steel such as low yield strength, continuous yielding behavior, and highly uniform total elongation are also as detrimental properties [4]. Beside the ferrite grain size [6], martensite volume fraction and its morphology affect the mechanical properties of duplex phase steels [5]. Alaneme [7] noted that a good combination of strength and ductility can be achieved by intercritical heat treatment to transform low carbon steels to duplex microstructures of ferrite and martensite on the right level. Increasing in heating time at low temperature decreases the microstructural strapping in cold-rolled dual phase steel, while in the intercritical range, the formation of ferrite was observed more by quenching. The present

* To whom any correspondence should be addressed.

work examines the microstructural changes in cold-rolled low carbon steel during intercritical annealing with a focus on the formation of duplex ferrite-martensite phase.

2. Experimental procedure

The material for the study is a hot-rolled low carbon steel plate with 100x75x3.40 mm in dimension. The element composition is as follows: C (0.155); Si (0.18); Mn (1.05); P (0.056); S (0.0065); Cr (0.011); Ni (0.008); Mo (0.0025); V (0.0036); Nb (0.0019); Ti (0.0012); W (0.0041), in wt%. The plates were initially subjected to intercritical annealing and quenching [8-9]. The plates with initial diameter of 3.40 mm were cold-rolled to approximately 41% after intercritical treatment was performed. Before intercritical treatment, the lower critical temperature (A_1) and the upper critical temperature (A_3) for the test material were determined following empirical relations in accordance with Andrews [10]. The plates were heated at intercritical temperatures of 775°C, 800°C, and 825°C for 10, 15 and 20 minutes, respectively, followed by water quenching. After quenching, the samples were tempered at 200°C for 20 minutes and then cooled in the air. A micro Vickers hardness was used to obtain the hardnes value of cold-rolled steel. The microstructures of samples were evaluated by optical microscopy. Martensite volume fraction was calculated based on ASTM E562-83 standard.

3. Result and discussion

Figure 1 shows the microstructure of cold-rolled plate before and after quenching. The cold-rolled plate before quenching was consisted of ferrite and pearlite phases (figure 1 (a)). Volume fraction of the phase was obtained by using point counting calculations.

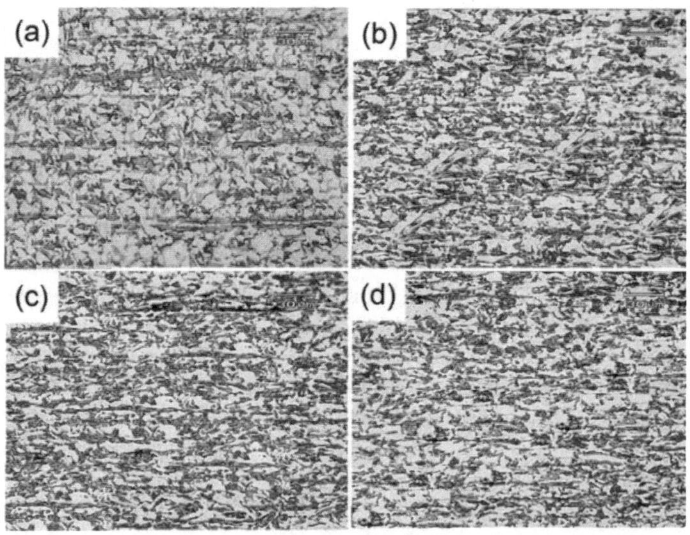

Figure 1. Microstructure of cold-rolled plate (a) before and after heat treatment at 775 °C for (b) 10, (c) 15 and (d) 20 minutes, water-quenched, 41% cold-rolled.

Volume fraction of ferrite and pearlite were 79.7% and 20.3%, respectively. After quenching, ferrite, pearlite and martensite phase were formed which can be shown in figure 1(b-d). Martensite volume fractions for heating time of 10, 15 and 20 minutes were 26.5%, 29.2% and 31.8%, respectively. The effect of intercritical temperature on martensite volume fraction at several heating time are shown in figure 2. Intercritical temperature has significant effect on increasing the hardenability compared to that of heating time. Martensite volume fraction will significantly increases with increasing temperature. When the hot-rolled plates were heated at intercritical annealing point, austenite start to nucleate and grow [11]. Austenite is unstable phase and it will be transformed during cooling.

Eutectoid transformation from austenite to ferrite, cementite and pearlite is begun in the A_1 temperature.

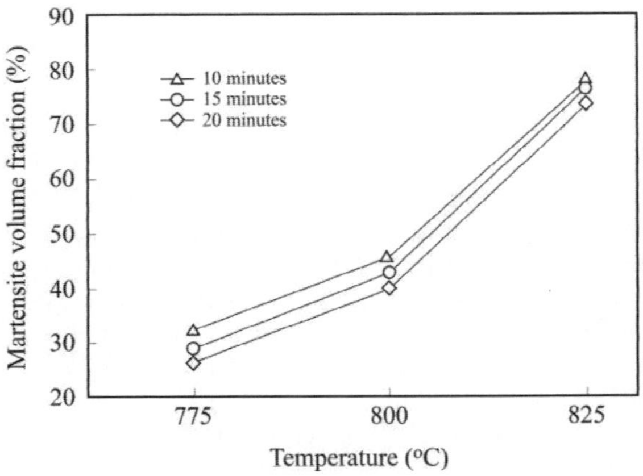

Figure 2. Effect of temperature on martensite volume fraction of specimens heated at various holding time.

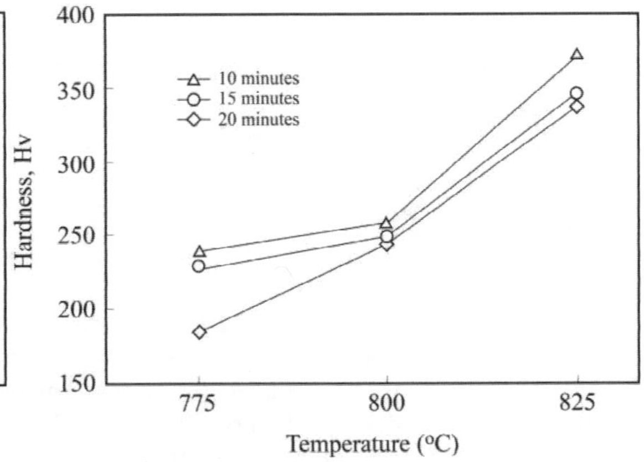

Figure 3. Effect of temperatures on hardness at various heating time.

When the temperature is raised above the A_1, carbon atom easily dissolved into austenite [12]. Therefore, increasing heating temperature increase the volume fraction of austenite after rapid cooling. Increasing in holding time will promote the formation of austenite. The heating time affected to the phase formation, because the holding time provides an opportunity for atoms to diffuse thus homogeneously into austenite [11]. Austenite is an unstable phase and it can be transformed easily into another phase during cooling process.

Figure 3 shows the hardness value of specimens. The hardness of hot-rolled plate before heat treatment was 135.6 HVN. Heating was carried out on the ferrite-austenite region. Increase in the volume fraction of martensite caused increase in hardness. The difference in phase formations; ferrite-pearlite and ferrite-martensite, respectively, was caused by the difference of cooling rate between specimens before and after heat treatment. Ferrite and pearlite were appeared in low carbon steel. Carbon content determined hardness of the cold-rolled plate which is affected by ferrite and pearlite appearance. After quenching, the final structure was ferrite and martensite containing saturated carbon atoms. The formation of martensite can cause on rising in hardness value. The increased cooling rate will be accompanied by an increase in heating temperature so that martensite can be easily formed. Increase in the number of martensites will be followed by an increase in the value of violence. Increase in martensite volume fraction cause the increasing of hardness value [13]. The strengthening mechanism of duplex-phase steel is caused by martensite structure as a strong constituent in the soft ferrite matrix. High tensile strength is generated by dislocation density, also has an influence to the strengthening mechanism in a high rate of work hardening duplex phase steel. The presence of high density of dislocation can be obstacle for the movement of the other dislocations.

Formation kinetics of austenite-martensite transformation during the heat treatment point of view intercritical annealing has long been examined. The transformation kinetics of austenite in duplex phase steel is controlled by diffusion growth. The transformation of austenite during intercritical annealing can be calculated using Johnson-Mehl-Avrami-Kolomogrov (JMAK) equation [14].

By modifying the JMAK equation, the grain growth rate constant (K) and the Avrami exponent (n) can be determined. K and n value is obtained from the slope and intercept which is resulted from linear equation relationship between ln [ln (1/(1-f_γ/fe)] versus ln t, as can be shown in figure 4. From figure 4, the n and K can be obtained, as listed in Table 1.

Table 1. n and K value at 775°C, 800°C and 825°C

Temp. (°C)	Equilibrium Austenite volume fraction (f_e)	Avrami's exponent (n)	Rate constant (K)
775	0,2808	0,461	0,15
800	0,4265	0,598	0,41
825	0,7692	0,281	0,52

Rate constant (K) is a value that depends on the temperature and velocity related with grain growth and frequency of nucleation. The value of K will increase if temperature of intercritical annealing is high. [15]. The Avrami exponent (n) depends on the nucleation and growth processes. The grain nucleation rate has a significant effect on the value of n [14]. From the data of grain growth rate constant (K), the value of activation energy (Q) and constant (A) can be calculated [15]. Figure 5 shows the value of activation energy (Q) and constant (A) were obtained from the slope and intercept resulting from linear equation between ln K and 1/T. From this correlation, activation energy (Q) and constant (A) will be 267kJ/mol and 2.97 x 1012, respectively.

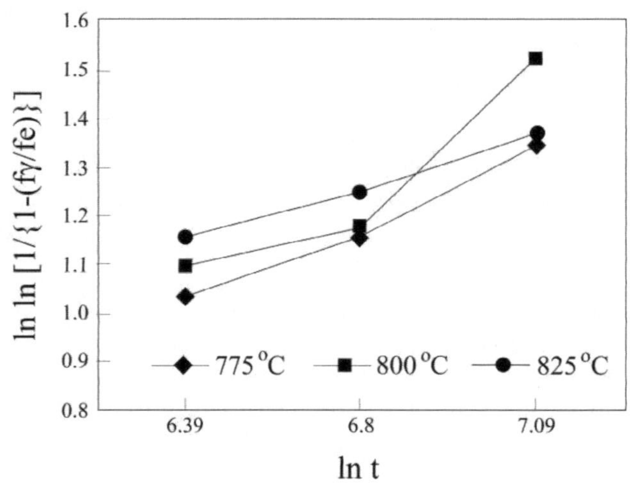

Figure 4. The relationship between ln [ln (1 / (1 - fγ/fe)] and ln t from experiment.

4. Conclusion

After the the chemical composition of the investigated steel allows to manufacture the products with the ferritic–martensitic structure by under hardening from the austenite and ferrite area. The coexistence of harder (martensite) and softer (ferrite) phase in the microstructure of duplex phase steel produce an excellent combination between strength and ductility. The promising hardness was obtained when treatment was performed at 825°C in comparison to 800°C and 777°C. The optimum value was 373 HVN with 78% of martensite. At the optimum point (at volume fraction of martensite approaching 20%), the value of hardness was 184 HVN. Rate constant will increase with increasing heating temperature. In this work, activation energy (Q) of specimen was 267 kJ/mol.

5. References

[1] Buriková K and Rosenberg G 2009 *Metal* **5** 19
[2] Lis J, Lis A K and Kolan C 2005 *Achiev. Mech. Mater. Eng.* **162-163** 350

[3] Tavares S, Pedroza P D and Teodosio J R 1999 *Mechanical properties of a quenched and tempered dual phase steels* **40** 887-892
[4] Ahmad E, Manzoor T, Ali K L and Akhter J I 2000 *Script. Mater.* **9** 306-310
[5] Erdogan M and Priestner R 2002 *J. Mater. Sci. Tech.* **18** 369
[6] Sun S and Pugh M 2002 *Mater. Sci Eng. A* **335** 298
[7] Alaneme K K 2010 *Mater. Res.* **13** 113
[8] Mineta S, Alfirano, Namba S, Yoneda T, Ueda K and Narushima T 2010 *Mater. Sci. For.* **654-656** 2176
[9] Alfirano, Mineta S, Namba S, Yoneda T, Ueda K and Narushima T 2010 *Mater. Sci. For.* **654-656** 2180
[10] Andrews K W 1965 *J. Iron Steel Inst.* **203** 721
[11] Calcagnotto M, Ponge D and Raabe D 2012 *Metal. Mater. Trans. A* **43** 37
[12] Panda A K, Ray P K and Ganguly R I 2000 *Mater. Sci. Tech.* **16** 648
[13] Francisca G C, Andrea G J, Carlos C and Carlos G A 2006 *Mater. Trans.* **47** 2269
[14] Waterschoot T, Verbeken K and De Cooman B C 2006 *ISIJ Int.* **46** 138
[15] Alfirano, Samdan W and Maulud H 2014 *Appl. Mech. Mater* **493** 721

The feasibility of desorption on Zeolite-water pair using dry gas

E Oktariani[1], K Nakashima[2], A Noda[2], B Xue[2], K Tahara[2], K Nakaso[2], and J Fukai[2]

[1]Department of Polymeric Chemical Engineering, *Polytechnic of STMI Jakarta*, Ministry of Industry, DKI Jakarta 10510, Indonesia

[2]Department of Chemical Engineering, Faculty of Engineering, *Kyushu University*, Fukuoka 819-0395, Japan

E-mail: erfina@kemenperin.go.id

Abstract. The increase in temperature, reduction in partial pressure, reduction in concentration, purging with an inert fluid, and displacement with a more strongly adsorbing species are the basic things that occur in the practical method of desorption. In this study, dry gas at constant temperature and pressure was employed as the aid to reduce the partial pressure in the water desorption on the zeolite 13X. The objective of this study is to confirm the feasibility of desorption using dry gas experimentally and numerically. The implication of heat and mass transfers were numerically investigated to find the most influential. The results of numerical simulation agree with the experimental ones for the distribution of local temperature and average water adsorbed in the packed bed.

1. Introduction

Adsorption heat pump systems using zeolite-water pairs have been developed since long time ago [1]. Direct contact system between water as working fluid and zeolite as adsorbent has been proposed to develop the adsorption heat pump system for steam generation [2]. The reason for proposing this direct contact system was described in the previous study [2,3]. The possibility for steam generation using this system has been confirmed experimentally although it still could not reach the prediction/ theoretical value. Steam generation is on the adsorption stage. For regeneration of saturated zeolite after adsorption process, inert gas is chosen to strip/release the water content which is believed in vapor/gas phase in the zeolite. From the experimental study before [1], inert gas such as dry nitrogen and dry air can regenerate zeolite for almost 80-90%. But there is problem in desorption study such as long time needed to release/remove water in the zeolite particle in the packed bed which means impractical to be applied in the industries. Long time needed for desorption accomplish means reaction rate occurs slowly.

Study about the solution to increase desorption rate was found in some report used microwave to shorten the time and also could decrease the energy level needed to remove water content in the zeolite [4,5]. Methods for regeneration of saturated zeolite have been investigated in various types [6]. General problem in desorption /regeneration process of zeolite is low reaction rate that may be because of high energy required to desorb water vapor from zeolite. Practical methods of desorption and regeneration including one, or more usually a combination, of the following such as increase in

the temperature, reduction in partial pressure, reduction in concentration, purging with an inert fluid, displacement with a more strongly adsorbing species, and change of chemical condition such as pH. As a variable for changing thermodynamic potential, a change in temperature is much more effective than a change in pressure. However, the final choice of regeneration method(s) depends upon technical and economic considerations. The most common methods are changes in temperature (thermal swing adsorption) and changes in pressure (pressure swing adsorption). The general advantages and disadvantages of each method together with some process examples.
The main objective of whole study is to regenerate steam from waste water in the industries such as petrochemical industries using the dry inert gas in medium/low level energy (noticed by the temperature degree). Because in the plant, there are plenty amount of gas at around 150 °C exhausted from the cracking furnace. It is wiser to use this gas for the regeneration of saturated zeolite after it producing the steam. In terms of the type of zeolite 13X is hydrophilic adsorbent, thermal swing is better method for desorption [7]. The purpose of this study is to find the best condition of desorption process of zeolite 13X to release the water vapor using dry gas stripping. Therefore the gas condition as the stripper was investigated experimentally and numerically.

2. RESEARCH METHODS

2.1 Experiment

Similar with the previous study [2], zeolite 13X was investigated in this study. Two different particle shape of 13X were examined. Initially, zeolite was calcined in furnace at 350 °C for 4.5 hours. Then the dry zeolite were put in a box at high humidity 98% until reached equilibrium condition to get saturated zeolite with water vapor. The mass fraction of water adsorbed in 13X at 25°C under atmospheric pressure is 0.271 kg-water/kg-dry zeolite as theoretical value. In the experiment, the amount of water adsorbed was also around 0.27 kg-water/kg-dry zeolite.
Figure 1 shows the schematic diagram of experimental apparatus. Saturated zeolite was packed in the reactor and kept at constant temperature. Water content in the zeolite particles at 110 °C at initial condition was 0.27 kg water/kg dry zeolite without free water. Free water was obtained from the difference between water concentration in the zeolite after experiment subtracted by the the maximum value of water can be adsorbed in the zeolite (for 13X was 0.27 kg water/kg dry zeolite). Dry N_2 at temperature 80 and 140 °C with various flow rates at 10, 30 and 50 L/min were then introduced to the reactor. The mass of water released was obtained by measuring the mass of water vapor in the air at outlet of the reactor.

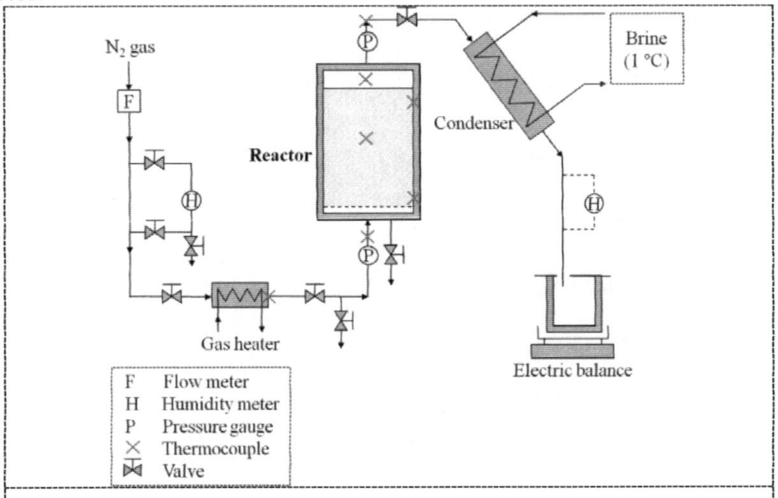

Figure 1. Schematic diagram of experimental apparatus.

2.2 Mathematical Model

Phenomenon during desorption using N_2 gas in packed bed was studied using a two-dimensional cylindrical model (r-z axes). Fig.2 shows scheme of module of the packed bed from the experiment that would be numerically calculated.

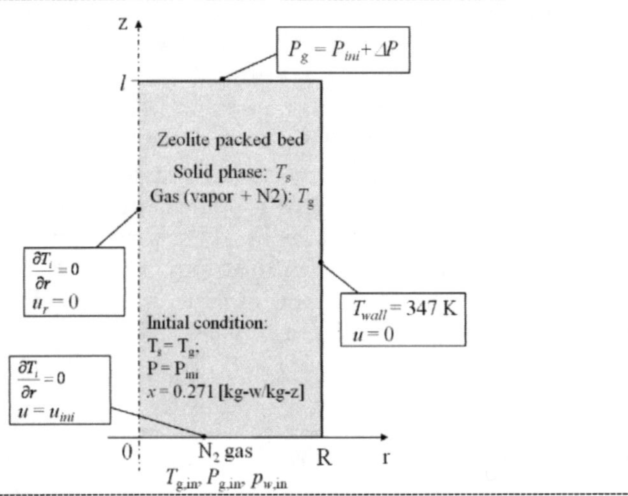

Figure 2. Computational domain, initial and boundary condition of calculation.

The main assumptions in the modeling are:

(a) All the zeolite particles have uniform properties such as geometry and size
(b) The gas velocity (including vapor) in the packed bed is determined using Darcy's equation
(c) The gaseous phase behaves as an ideal gas
(d) The particle is covered by laminar film, bounding the particle from the gas fluid.
(e) Macropore conducts the transportation of gas molecule from inside to the surface of particle.
(f) Molecules of water vapor are only adsorbed on the surface of zeolite crystal, therefore effect of micropore diffusion is not considered.

The governing equations of the studied system consist of energy balance of solid and gas (water vapor and N_2 gas) and mass balance in the solid related to the gas in the pores. Model equations of energy balance, mass balance, momentum, and kinetic model are shown in the following and the computational domain including initial and boundary conditions is shown in Fig. 3. The temperature of gas and solid is solved separately.

- Energy balance for solid phase (the zeolite):

$$(1-\varepsilon_t)\rho_s C_{ps}\frac{\partial T_s}{\partial t} = (1-\varepsilon_t)\nabla\cdot k_s \nabla T + (1-\varepsilon_t)Q_0\frac{dv}{dt} + S_{sg}h_{sg}(T_s - T_g) \tag{1}$$

- Energy balance for gas phase (N_2 and vapor):

$$\varepsilon_t \rho_g C_{pg}\frac{\partial T_g}{\partial t} + \rho_g C_{pg} u_g \nabla T = \varepsilon_t \nabla\cdot k_g \nabla T + S_{sg}h_{sg}(T_s - T_g) \tag{2}$$

- Mass balance for the zeolite:

$$(1-\varepsilon_t)\frac{\partial}{\partial t}\rho_s = m \tag{3}$$

- Mass balance for gas phase:

IOP Conf. Series: Materials Science and Engineering **344** (2018) 012002 doi:10.1088/1757-899X/344/1/012002

$$\frac{\partial}{\partial t}\varepsilon_t\rho_g + \nabla\cdot(\rho_g u) = -m \tag{4}$$

where m is the desorption rate of water desorbed from zeolite which is defined as the following:

$$m = (1-\varepsilon_t)\Delta\rho_s\frac{\partial v}{\partial t} \tag{5}$$

Gas diffusive velocity u is calculated using Darcy's equation as the momentum equation:

$$u = -\frac{K}{\mu}\nabla P \tag{6}$$

with K as the permeability calculating using Kozeny-Carman's:

$$K = \frac{1}{5}\frac{\varepsilon_t^{\,3}}{(1-\varepsilon_t)^2}\frac{1}{S_p^{\,2}} \tag{7}$$

The specific surface area Sp depends on the particle shape is determined by $6/d_p$ and $4/d_p$, for spherical and cylindrical particles, respectively.
Reaction rate model is expressed as below:

$$\frac{\partial v}{\partial t} = k\left(v_{eq} - v\right) \tag{8}$$

where v is the reacted fraction defined from the water content in the zeolite based on thermal analysis (TG/DTA) measurement. Reacted fraction is obtained from:

$$v = \frac{x_0 - x}{x_0 - x_f} \tag{9}$$

Reaction rate controller k is defined using Linear Driving Force (LDF) model because diffusion on the surface (external film) of particle and macropores influence the desorption rate.

$$k = k_{LDF} \tag{10}$$

$$\frac{1}{k_{LDF}} = \frac{\wedge r_p}{3(1-\varepsilon_b)k_f} + \frac{\wedge r_p^{\,2}}{15(1-\varepsilon_b)\varepsilon_p D_p} \tag{11}$$

Therefore, the rate controller consists of effect of external film and macropores region, respectively. $\wedge$ is partition ration that represent the fraction of water vapor as adsorbate that is exactly adsorbed in the zeolite particle, expressed as follow:

$$\wedge = \frac{\rho_b x}{\rho_g} \tag{12}$$

k_f as the mass transfer coefficient on the external film is defined using Wakao equation as:

$$k_f = \left(2.0 + 1.1\mathrm{Sc}^{1/3}\,\mathrm{Re}^{0.6}\right)\frac{D_{AB}}{r_p} \tag{13}$$

The gas diffusivity is calculated assuming both Knudsen and molecular diffusion:

$$\frac{1}{D_p} = \tau_p\left(\frac{1}{D_K} + \frac{1}{D_{AB}}\right) \tag{14}$$

with

$$D_{AB} = 1.8583\times10^{-3}\frac{\sqrt{T^3/MM_g}}{p\sigma^2\Omega_D} \tag{15}$$

and

$$D_K = 4.85 d_{pore} \sqrt{\frac{T}{MM_g}} \tag{16}$$

where τ_p as the particle tortuosity is determined as the inverse of particle pore ($1/\varepsilon_p$) based on Wakao while the molar mass of gas (nitrogen and vapor) MM_g, defined as:

$$\frac{1}{MM_g} = \left(\frac{1}{MM_{N_2}} + \frac{1}{MM_{H_2O}} \right) \tag{17}$$

Enthalpy of desorption (ΔH), heat capacity (C_{ps}) obtained from DSC measurement and such as frequency factor (A) and activation energy (E_a) obtained from TG/DTA measurement. Kunii and Smith's equation is used to get λ_{eff}. The total number of mesh points in the computational domain are 80×15 with the time step is 10^{-3} s.

The governing equations above are numerically solved using finite element method (FEM). The followings are the discretization of the governing equations which are solved for mass and heat balances separately.

3. Results and discussion

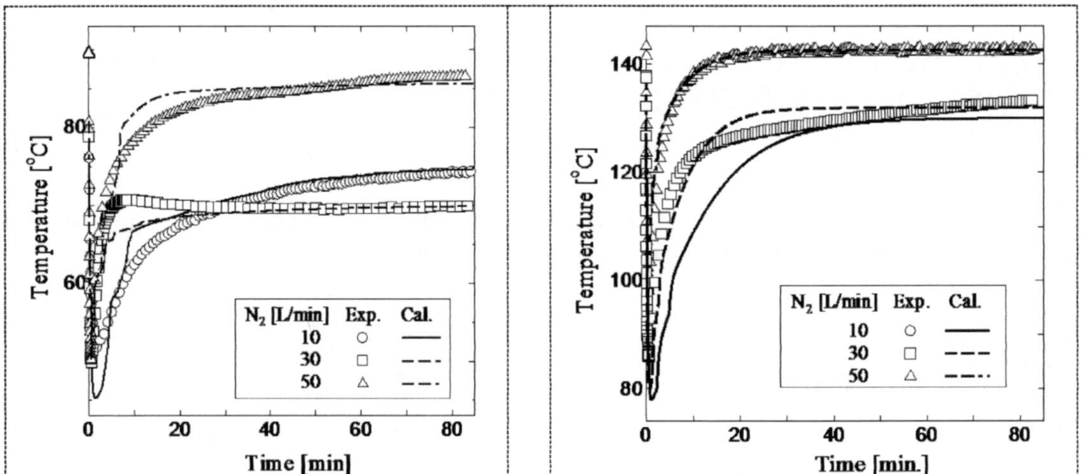

a. Inlet temperature of N_2 at 80°C | b. Inlet temperature of N_2 at 140°C

Figure 3. Comparison between experimental and calculation results for spherical particles ϕ2.1 mm: local bed temperatures.

Good agreements of temperature variations with time were obtained for desorption using spherical particles which the flow-rate of Nitrogen gas varies at 10, 30 and 50 L/min at temperatures 80 and 140°C (Fig. 3). According to the temperature variations in the center of the bed, flow-rate of Nitrogen gas has big influence in desorption. This is confirmed from the peak of temperature distribution becomes narrower when flow-rate of inlet gas increased. Higher flow-rate will influence the distribution because it can reduce the pressure drop in the bed (Eq. 6) and increase the heat transfer coefficient (Eqs. (1) and (2)).

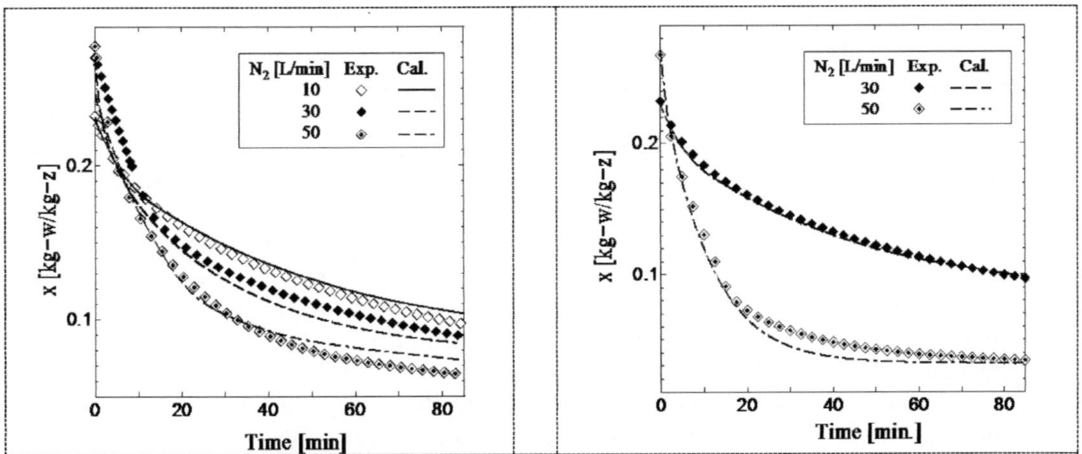

a. Inlet temperature of N_2 at 80°C | b. Inlet temperature of N_2 at 140°C

Figure 4. Comparison between experimental and calculation results for spherical particles ϕ2.1 mm: average water adsorbed in the zeolite [kg-water/kg-dry zeolite].

Fig. 4 shows water adsorbed variations with time between experiment and calculation confirmed good agreement for various flow-rate and temperature of Nitrogen gas. Initial value of water adsorbed in the zeolite is not uniform for each desorption condition due to difficulty to get uniform temperature in the preparation of saturated zeolite. The amount of water released from zeolite depends on the flow-rate and temperature of Nitrogen gas. Higher flow rate and temperature, more water can be desorbed. Furthermore, time required to release 80% of water adsorbed in zeolite can be shortened if the flow rate increased. Temperature of inlet gas also has effect in the shortening desorption time. This implies desorption rate can be increased if temperature and flow-rate of inlet gas increased.

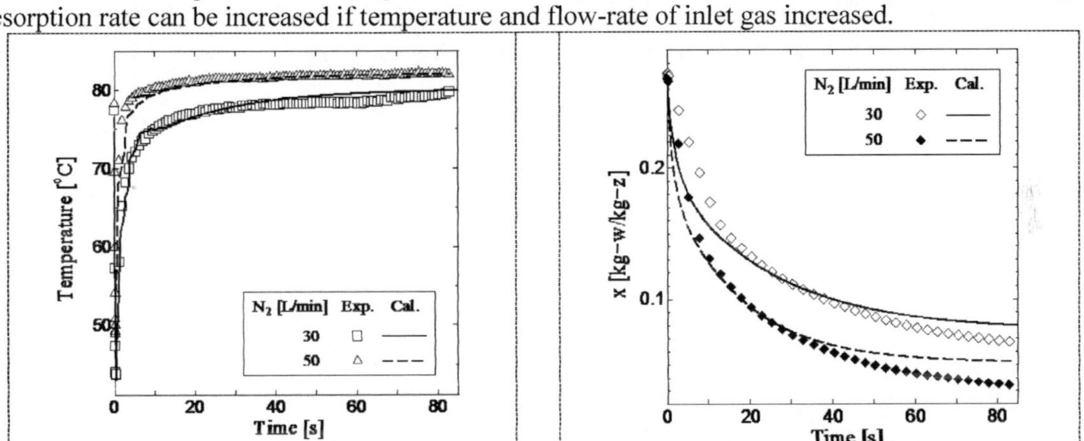

a. Local bed temperature in the center | b. Average water adsorbed

Figure 5. Comparison between experimental and calculation results for cylindrical particles ϕ1.6 mm, temperature of inlet gas at 80°C.

Desorption on cylindrical particles of zeolite 13X was investigated using Nitrogen gas at 80°C with flow-rate 30 and 50 L/min. Fig. 5 shows a good agreement of temperature variations with time can be obtained for desorption using spherical particles which the flow-rate of Nitrogen gas varies at 30 and 50 L/min at 80°C. The temperature distribution becomes narrower when flow-rate of inlet gas increased similar tendency with using spherical particles. More amount of water adsorbed can be released if higher flow rate used to desorb zeolite. This is confirmed experimentally and numerically through Figure 5-8. However, theoretical calculation predicted the amount of water released from zeolite is 0.015 [kg-w/kg-z] higher than experimental one.

4. Conclusion

This study could confirm the feasibility of desorption using dry gas where the results were summarized as below:

1. According to the overall heat balance of desorption, the drying time is theoretically calculated. From the comparison of drying time between experiments and calculations, the relative errors were within 9%. The relative errors show that heat transfer was dominant in desorption.
2. The results of numerical simulation agree with the experimental ones for the distribution of local temperature and average water adsorbed in the packed bed.

From the calculations, there was limitation in the operating condition of desorption. There was almost no different when the gas flow rate was more than 50 L/min and the gas temperature was more than 100 °C.

Acknowledgment: This work was supported in part by.Global Center of Excellence in Novel Carbon Resource Sciences, Kyushu University and JSPS Grant in aids for young scientist.

5. References

[1] Demir H, Mobedi M, Ulku S 2008 *Renewable and Sustainable Energy Reviews.* **12** 2381
[2] Oktariani E, Noda A, Nakashima K, Tahara K, Xue B, Nakaso K and Fukai J 2012 *International Journal of Energy Research* **36** 1077
[3] Oktariani E, Noda A, Nakashima K, Tahara K, Xue B, Wijayanta AT, Nakaso K and Fukai J 2012 *Journal of Chemical Engineering of Japan* **45** 355
[4] Watanabe F, Sumitani K, Kashiwagi T, Huang H, Hasatani M and Kobayashi N 2009 *Kagaku Kogaku Ronbunshu* **35** 431
[5] Saitake M, Kubota M, Watanabe F, Matsuda H 2007 *Kagaku Kogaku Ronbunshu* **33** 53
[6] Kärger J, Ruthven D M 1993 *Berichte der Bunsengesellschaft fur physikalische Chemie* **97** 146

IOP Conf. Series: Materials Science and Engineering **344** (2018) 012003 doi:10.1088/1757-899X/344/1/012003

Surface deformation monitoring of Sinabung volcano using multi temporal InSAR method and GIS analysis for affected area assessment

A Aditiya[14]**, Y Aoki**[2] and **R D Anugrah**[3]
[1] Civil Engineering Department, The University of Tokyo, Tokyo 151-0063, Japan
[2] Earthquake Research Institute, The University of Tokyo, Tokyo 151-0063, Japan
[3] Urban Engineering Department, The University of Tokyo, Tokyo 151-0063, Japan
[4] Geospatial Information Agency of Indonesia, Jl. Raya Jakarta-Bogor Km. 46, Cibinong, Indonesia

E-mail: arif@iis.u-tokyo.ac.jp, yaoki@eri.u-tokyo.ac.jp, ranie@env.t.u-tokyo.ac.jp

Abstract. Sinabung Volcano which located in northern part of Sumatera island is part of a hundred active volcano in Indonesia. Surface deformation is detected over Sinabung Volcano and surrounded area since the first eruption in 2010 after 400 years long rest. We present multi temporal Interferometric Synthetic Aperture Radar (InSAR) time-series method of ALOS-2 L-band SAR data acquired from December 2014 to July 2017 to reveal surface deformation with high spatial resolution. The method includes focusing the SAR data, generating interferogram and phase unwrapping using SNAPHU tools. The result reveal significant deformation over Sinabung Volcano areas at rates up to 10 cm during observation period and the highest deformation occurs in western part which is trajectory of lava. We concluded the observed deformation primarily caused by volcanic activity respectively after long period of rest. In addition, Geographic Information System (GIS) analysis produces disaster affected areas of Sinabung eruption. GIS is reliable technique to estimate the impact of the hazard scenario to the exposure data and develop scenarios of disaster impacts to inform their contingency and emergency plan. The GIS results include the estimated affected area divided into 3 zones based on pyroclastic lava flow and pyroclastic fall (incandescent rock and ash). The highest impact is occurred in zone II due to many settlements are scattered in this zone. This information will be support stakeholders to take emergency preparation for disaster reduction. The continuation of this high rate of decline tends to endanger the population in next periods.

1. Introduction

Sinabung volcano which located in northern part of Sumatera island is part of a hundred active volcano in Indonesia called by ring of fire. Therefore some potential impact of disasters caused by volcano and tectonic activities are frequently occurred such as eruption, landslide and earthquake in surround area. Reported by the Center for Volcanology and Geological Hazard Mitigation (CVGHM) of Indonesia, Sinabung volcano since August 2010 showed its activities until recent days (Fig.1). It has been observed thin white smoke reaches a height of 50-1000 m from the mount. Visually and through the seismograph has been observed 1 eruption with the ash column height reaching 2000 m above the mount, leaning to the east. Observed lava falling as far as 500-1000 m to the southeast and east [1].

In order to observe the pattern of suface deformation occured in Sinabung volcano during 2014 to 2017 period, several geodetic methods such as GPS survey, InSAR, leveling, electronic istance measurement (EDM), and tiltmeter can be conducted. However various methods are available, each

[1] To whom any correspondence should be addressed.

with advantages and disadvantages in terms of cost, complexity, reliability, power requirements, sensitivity, and ease of use. The greatest advantage of InSAR method in a crisis is that their data is continuous observed to the ground without direct contact during a few to ten of days, wide area coverage and no need any instrument in the ground. Meanwhile other methods have limited coverage, need to close the object and power requirement.

Spaceborne Synthetic aperture radar (SAR) interferometry is a geodetic technique based space that provides high-resolution measurements of the ground displacement associated with various geophysical phenomenons such as earthquake [2], volcano [3] [4] [5], urban subsidence [6], landslide [7], atmospheric researches [8], ocean applications [9].In the application of volcano monitoring, InSAR technique has demonstrated its high performance to measure surface displacements in different conditions and scenarios [10]. The advanced InSAR time series processing method referred to as Small BAseline Subset (SBAS) [11] that allows studying both the spatial and temporal variability of the surface displacements. It has proven to optimize for different models of SAR signal scattering.

This study presents surface deformations due to activities of Sinabung eruption are detectable with InSAR approach and to estimate disaster affected assessment by using GIS technique. Disaster assessment aims to mitigate the impact of eruption by consider number of houses and population when a disaster occurs. The main objective this research is to obtain rates of deformation and to assess affected area to reduce risk of disaster in the term of towards civilian resilience.

Figure 1. Recent of Sinabung eruption activities from 2014-2017

2. Methodology

2.1. Study Area and Data Sets

Sinabung volcano (2460 m) is located in the Karo plateau of Karo Regency, North Sumatra, Indonesia, 40 km from the Lake Toba supervolcano and 90 km from Medan city. The volcano was claasified as a Pleistocene-to-Holocene stratovolcano. Recent documented events include an eruption in the early hours of 29 August 2010 after long rest period 400 years ago. The first eruptions were following by eruptions in September, November 2013, February and October 2014, April 2015, May 2016, and August 2017. According to the evacuation report, 27.671 people have been evacuated during the eruption in 2014. Therefore this study has been conducted to support local emergency response plan.

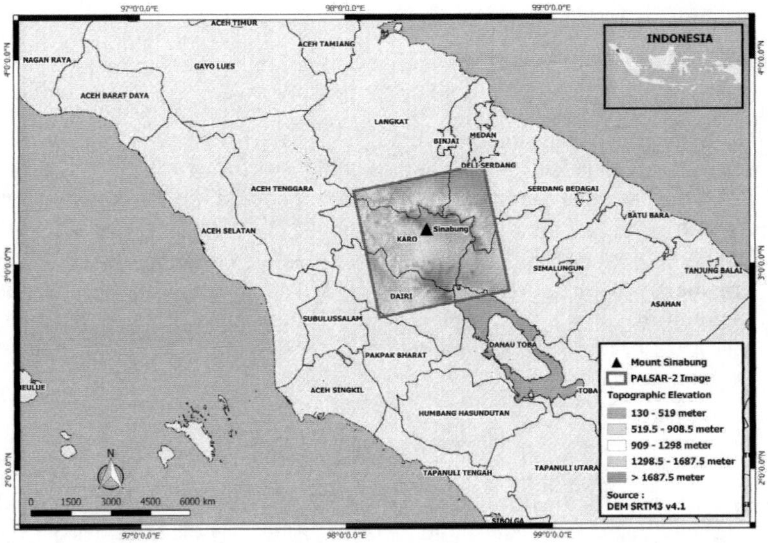

Figure 2. Geographic location of Sinabung Volcano, North Sumatera of Indonesia and red rectangle of PALSAR-2 imagery

We used the PALSAR-2 (Phased Array Type L-band Synthetic Aperture Radar) sensor onboard of the Advanced Land Observing Satellite (ALOS-2) images in this study and acquired from December 2014 to July 2017. Table 1 shows details of cover ranges PALSAR-2 data that used to observe Sinabung volcano. These data sets were acquired in the ascending orbit with an off-nadir angle 28.2°. Observation parameters for all the images were as mode Strip Map and right looking direction. All baseseline of pairs images less than 350 m , remind one pairs had a baseline of 318 m. The coherence was good performed at volcano summit and its surround, meanwhile less coherence in vegetation area.

Table 1. List of ALOS-2/PALSAR-2 data

Satellite	Orbit	$B_{PERP\,(m)}$	Track	B_{TEMP}	Nadir Angle	Acquisition Date
ALOS-2	Asc.	0 (M)	152	0	28.2°	20141202
ALOS-2	Asc.	-125.1	152	294	28.2°	20150922
ALOS-2	Asc.	-318.9	152	364	28.2°	20151201
ALOS-2	Asc.	-123.1	152	546	28.2°	20160531
ALOS-2	Asc.	0	152	588	28.2°	20160712
ALOS-2	Asc.	-4.4	152	658	28.2°	20160920
ALOS-2	Asc.	-34.6	152	798	28.2°	20170207
ALOS-2	Asc.	-258.2	152	952	28.2°	20170711

These data were used to generate interferogram and to obtain rates of surface deformation. Precise orbit data which is delivered from the product header was also used to remove the reference phase from the differential interferograms. A Shuttle Radar Topography Mission (SRTM) version 4.1 was employed to eliminate the topographic phase.

2.2. Multi-temporal InSAR

SAR is microwave remote sensing technique using at least two or more images acquired at different times to generate displacement maps to detect surface changes over a specific area. Furthermore multi temporal InSAR is advanced methods involving the parallel processing of several SAR acquisitions in time increase **[12]**. In the interferometric SAR data processing, the interferograms are generated by combining two complex SAR images **[13]**. InSAR measures the change in path length in the satellite line-of-sight (LOS) between observations. Generally for InSAR approach two SAR images from slightly different orbit configurations and at different times are combined to exploit the phase difference of the signals LOS (line of sight). By assuming that the scattering phase is the same in both images, the interferometric phase ϕ is a very sensitive measure of the range difference $R_2 - R_1$ i.e :

$$\phi = \phi_1 - \phi_2 = \frac{4\pi}{\lambda}(R2 - R1)$$

Here, ϕ_1 and ϕ_2 are the phases of the first and second SAR images, respectively; R_1 is the distance from the SAR to the scatterer by the first acquisition; R_2 is the distance by the second acquisition; and , λ is the wavelength, as for L-band ALOS-2 data λ is 23.6 cm. Many factors contribute to changes in path length, but with appropriate removal of topographic effects, and if atmospheric and ionospheric effects are small and/or can be isolated, path length changes correspond to deformation of the Earth's surface. It can be expressed as below.

$$\phi_{int} = \phi_{topo} + \phi_{defo} + \phi_{orb} + \phi_{atm} + \phi_{scatt} + \phi_{noise}$$

where ϕ_{int} is interferometric phase, ϕ_{topo} is topographic phase, ϕ_{defo} is deformation phase due to the deformation in the radar line of slight, ϕ_{orb} is deterministic flat earth phase and the residual phase signal due to orbit in determination, ϕ_{atm} is atmospheric phase, ϕ_{scatt} is phase due to a temporal and spatial change in the scatter characteristics of the earth surface between the two observation times, and ϕ_{noise} is phase degradation factors, caused by e.g., thermal noise, coregistration noise and interpolation noise. We applied SBAS method on an appropriate combination of differential interferograms produced by data pairs characterized by a small orbital separation (baseline) in order to reduce the spatial decorrelation phenomena [11]. Cumulative phase can be obtained by solving a linear least squares problem. However, this can produce different subsets of InSAR pairs connected in time and separated by large baselines. SBAS method allow to obtain surface deformation and to analyze their space time characteristics.

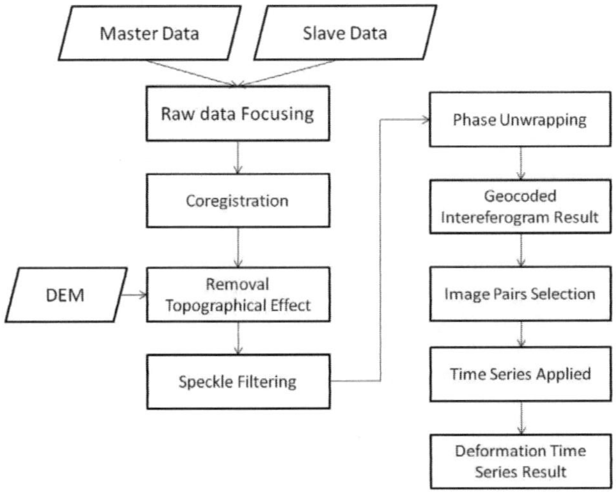

Figure 2. Processing Diagram of multi temporal InSAR

2.3. GIS Assessment

Geographic Information System (GIS) technique is a set of tools in a computer based information system that can store, retrieve, create a model, manipulate, transform, analyse, share, and display the data which are referenced to the Earth spatially for a specific purpose [14]. One of the most important application of GIS for disaster is establishing data and analytical modelling, support the decision making for pre-impact planning, post event response, and the mitigation process [15]. GIS applied to identify and prioritize the prone area to volcano hazard in Sinabung area, North Sumatera province. In this assessment, two types of data are used: (1) hazard data defined as disaster prone areas of Sinabung volcano, and (2) exposure data defined as the number of building and population.

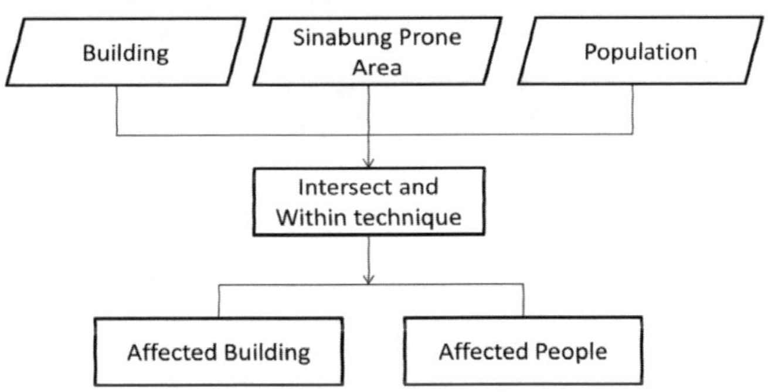

Figure 3. Diagram of GIS technique for Affected Disaster Estimation

The hazard data obtained from National Agency for Disaster Management (BNPB) which is divided into 3 levels of disaster vulnerability from high to low namely (1) disaster prone area 3 is a very potential area affected by hot clouds, lava flows, lava falling, stone pouring, and poisonous gas, (2) disaster prone area 2 is a potential area of hot clouds, lava flows, stones and lava storms, heavy ash rains, hot mud, lava flows, and toxic gases; and (3) disaster prone areas 1 is a potentially lava-stricken area, crushed material fall in the form of rain ash, and water with high acidity. While the exposure data obtained from OpenStreetMap for building and population from WorldPop which open population data for almost all of Africa, Asia and America with high resolution data approximately 100 m cell size. In addition each data layer between hazard and building/population is processed using overlay technique (intersect and within tool) in GIS tools [16] and the number of affected buildings and the number of affected populations based on the level of disaster prone areas of Sinabung and visualized into the map.

3. Result and Discussion

We generated 28 interferograms and processed time series data for a few points of Sinabung volcano by using GIAnT program [17]. Thereafter we computed all interferograms and the wrapped phase was corrected for spatially-uncorrelated look angle error and noise associated with the master image. Small Baseline Subset (SBAS) technique [11] was employed to reduce spatial decorrelation. Atmospheric artifacts are reduced through spatial-temporal filtering after the mean deformation is obtained. Iterative processing can further improve the time-series deformation estimates. By using this algorithm each deformation according to time period is able to calculate more precisely by minimizing all error components.

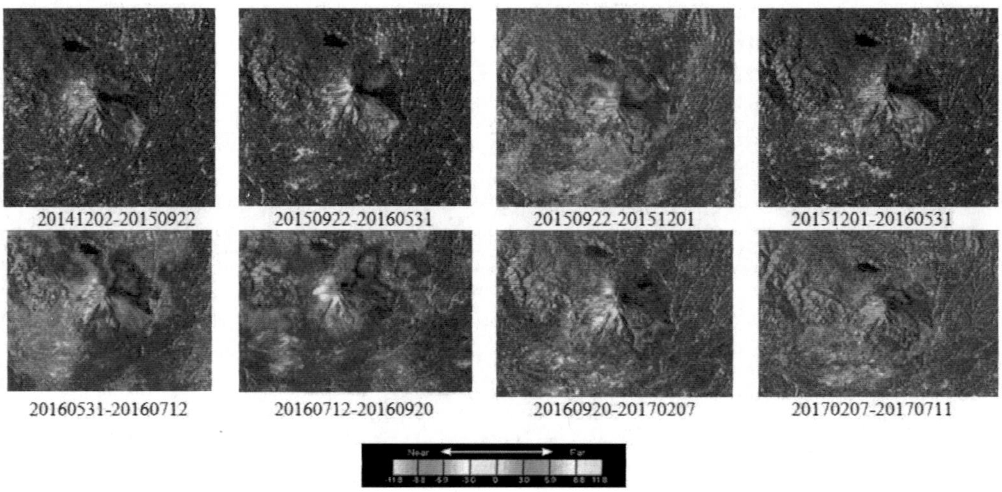

Figure 4. Interferograms of Sinabung volcano

Figure 5 shows point A and B were located in western part of Sinabung summit have been surface deformation during obervation period. The surface deformation in point A and B reach up to 6-7 cm/ year. Meanwhile point C which is located in northern of Sinabung summit has small deformation 1.9 cm/ year and point D 0.8 cm/ year. Point G has significant deformed up to 16 cm in almost 3 years and followed by point F where is near Sinabung summit approximately 9 cm.

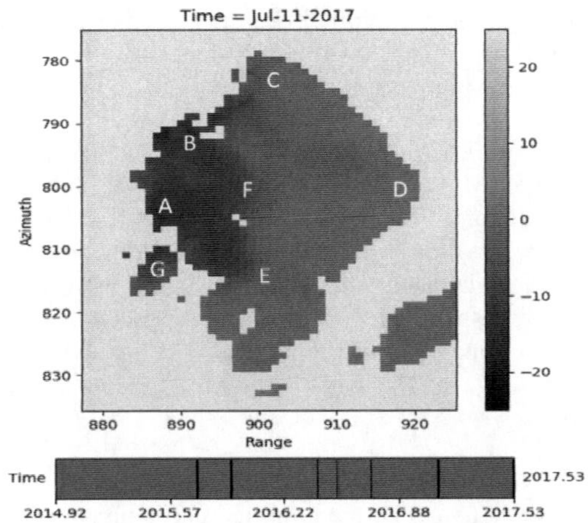

Figure 5. Observation Point at Sinabung Volcano

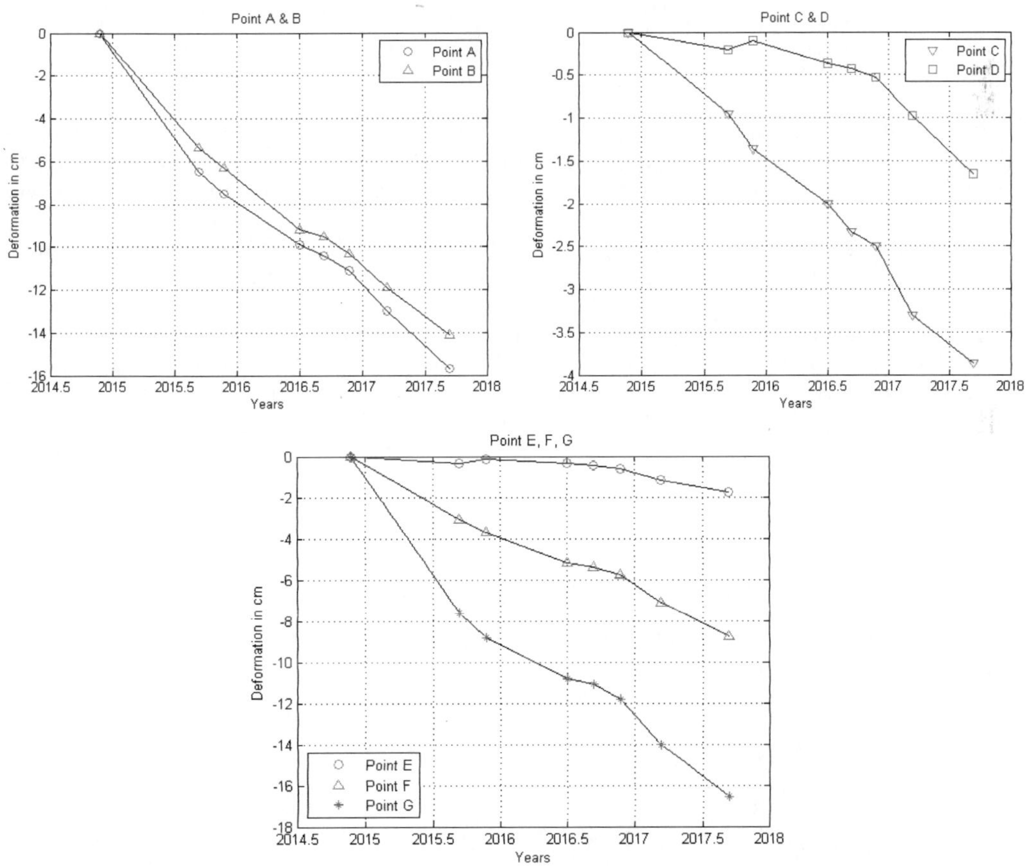

Figure 6. Time Series InSAR Result

The deformation was likely due to the deflation of magma from a deep source to the shallow reservoir. This phenomenon may have been caused by the deppressurization of the magma chamber due to periodic volcanic degassing at the Sinabung summit.

In term of disaster management GIS is used to support pre-impact planning, post event response, and the mitigation process. There are two types of output that are made such as a map showing the number of buildings affected and population affected by the eruption of Sinabung volcano based on the level of disaster prone area. This information could be used as reference for the development of volcano mitigation plans in reducing the damages and losses. It could be used to improve awareness and preparedness of the community, especially those located in volcano prone areas. Table 2 shows the number of buildings and affected persons in prone areas 1, 2 and 3. The disaster-prone area 2 has potential to suffer huge losses due to the large number of residents who live and the number of building affected by eruption.

Table 2. The Estimation Results of Buildings and Populations affected by the eruption of Sinabung

Disaster Zone	Building affected	People affected
Disaster prone area 3	4800	7600
Disaster prone area 2	5700	7100
Disaster prone area 1	1100	1500
Total	11,600	16,200

The total buildings were affected by 11,600 buildings and the highest number of affected buildings were in disaster prone areas 2 as many as 5700 buildings. While total people were affected by 16,200 people and the highest number of affected population is in disaster prone areas 3 as many as 7600 people, then the lowest number of affected population is in disaster prone areas 1 as many as 1600 people. This is because that in disaster prone areas 2 and 3 is a residential area with high population. The following is a table showing the estimated number of buildings and populations affected.

Figure 7 shows the map of buildings and people affected of Sinabung eruption based on criteria refers to the National Disaster Management Agency. This method directly related to the geographical conditions, social and economic communities living around the volcano. By adopted the result obtained from this method can be mitigate disaster impact.

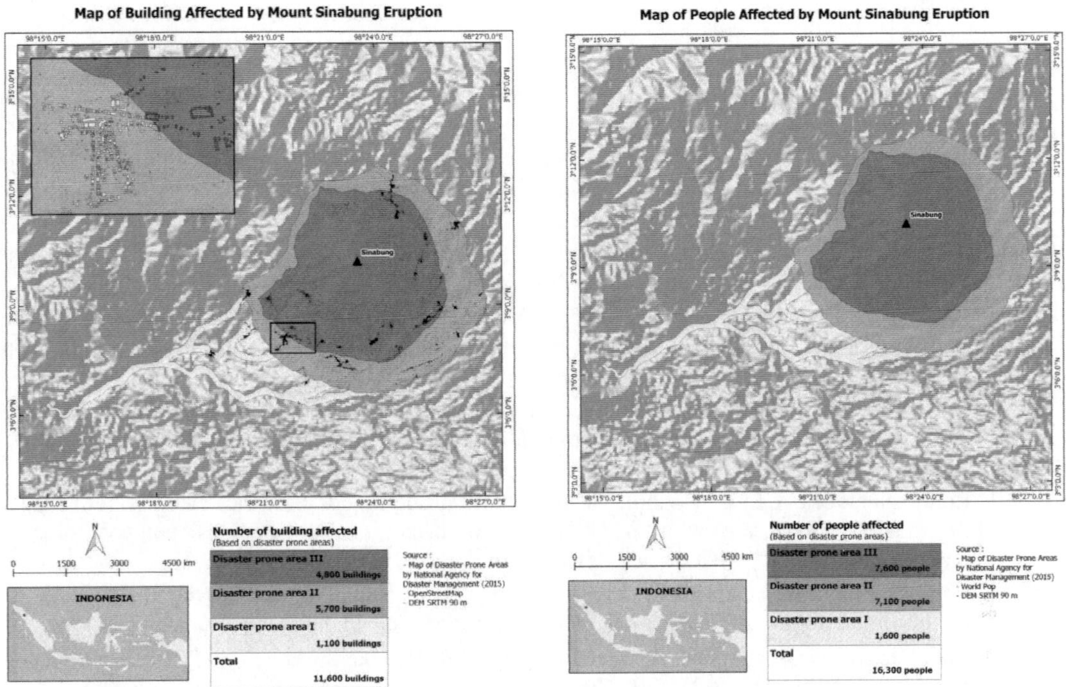

Figure 7. GIS Result of Building (left) and People Affected (right)

4. Conclusion

This work has presented an analysis of the surface deformation phenomena in Sinabung volcano. The advanced multi-temporal InSAR technique has been applied to this site by using eight PALSAR-2 images acquired from December 2014 to July 2017. We identified a few locations have deformed rates up to 15 cm during observation period. The average subsidence velocity map has been retrieved by Small Baseline approach to reduce spatial decorrelation characteristics among image itself. A rapid period of surface defromation occured during the 2,5 years of the post-eruption period. The greatest surface deformation was measured up to 6-7 cm/ year at the western part of Sinabung summit. We concluded the observed deformation primarily caused by volcanic activity respectively after long period of rest. The surface deformation at Sinabung volcano was correlated greater locally. There could also be possible existence of other causes due to geological factors in almost all deformation areas such as tectonic processes in Sumatra island. Future investigation can improve the findings of this study by utilizing other techniques. In addition the results of GIS analysis show that disaster-prone area 2 has potential to suffer huge losses than disaster prone area 1 and 3 due to the large number of population and building affected by eruption. This information could be used by local government and stake holders to mitigate disaster impact and people evacuation needed.

Acknowledgement

We thank to Indonesia Endowment Fund for Education (LPDP) for financial study support. These data were made available by PALSAR Interferometry Consortium to Study our Evolving Land surface (PIXEL) under research contract with the Earthquake Research Institute of the University of Tokyo.

References

[1] CVGHM. (2017, August) Magma Indonesia. [Online]. https://magma.vsi.esdm.go.id/

[2] D. Massonnet et al., "The displacement field of the Landers earthquake mapped by Radar interferometry," *Nature*, pp. 138-142, 1993.

[3] Y. Aoki and T. P. Sidiq, "Ground deformation associated with the eruption of Lumpur Sidoarjo mud volcano, east Java, Indonesia," *Journal of Volcanology and Geothermal Research*, pp. 96-102, 2014.

[4] Y. Aoki, "Space Geodetic Tools Provide Early Warnings for Earthquakes and Volcanic Eruptions," *J. Geophys. Res.*, pp. 3241-3244, April 2014.

[5] E. Chaussard, F. Amelung, and Y. Aoki, "Characterization of open and closed volcanic systems in Indonesia and Mexico using InSAR time series," *J. Geophys. Res. Solid Earth*, vol. 118, pp. 3957–3969, 2013.

[6] Estelle Chaussard, Falk Amelung, Hasanudin Abidin, and Sang-Hoon Hong, "Sinking cities in Indonesia: ALOS PALSAR detects rapid subsidence due to groundwater and gas extraction," *Remote Sensing of Environment*, vol. 128, pp. 150-161, 2013.

[7] Romy Schlögel, Cécile Doubre, Jean-Philippe Malet, and Frédéric Masson, "Landslide deformation monitoring with ALOS/PALSAR imagery : A D-InSAR geomorphological interpretation method," *Geomorphology*, pp. 314-330, 2014.

[8] Michael Jehle and David Small, "Measurement of Ionospheric TEC in Spaceborne SAR Data," *IEEE Transactions on Geoscience and Remote Sensing*, vol. 48, pp. 2460-2468, June 2010.

[9] Rudi Gens, "Oceanographic Applications of SAR Remote Sensing," *GIScience & Remote Sensing,* vol. 45, no. 3, pp. 275-305, 2008.

[10] Yan Zhan and Patricia M. Gregg, "Data Assimilation Strategies for Volcano Geodesy," *J. Volcanol. Geotherm. Res.*, 2017.

[11] Paolo Berardino, Gianfranco Fornaro, Riccardo Lanari, and Eugenio Sansosti, "A New Algorithm for Surface Deformation Monitoring Based on Small Baseline Differential SAR Interferograms," *IEEE Transactions on Geoscience and Remote Sensing*, pp. 2375-2382, 2002.

[12] A. Hooper, "A Mutli-temporal InSAR Method Incorporating both Persistent Scatterer and Small Baseline Approaches," *Gephys. Res. Lett.*, vol. 35, 2008.

[13] R F Hanssen, *Radar Interferometry : Data Interpretation and Error Analysis*. Dordrecht: Kluwer Academic, 2001.

[14] I. Heywood, S. Cornelius, and S Carver, *An introduction to geographical information systems.*: Pearson Prentice Hall, 2006.

[15] K. Chen, R. Blong, and C. Jacobson, "Towards integrated approach to natural hazard risk assessment using GIS: With references bushfires," *Environmental Management*, pp. 546-560, 2003.

[16] I. R. Pranantyo, M. Fadmastuti, and F. Chandra, "InaSAFE Applications in Disaster Preparedness," in *4th International Symposium on Earthquake and Disaster Mitigation 2014 (ISEDM 2014)*, Bandung, 2015.

[17] P. S. Agram et al., "New Radar Interferometric Time Series Analysis Toolbox Released," *Eos Trans. AGU*, vol. 97, no. 7, pp. 69-70, February 2013.

IOP Conf. Series: Materials Science and Engineering **344** (2018) 012004 doi:10.1088/1757-899X/344/1/012004

The machined surface of magnesium AZ31 after rotary turning at air cooling condition

G Akhyar [1*], B Purnomo [1], A Hamni [1], S Harun [1], Y Burhanuddin[1]
Department of Mechanical Engineering, Faculty of Engineering, University of Lampung. Jln. Prof.Sumantri Brojonegoro No.1 Gedung H FT Lt.2 Bandar Lampung

Email: gusri.akhyar@eng.unila.ac.id

Abstract. Magnesium is a lightweight metal that is widely used as an alternative to iron and steel. Magnesium has been applied in the automotive industry to reduce the weight of a component, but the machining process has the disadvantage that magnesium is highly flammable because it has a low flash point. High temperature can cause the cutting tool wear and contributes to the quality of the surface roughness. The purpose of this study is to obtain the value of surface roughness and implement methods of rotary cutting tool and air cooling output vortex tube cooler to minimize the surface roughness values. Machining parameters that is turning using rotary cutting tool at speed the workpiece of (Vw) 50, 120, 160 m/min, cutting speed of rotary tool of (Vt) 25, 50, 75 m/min, feed rate of (f) 0.1, 0.15, 0.2 mm/rev, and depth of cut of 0.3 mm. Type of tool used is a carbide tool diameter of 16 mm and air cooling pressure of 6 bar. The results show the average value of the lowest surface roughness on the speed the workpiece of 80 m/min, cutting speed of rotary tool of 50 m/min, feed rate of 0.2 mm/rev, and depth of cut of 0.3 mm. While the average value of the highest surface roughness on the speed the workpiece of 160 m/min, cutting speed of rotary tool of 50 m/min, feed rate of 0.2 mm/rev, and depth of cut of 0.3 mm. The influence of machining parameters concluded the higher the speed of the workpiece the surface roughness value higher. Otherwise the higher cutting speed of rotary tool then the lower the surface roughness value. The observation on the surface of the rotary tool, it was found that no uniform tool wear which causes non-uniform surface roughness. The use of rotary cutting tool contributing to lower surface roughness values generated.

1. Introduction

Magnesium is a lightweight metal and has characteristics similar to aluminum. Magnesium can be used as an alternative to iron and steel because magnesium is abundant elements and elements that make up the eighth most 2 % of the earth's crust, and is the third most element dissolved in seawater [1]. On development, magnesium or its alloys has been widely applied in the automotive industry among others to lose weight because of a component is a metal that is light [2]. Magnesium machining process is known to have excellent cutting characteristics because it has a low specific cutting force, furious chips are short, relatively low tool wear, high surface quality and can be cut at cutting speeds and feeds are high [3]. Although magnesium has many advantages, but has the disadvantage that magnesium is highly flammable because it has a low flash point. At such a low flash point will be burning furiously, where the cutting temperature exceeds the melting point of the material, namely (400° C - 600° C) [4].

In addressing these issues, numerous studies have been conducted to find an effective method to reduce the cutting temperature. As stated by previous researchers that a high cutting temperatures will produce a high surface roughness values [5]. Because the machining process with high temperature will cause occurrence of wear on the cutting tool so that it can degrade the quality of the workpiece surface roughness [6].

In the machining process, generally the method that is widely used to lower the temperature of cutting is to use liquid. But in its development began to minimized use this liquid because it is very dangerous for health and the environment [3]. In the study conducted by Doni [7] which aims to analyze the surface roughness value by lowering the temperature of the cutting. The chosen method is to use a rotating cutting tool on a lathe machining processes without coolant and use a rotary tool tilt angle of 0° with respect to the workpiece. Rotary tool method used successfully obtain minimum surface roughness value of 0.62 μm and a maximum surface roughness values of 2.86 μm. In his research, Doni [7] recommends using low feed rate , because the higher of feed rate will generate greater roughness values. Research on machining magnesium is also done by Andriyansyah [8] which aims to determine the effect of cutting parameters on surface roughness values of magnesium using air cooling of vortex tube cooler with a temperature of 15° C. Machining process used is a milling machining. Research results show that the value of the minimum surface roughness of 0.35 μm and maximum roughness value of 1.50 μm.

of the research that has been described above concludes that the value of the surface roughness of magnesium in addition influenced by cutting parameters, including feed rate and the cutting speed, also influenced by the temperature of the cutting. Therefore in this study will be analyzed on a surface roughness values lathe machining process using a rotary cutting tool with a tilt angle of 10 °. In a previous study of this corner enabled to reduce power by 30% of the total power use and reduce the amount of cutting force [9] [10]. The cutting process will be given air cooling output of vortex tube cooler constant. By using cold air is expected to reduce the temperature of the cutting so as to reduce the rate of tool wear and improve tool life [11] [12]. So this method is expected to produce a better roughness values and can be used as a substitute for innovation in the machining process fluid.
The purpose of this study is to obtain the value of surface roughness and implement methods of cutting tool rotary tool and air cooling output of vortex tube cooler to minimize the surface roughness values.

2. Research Methodology

Laboratory of Manufacturing, Mechanical Engineering, Lampung University. Materials used in the study are magnesium alloy AZ31 (Al of 3% and Zinc of 1%). Here is an AZ31 magnesium physical table property.

Table 1. Physical properties of magnesium AZ31

Density [kg/mm^3]	1,77 x 10^{-6}
Young's Modulus [kN/mm^2]	45,000
Possion's ratio	0.35
Melting temperature [K]	891
Thermal Conductivity [w/(mK)]	77 + 0.096T
Heat specific capacity [J/(kgK)]	1000 + 0.666T
Heat coefficient [K^{-1}]	2.48 x 10^{-5}

The machining is done on conventional lathes brands PINACHO type S-90/200. By using a rotary cutting tool system as shown in Figure 1. The material of tool used is a carbide, diameter of 16 mm and the air cooling pressure of 6 bar. Here are the figure and specifications of the tool cutting system spins and vortex tube :

Figure 1. Rotary cutting tool system

Table 2. Specifications of rotary tool

Merk	AXUM590-A
Tool type	*Insert-Propeller*
Tool rotation speed	0-2000 rpm
Speed direction	CW/CCW
Diameter of insert	16 mm

Figure 2. Vortex tube used to produce air cooling.

Machining parameters that is turning using rotary cutting tool at speed the workpiece of (Vw), cutting speed of rotary tool of (Vt), feed rate (f), depth of cut (d). Can be seen in Table 3.

Table 3. Cutting parameters selected in turning process.

Speed of workpiece Vw, m/min			Feed rate *f*, mm/ref			Depth of cut d, mm	Speed of rotary tool, Vt m/menit			Temperature
80	120	160	0,10	0,15	0,20	0,3	25	50	75	0 °C. 6 Bar

Data collection roughness on every parameter done four times. This is done to get the maximum results. Measurement values using a surface roughness tester with an accuracy of 0.01 μm and surface profile of picture a using USB camera with a magnification of 600 x.

3. Results and Discussions

In Experimental trials of AZ31 magnesium alloys using conventional turning and rotary cutting tool system with air cooling. To determine the value of the surface roughness material magnesium alloy AZ31 has been carried out with a variety of machining parameters that is turning using rotary cutting tool at speed the workpiece of (Vw) 50, 120, 160 m/min, cutting speed of rotary tool of (Vt) 25, 50, 75 m/min, feed rate of (f) 0.1, 0.15, 0.2 mm/rev, and depth of cut of 0.3 mm. Here are the data obtained from the machining process with a variety of conditions, produces a variety of data as shown in Table 4-7:

Table 4. Measurement data value of roughness on speed the workpiece of 120 m/min, depth of cut of 0.3 mm, and cutting speed of rotary tool of 25 m/min.

No	Speed of workpiece (Vw) (m/min)	Feed rate (f) (mm/rev)	Depth of cut (d) (mm)	Speed of rotary tool (Vt) (m/min)	Surface roughness (μm) 1 *Ra*	2 *Ra*	3 *Ra*	Average (μm)	Time (t) (Minute)
1					1,22	1,32	1,27	**1,27**	05:28
2	120	0.1	0.3	25	1,37	1,28	1,14	**1,26**	10:53
3					1,24	1,09	1,24	**1,19**	16:20
4					1,24	1,02	1,02	**1,09**	21:47
1					0,89	1,15	0,78	**0,94**	02:43
2	120	0.2	0.3	25	1,29	0,82	0,9	**1,00**	05:27
3					1,22	1,32	1,27	**1,27**	05:28
4					1,37	1,28	1,14	**1,26**	10:53

Table 5. Measurement data value of roughness on speed the workpiece of 160 m/min, depth of cut of 0.3 mm, and cutting speed rotary tool of 50 m/min.

No	Speed of workpiece (Vw) (m/min)	Feed rate (f) (mm/rev)	Depth of cut (d) (mm)	Speed of rotary tool (Vt) (m/min)	Surface roughness (μm) 1 *Ra*	2 *Ra*	3 *Ra*	Average (μm)	Time (t) (Minute)
1	160	0.1	0.3	50	1,59	1,35	1,3	**1,41**	04:20
2					1,45	1,64	1,4	**1,50**	08:40
3					1,73	1,64	1,51	**1,63**	13:01
4					1,49	1,6	1,59	**1,56**	17:22
1	160	0.2	0.3	50	1,64	2,48	1,79	**1,97**	02:10
2					2,87	2,33	1,71	**2,30**	04:20
3					1,76	2,54	2,51	**2,27**	06:31
4					2,25	1,35	1,85	**1,82**	08:43

Table 6. Measurement data value of roughness on speed the workpiece of 80 m/min, feed rate of 0.15 mm/rev, and depth of cut 0.3 mm.

No	Speed of workpiece (Vw) (m/min)	Feed rate (f) (mm/rev)	Depth of cut (d) (mm)	Speed of rotary tool (Vt) (m/min)	Surface roughness (μm) 1 *Ra*	2 *Ra*	3 *Ra*	Average (μm)	Time (t) (Minute)
1	80	0.15	0.3	25	1,01	0,7	0,95	**0,89**	06:37
2					1,05	0,97	0,91	**0,98**	13:15
3					0,63	0,85	0,69	**0,72**	19:53
4					1,17	0,62	0,95	**0,91**	26:29
1	80	0.15	0.3	75	0,82	0,61	0,97	**0,80**	06:38
2					0,76	0,88	0,83	**0,82**	13:15
3					0,84	0,7	0,77	**0,77**	19:54
4					0,95	0,84	0,66	**0,82**	26:30

Table 7. Measurement data value raughness on the feed rate 0.15 mm/rev, depth of cut 0.3 mm, and cutting speed of rotary tool of 25 m/min

No	Speed of workpiece (Vw) (m/min)	Feed rate (f) (mm/rev)	Depth of cut (d) (mm)	Speed of rotary tool (Vt) (m/min)	Surface roughness (μm)			Average (μm)	Time (t) (Minute)
					1	2	3		
					Ra	*Ra*	*Ra*		
1	80	0.15	0.3	25	1,01	0,7	0,95	**0,89**	06:37
2					1,05	0,97	0,91	**0,98**	13:15
3					0,73	0,85	0,89	**0,82**	19:53
4					1,17	0,62	0,95	**0,91**	26:29
1	160	0.15	0.3	25	1,81	0,7	1,6	**1,37**	02:53
2					1,64	1,6	1,49	**1,58**	04:46
3					2,06	1,44	1,38	**1,63**	07:39
4					1,19	1,74	1,53	**1,49**	10:32

3.1. Feed rate and surface roughness value

The following is a graph showing the comparison surface roughness values between feed rate 0.1 mm/rev and 0.2 mm/rev at the depth of cut (d) 0.3 mm.

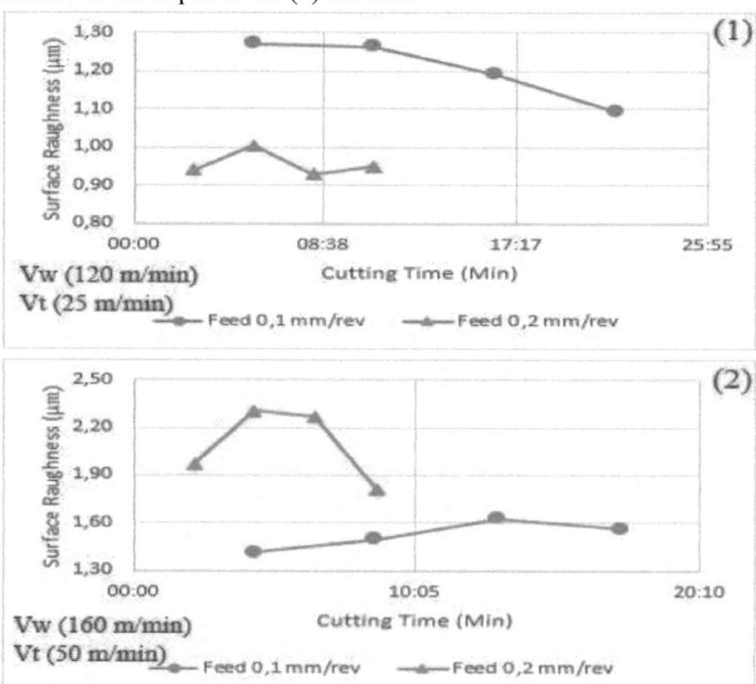

Figure 3. Comparison of surface roughness between feed rate 0.1 and 0.2 mm/rev.

Figure 3 is a comparison of the effect of feed rate on the value of surface roughness. Where in graph 1 surface roughness value was lower than in the feed rate of 0.2 mm/rev. It is inversely proportional to the statement earlier researchers [7]. Which states that the chip of feed is very influential on the surface roughness, where the greater the price of feed. the greater the level of coarseness. But in general, the value of the surface roughness is influenced by a variety of many factors. These factors include other parameters of the machining process, the condition of the tool, the workpiece and the cutting phenomenon [13] [14]. From the statement of course in this case the value of the feed is not completely affect the value of the resulting roughness.

Whereas in Figure 2 the surface roughness value was lowest in the feed rate 0.1 mm/rev. In both cases carried out observations of the surface profile, wherein the surface profile measurements to feed mark the resulting on cutting scars. Here is a comparison of the measured surface profile of the workpiece :

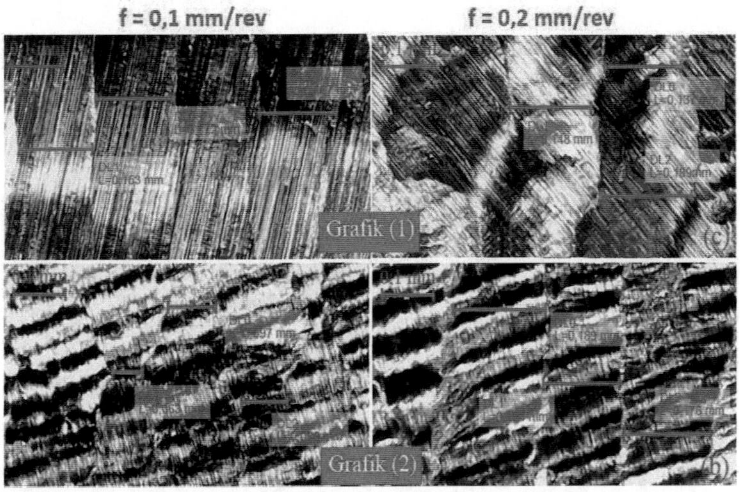

Figure 4. Comparison between profiles of feed rate (f) 0.1 and 0.2 mm/rev.

In Figure 4 obtained value feed mark on the first surface profile of feed rate 0.1 mm/rev of 0.164 mm and feed rate 0.2 mm/rev of 0.158 mm. While the second profile feed rate 0.1 mm/rev produce value feed mark of 0.082 mm and feed rate 0.2 mm/rev of 0.173 mm.

3.2. Cutting speed Rotary and surface roughness value

The following is a graph showing the comparison between the surface roughness value of the cutting speed of rotary tool of 25 m/min and 75 m/min with the depth of cut (d) 0.3 mm :

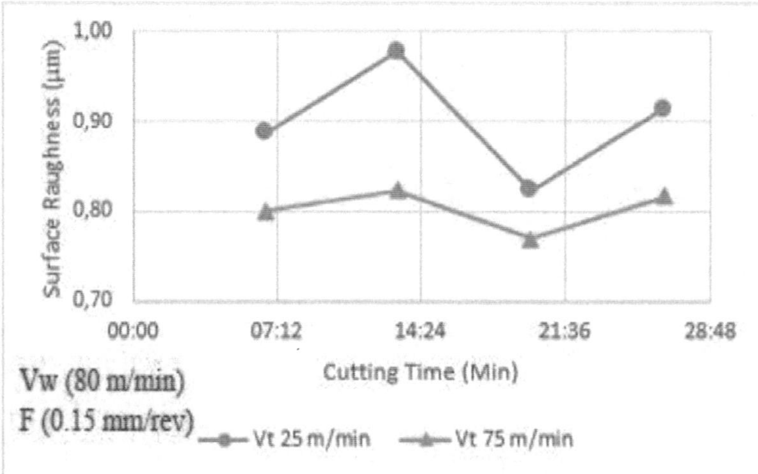

Figure 5 . Graph comparison of cutting speed of rotary tool (Vt) of the surface roughness values.

Figure 5 is a comparison of the effect of cutting speed rotary tool against the surface roughness values. Where is concluded that the value of the surface roughness was lowest in the cutting speed of rotary tool (Vt) 75 m/min and the surface roughness values were highest in cutting speed of rotary tool (Vt) 25 m/min. As already stated by Doni [7], the greater the cutting speed of the rotary cutting tool resulting roughness value would be lower. For a comparison of the surface profile can be seen in Figure 6.

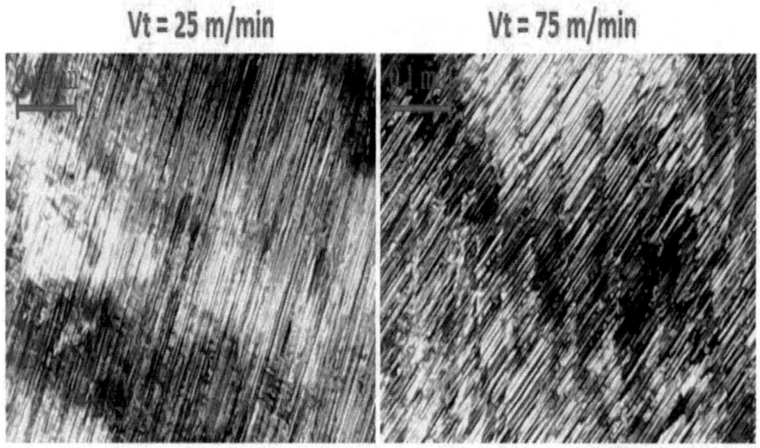

Figure 6. Comparison between the profile cutting speed of rotary tool 25 and 75 m/min.

3.3. Workpiece speed and surface roughness value

The following is a graph showing the comparison between the value of the surface roughness of the speed the workpiece of 80 m/min and 160 m/min with the depth of cut (d) 0.3 mm:

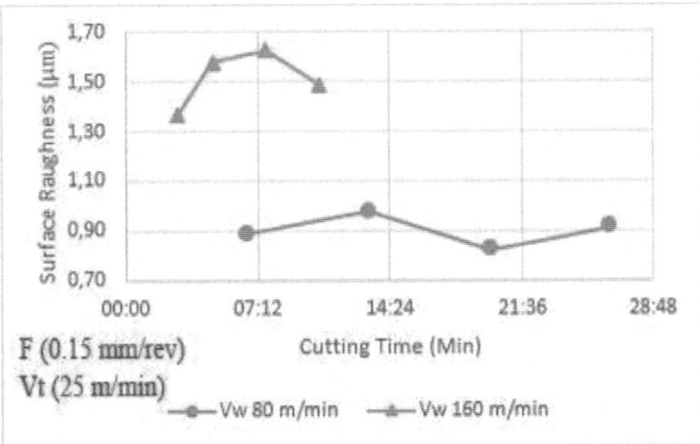

Figure 7. Graph speed the workpiece (Vw) of the surface roughness values.

Figure 7 is a comparison of the effect of speed of workpiece on the value of surface roughness. Where is concluded that the value of surface roughness was lowest in the speed of workpiece of 80 m/min and the highest surface roughness values occur at the speed of workpiece 160 m/min. As stated by Fariza [15] in his study, the greater the speed of workpiece then the resulting roughness value will be higher. For a comparison of the surface profile can be seen in Figure 8.

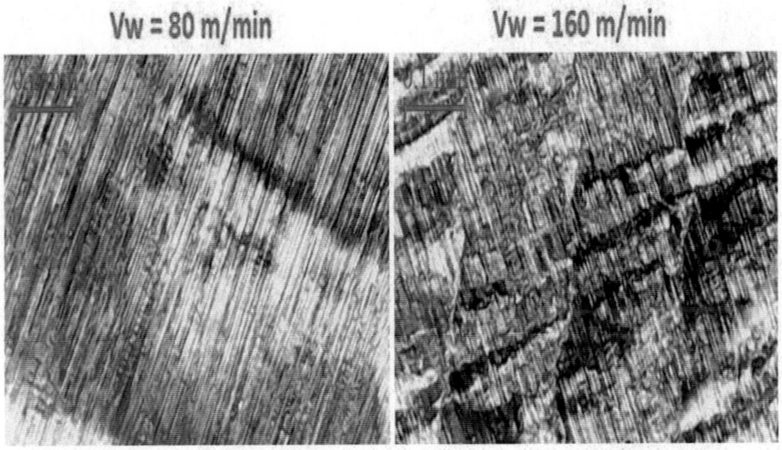

Figure 8. Comparison between the velocity profile speed of workpiece 80 and 160 m/min..

4. Conclusions

1. The average value of the lowest surface roughness obtained on the speed of the workpiece (VW) to 80 m/min, cutting speed of rotary tool (Vt) of 50 m/min with feed rate 0.2 mm/rev and depth of cut 0.3 mm. While the average value of the highest surface roughness obtained on the speed of the workpiece (Vw) of 160 mm/min, the cutting speed of rotary tool (Vt) of 50 m/min with feed rate 0.2 mm/rev and depth of cut 0.3 mm.

2. Values of surface roughness is not uniform throughout the machining process, this is due to tool wear is not uniform along the edge of the rotary cutting tool.
3. Values of surface roughness parameters are more influenced by the speed of the workpiece (Vw) and the cutting speed of rotary tool (Vt). Higher speed of the workpiece (Vw) then the surface roughness of the resulting higher. Instead higher cutting speed of rotary tool (Vt) then the value of the resulting surface roughness is lower.
4. Turning process using a rotary tool and air cooling can be implemented in a lathe machining process AZ31 magnesium material for producing a lower roughness values compared to using a silent tool.

References

[1] Ibrahim G A 2014 Analisa Kekasaran Permukaan pada Pemesinan Paduan Magnesium. Jurusan Teknik Mesin. Universitas Lampung. Bandar Lampung
[2] Blawert C, Hort N and Kainer K U 2004 Automotive Applications Of Magnesium And Its Alloys Trans Indian Inst. Met. Vol.57 No. 4 pp. 397- 408
[3] Harun S 2012 Peningkatan Produktifitas dan Pengendalian Suhu Pengapian Pemesinan Magnesium Dengan Sistem Pahat Putar (Rotary Tool System) dan Pendingin Udara (Air Cooling) Universitas Lampung. Bandar Lampung
[4] Mahrudi H, and Burhanuddin Y 2013 Rancang Bangun Aplikasi Thermovision Untuk Pemetaan Distribusi Suhu Dan Permulaan Penyalaan Magnesium Pada Pembubutan Kecepatan Tinggi Jurusan Teknik Mesin Universitas Lampung, Bandar Lampung.
[5] Bruni C, Forcellese A, Gabrielli F, and Simoncini M 2004 Effect Of Temperature, Strain Rate And Fibre Orientation On The Plastic Flow Behaviour And Formability Of AZ31 Magnesium Alloy Department of Mechanics Università Politecnica delle Marche Via Brecce Bianche Ancona 60131 Italy
[6] Su Y, He H, Li L, Iqbal A, Xiao, M H Xu, S Qiu, B G 2007 Refrigerated Cooling Air Cutting Of Difficult-To-Cut Materials International Journal of Machine Tools & Manufacture 47 927 – 933
[7] Doni A R 2015 Analisa Nilai Kekasaran Permukaan Paduan Magnesium AZ31 Yang Dibubut Menggunakan Pahat Potong Berputar. Tugas Akhir Universitas Lampung.
[8] Andriyansyah 2014 Pengaruh Parameter Pemotongan Terhadap Kekasaran permukaan Dalam Pengefreisan Magnesium Tersuplai Udara Dingin. Tugas Akhir. Universitas Lampung.
[9] Akhmad Isnain P, Gusri A I, and Burhanuddin Y 2013 Unjuk Kerja Vortex Tube Cooler Pada Pemesinan Baja St41. Jurnal FEMA. Universitas Lampung
[10] Paryanto R, an Tony S U 2012 Effect Of Air Jet Cooling On Surface Roughness And Tool Wear Jurnal Teknosains Universitas Diponegoro
[11] Stejernstoft T 2004 Machining of Some Difficult-to-Cut Materials with Rotary Cutting Tools Stockholm The Royal Institute of Technology KTH.
[12] Novriadi D 2016 Rancang Bangun Sistem Pahat Putar Modular (Modular Rotary Tool System) Untuk Pemesinan Alat Kesehatan Ortopedi Universitas Lampung Bandar Lampung
[13] Benardos P G, and Vosniakos G C 2003 Predicting surface roughness in machining: a review International Journal of Machine Tools & Manufacture 43 833 – 844.
[14] Salgado D R, Alonso F J, Cambero I, and Marcelo A 2009 In-process surface roughness prediction system using cutting vibrations in turning International Journal of Advance Manufacturing Technology 43 40 – 51.
[15] Fariza F 2016 Evaluasi Dan Analisa Kinerja Sistem Pahat Putar Modular Untuk Pemesinan Peralatan Kesehatan Ortopedi Berbasis Material Titanium 6al-4v ELI Tugas Akhir. Universitas Lampung

Amino acid and fatty acid compositions of Rusip from fermented Anchovy fish (*Stolephorussp*)

D Koesoemawardani[1], **S Hidayati** and **Subeki**

Department of Agricultural Technology Faculty of Agriculture Lampung University
Jl. Prof. Sumantri Brojonegoro No. 1 Bandar Lampung
Lampung 35145 Indonesia, Phone (0721)-781823, fax (0721)-702767

Email : dyahthp@gmail.com

Abstract. Rusip is a typical food of Bangka Belitung Indonesia made from fermented anchovy. This study aims to determine the properties of chemistry, microbiology, composition of amino acids and fatty acids from fermented fish spontaneously and non spontaneously. Spontaneous rusip treatment is done by anchovy fish (*Stolephorus*sp) after cleaning and added salt 25% (w/w) and palm sugar 10% (w/w). While, non-spontaneous rusip is done by adding a culture mixture of *Streptococcus, Leuconostoc,* and *Lactobacillus* bacteria 2% (w/v). The materials are then incubated for 2 weeks. The data obtained were then performed t-test at the level of 5%. Spontaneous and non-spontaneous rusip fermentation process showed significant differences in total acid, reducing sugar, salt content, TVN, total lactic acid bacteria, total mold, and total microbial. The dominant amino acid content of spontaneous and non-spontaneous rusip are glutamic acid and aspartic acid, while the dominant fatty acids in spontaneous and non-spontaneous rusip are docosahexaenoic acid, palmitic acid, oleic acid, arachidonic acid, stearic acid, eicosapentaenoic acid, palmitoleic acid, and myristic acid.

1. Introduction

Rusip is typical of Bangka Belitung Indonesian food made from fermented anchovy [18] [1]. Besides fish, other ingredients are added in the manufacture of salt rusipare 25% and 10% of palm sugar, then fermented for 1-2 weeks. Rusip have a light brown color to dark gray, sweet, sour and salty, typical aroma of fish, and generally consumed as a condiment or mixed dishes [18].[17]states that the value of TVN rusip still high, so it needs further in detail processing with the addition of lactic acid bacteria culture. The results showed changes in microbiological and chemical properties rusip during the fermentation process. The fermentation process is non-spontaneous rusip have chemical and microbiological properties better than spontaneous fermentation [19]. Changes in chemistry, microbiology, amino acids, fatty acids, as well as sensory may occur during the process of fermentation of fish due to processing and differential treatment[3],[8],[15],[16],[21], [33],[35].Lactic acid bacteria have a big impact on the nutritional quality and shelf life of the products of fermentation [2].This study aims to determine the nature of chemistry, microbiology, as well as the composition of amino acids and fatty acids rusip from fermented anchovy spontaneous and non-spontaneous.

[1] To whom any correspondence should be addressed.

2. Materials and methods

Spontaneous rusip processing is done by anchovy *(Stolephorussp)* after cleaning and then added salt as much as 25% (w / w) and palm sugar 10% (w / w). While the non-spontaneous rusip done by adding mixed cultures of bacteria *Streptococcus, Leuconostoc,* and *Lactobacillus* as much as 2% (w/v). The next ingredient incubated for 2 weeks [19].

Observations chemical and microbiological include: total acid [4], protein content [4], fat [4], reduced sugar [9], the water content [4], salinity [7], pH [4], TVN [4], total lactic acid bacteria [11], total microbial [11], total mold [11], the amino acid composition (HPLC), and the fatty acid composition (GC-MS). The data were then performed t test at 5% level.

3. Results and discussion

Spontaneous and non-spontaneous rusip production was performed with the same incubation period of 2 weeksThe observation of chemical and microbiological properties of spontaneous and non-spontaneous rusip can be seen in Table 1.

Table 1. Characteristics of chemical and microbiological properties of Rusip

parameters	Rusip	
	Spontaneous	Non spontaneous
Total acid (%)	2.28[a]	3.60[b]
protein content (%)	14,45[a]	13,64[a]
Fat content (%)	0,38[a]	0,44[a]
Sugar reduction (%)	6.42[a]	5.01[b]
Moisture content (%)	66,52[a]	67,27[a]
Salinity (%)	24,94[a]	19,49[b]
pH	5,61[a]	5,74[a]
TVN (mg N / 100 g)	31.76[a]	20.19[b]
Total lactic acid bacteria (log cfu /g)	9.66[a]	12.09[b]
Total mold (log cfu /g)	9.66[a]	5.81[b]
Total microbe (log cfu /g)	9.84[a]	8.42[a]

Description: Figures in row followed by the same letters mean no significant difference in the level of 5%

The results showed that the protein content of spontaneous rusip(14.45%) and non-spontaneous (13.64%) more than in the previous study by 12.81% [15]. Table 1 shows the differences in the characteristics of the spontaneous and non-spontaneous rusip due to differences in the total amount of lactic acid bacteria in rusip. Rusip non spontaneous containing lactic acid bacteria 12:09 log cfu / g higher than the spontaneous rusip 9.66 log cfu/g. This is because the process of making non spontaneous rusip added mixed cultures of lactic acid bacteria *Streptococcus,*Leuconostoc,and*Lactobacillus.*Thus, the lactic acid bacteria dominate the growth during the fermentation process rusip[19]. This is in line with research [1] which states that the fermentation rusip with the addition of mixed cultures of lactic acid bacteria produce a total lactic acid bacteria higher compared to single culture. the high total lactic acid bacteria fermentation rusip give different characteristics of the resulting product as the value of total acid, salinity, TVN, total fungi, and total microbes.

Rusip non spontaneous containing acid total 3.60% higher than rusip spontaneous 2.28%. This is because the mixed culture of lactic acid bacteria added to the fermentation process can increase the growth rate and produce lactic acid through heterofermentative and homofermentative processes [20]One of the lactic acid bacteria added to the process of making rusip is Leuconostocsp which is a heterofermentative group of bacteria [23]. Leuconostocsp bacteria can grow well for 15 days of fermentation process [22]. Addition of lactic acid bacteria culture to the sindol which is a fish

fermentation product from India gives changes of organoleptic, physicochemical, and microbiological properties of the resulting product [5].
Non-spontaneous rusip has a salt content 19.49% lower than spontaneous rusip 24.94%. The dominant growth of lactic acid bacteria since the beginning of the fermentation process is able to suppress the growth of bacterial decomposition and pathogens [30].Thus, the use of salt by lactic acid bacteria in spontaneous rusip more than spontaneous rusip[20].The high lactic acid bacteria in non spontaneousrusip also affects the decrease of TVN, total mold, and total microbial value[19] [20].The addition of cultures of lactic acid bacteria to fermentation products will produce metabolite products that are antimicrobial, increase sensory, nutrient, and shelf life of the resulting product[2], [31], [32].

Table 2.Amino acid compositions ofRusip

Amino acids (%)	Rusip	
	Spontaneous	Non spontaneous
aspartic acid	1,39 a	1.41 a
glutamateacid	2,07 a	2.06 a
serine	0,56 a	0 , 56 a
histidine	0,31 a	0.28 a
glycine	0,57 a	0.61 a
threonine	0.55 a	0.51 a
arginine	0.55 a	0.53 a
alanine	0.91 a	0.90 a
tyrosine	0.43 a	0.45 a
methionine	0.33 a	0.07 b
valine	0.66 a	0.66 a

Description: Figures in row followed by the same letters mean no significant difference in the level of 5%

Table 2 shows that the amino acid composition of spontaneous and non spontaneousrusip is complete both essential and non essential amino acids. The results showed that amino acid composition between non spontaneous and spontaneous rusip was not significantly except for methionine and lysine. Dominant amino acid in rusip is glutamate acid and aspartate acid. [15] mentioned that glutamate acid, aspartate acid, and lysine are the dominant amino acids in the rusip with higher protein digestibility values than budu.[12]stated that the addition of mixed cultures to the fish fermentation can increase the amino acid composition. The composition of amino acids is also similar to other fish fermentation products such as oyster sauce[3], fish sauce [14] and[6], sauce oyster [13],fermented shrimp [10],budu and Rusip[15].
The dominant amino acid composition in fish fermentation product is glutamate acid. The difference in the number of each amino acid is caused by the difference in the free amino acid balance produced by autolysis and microbial activity [14]. Amino acids significantly affect the taste of Yu-lu fish fermentation products [25],[14]. Glutamate acid gives the aroma of meaty[14]. Amino acids glycine, alanine, serine, and threonine give a sweet taste, while valine, phenylalanine, and histidine give a bitter taste [24].[27]states that glutamic acid is closely related to "umami", which is produced as a relatively stable amino acid in the absence of secondary decomposition during fermentation.

Table 3.Fatty acidcompositions of Rusip

Fatty acids (%)	Rusip	
	Spontaneous	Non spontaneous
caprylic acid, C8: 0	0.05 a	0.05 a
capric acid, C10: 0	0.03 a	0.04 a
lauric acid, C12: 0	0.34 a	0.30 a
acid tridecanoat, C13: 0	0.11 a	0.11 a
acid myristate, C14: 0	3.21 a	2.87 a
acid pentadekanoat, C15: 0	0.61 a	0.56 a
acid palmitate, C16: 0	10.95 a	9.62 a
acids palmitoleic, C16: 1	3.26 a	2.89 a
acid heptadekanoat, C17: 1	0.82 a	0.71 a
acid cis-10-heptadecanoat, C17: 1	0.16 a	0.15 a
stearic acid, C18: 0	4.12 a	3.52 a
acid elaidic, C18: 2n6c	0.08 a	0.06 a
oleic acid, C18: 1n9c	4.66 a	4 , 15 a
linoleic acid, C18: 2n6c	0.67 a	0.61 a
arakidat acids, C20: 0	0.31 a	0.26 a
g-linolenic acid, C18: 3n6	0.15 a	0.13 a
acid cis-11-eicosenoat, C20: 1	0.08 a	0.07 a
acids linolenic, C18: 3n3	0 , 40 a	0.25 a
heneicosanoat acids, C21: 0	0.09 a	0.08 a
acid cis-11,14- eicosedienoat, C20: 2	0.10 a	0.10 a
behenic acid, C22: 0	0.43 a	0.37 a
acid cis-8,11,14-eicosetrienoat, C20: 3n6	0.13 a	0.13 a
erucat acid, C22: 1N9	0.02 a	0.03 a
acid cis-11,14,17-eicosetrienoat, C20: 3n3	0.04 a	0.05 a
arachidonic acid, C20:4n6,	4.53 a	4.01 a
tricosanoat acid C23: 0	0.08 a	0.12 a
acid lignocerat, C24: 0	0.79 a	0.74 a
acid cis-5,8,11,14,17-eicosapentaenoic, C20: 5n3	4.05 a	3.37 a
nervonat acid, C24: 1	0.23 a	0.19 a
cis-4,7,10,13,16 Acid, 19-docosahexaenoat, C22: 6n3	16.62 a	14.77 b

Table 3 shows that fatty acids dominant in rusip are docosahexaenoic acid, palmitic acid, oleic acid, arachidonic acid, stearic acid, eicosapentaenoate, palmitoleic acid, and myristic acid. The results showed that the composition of spontaneous and non-spontaneous fatty acids was not significantly different except docosahexaenoic acid. The content of docosahexaenoic acid in spontaneous rusip is 16,62% higher than non spontaneous 14,77%. The high content of docosahexaenoic acid causes a distinctive aroma in fish sauce[6].[34] mentioned that the content of docosahexaenoic acid in the bekasem from anchovy fish is lower than the fresh anchovy. The fatty acid docosahexaenoate decreases after fermentation by addition of bacterial culture of lactic acid. This can happen because of the decrease in salt concentration in the rusip. Non-spontaneous rusip has a lower salinity than spontaneous rusip due to the growth of lactic acid bacteria utilizing salt from 25% to 19.49[20].

The fatty acid with one double bond (MUFA) or more than one double bond (PUFA) is an essential fatty acid commonly found in fish and is needed by the body [29].[26]states that free radicals can cause cellular damage in the body, so antioxidant compounds are needed. Antioxidant compounds can prevent damage by inhibiting the production of major catalysts in lipid peroxidation. The Maillard Reaction products (MRPs) formed during the fermentation process can increase antioxidant activity [28].[3]states that shrimp paste fermented with 25% salt addition for 36 days indicates antioxidant activity. Rusip is a fermented product of fish fyang is also likely to show the antioxidant activity obtained from fatty acid content.

4. Conclusions

The fermentation process rusip spontaneous and non-spontaneous show a real difference to the total acid, reducing sugar, salt content, TVN, total lactic acid bacteria, total fungi, and total microbes. The amino acid content dominant in rusip spontaneous and non-spontaneous are glutamic acid and aspartic acid, while the fatty acids predominantly on rusip spontaneous and non-spontaneous are docosahexaenoatacid, palmitic acid, asam oleic asamarachidonic, stearic acid, eicosapentaenoic acid, palmitoleic acid and myristic acid.

5. References

[1] Afriani 2008 Quality and potency of curd (dadih) as source additional income of buffalo farmer in Kerinci*JIIP***11** (3) 144

[2] Adesokan I A, Odetoyinbo B B, Ekanola Y A, Avanrenren R E and Fakorede S 2011 Production of Nigerian Nono using lactic starter cultures Pakistan Journal of Nutrition 10 (3) 203.

[3] Anggo A D, Ma'ruf W F, Swastawati F and Rianingsih L 2015 Changes of amino and fatty acids in anchovy (*Stolephorus*sp) fermented fish paste with different fermentation periods *Procedia Environmental Sciences***23** 58.

[4] Association of Official Analytical Chemist 2005*Official Methods of Analysis of The Association of Official Analytical Chemist. Arlington* (Virginia USA Association of Official Analytical Chemists Inc) p 1141.

[5] Das M K 2014 Starter culture development of sindol-a fermented fish product of Assam. International Journal of Agricultural Science and Research **4** (2) 107.

[6] Dincer T, Cakli S, Kilinc B and Tolasa S 2010 Amino acid and fatty acid composition of fish sauce *Journal of Animal and Veterinary Advance***9** (2) 311.

[7] FortunyM, SilvaE B,FilhoA C, MeloR LF V,NeleM, CoutinhoR CC, SantosA F 2008 Measuring Salinity in crude oils: Evaluation of methods and an improved procedure*Fuel***87** **(7)** 1241

[8] Giri A, Nasu M, and Ohshima T 2012 Bioactive properties of Japanese fermented fish paste, fish miso, using koji inoculated with *Aspergillusoryzae International Journal of Nutrition and Food Sciences***1**(1)13.

[9] Gusakov A V, Kondratyeva E G and Sinitsyn A P 2011 Comparison of TwoMethods for Assaying Reducing Sugars in the Determination of Carbohydrase Activities *International Journal of Analytical Chemistry* 1

[10] Hajeb P and Jinap S 2012 Fermented shrimp products as source of umami in Southeast Asia *J Nutr Food Sci*1.

[11] Harrigan W. Fand McCance M E 1966 *Laboratory Methods in Microbiology* (London and New York: Academic Press) p 357

[12] Ijong F G, and Ohta Y1995 Characteristics of bakasang fermented with lactic acid bacteria-mixed culture *Journal of the Faculty of Applied Biological Science***34** (3)1.

[13] Je J-Y, Park P-J, Jung W-K, and Kim S-K 2005 Amino acid changes in fermented oyster (*Crassostreagigas*) sauce with different fermentation periods *Food Chemistry***91**(1)15.

[14] Jiang J-J,ZengQ-X, Zhu Z-W, Zhang L-Y 2007 Chemical and sensory changes associated Yu-lu fermentation process – a traditional Chinese fish sauce *Food Chemistry***104** (4)1629.

[15] Khairi I N, Huda N, Abdullah W N W and Al-Karkhi A F M 2014 Protein quality of fish fermented product: Budu and Rusip. *Asia Pacific Journal of Sustainable Agriculture Food and Energy***2** (2) 17.

[16] Kim Y-B, Choi Y-S, Ku S-K, Jang D-J, Hasnah and Moon K B 2014 Comparison of quality characteristics between belacan from Brunei Darussalam and Korean shrimp paste *Journal of Ethnic Foods***1** 19.

[17] Koesoemawardani D[a] 2007 *Conference proceedings of results of research and community service*(Lampung University 6-7 September 2007) Research institutions Lampung University p 304

[18] KoesoemawardaniD[b] 2007 Sensory analysis Rusip from Sungai Liat-Bangka *JurnalTeknologidanIndustriHasilPertanian***12** (2) p 36.

[19] Koesoemawardani D, Rizal S, and Tauhid M 2013 Microbiological and chemical changes of Rusip during fermentation *Agritech***33** (3) 265.

[20] Koesoemawardani D, Yuliana N and Sari M 2015 *Proceedings of the national seminar and exhibition of food products 2015 "Innovation for Strengthening the role of industrial technology Toward Fulfillment acceleration of national food"* (Semarang 20-21 October 2015) The association of Indonesian food technologists PATPI p 593.

[21] Koo O K, Lee S J, Chung K R, Jang D J, Yang H J and Kwon D Y 2016 Korean traditional fermented fish products: jeotgal*Journal of Ethnic***3** 107.

[22] Kurniati Y 2006 *Eassy*Fakulty of Agriculture Lampung University (Bandar Lampung: Lampung University)

[23] Lahtinen S, Ouwehand A C, Salminen S, Wright A V 2011 *Lactic Acid Bacteria: Microbiological and Functional Aspects* (London:Taylor and Francis CRC Press) p 798.

[24] Liu, Pei-zhi 1989 The umami of Yu-lu*Food Science(Chinese)***4** 37.

[25] Lopetcharat K, Choi Y J, Park J W and Daeschel M A 2001 Fish Sauce Products and Manufacturing *Food Reviews International***17** 65.

[26] Mohammed M T, Kadhim S M, Jassimand and Abbas S I 2015 Free radicals and human health. *International Journal of Innovation Sciences and Research***4** (6) 218.

[27] Mizutani, T., Kimizuka, A., Ruddle, K., Ishige, N. 1992. Chemical components of fermented fish products.Journal of Food Composition and Analysis. Vol. 5 (2):152-159

[28] Noguchi N and Niki E 1999 *Chemistry of active oxygen species and antioxidants* in *A M Papas, Antioxidant status, diet, nutrition and health* (Florida: CRC Press) p 3

[29] Newton D E 2007 Food Chemistry (New York:An imprint of Infobase Publishing) p 225

[30] Rosma A, Afiza TS, Wan Nadiah, WA, Liong MT and Gulam R R A 2009 Short communication microbiological, histamine and 3-MCPD contents of Malaysian unprocessed 'budu'. *International Food Research Journal***16** 589.

[31] Savadogo A, Ouattara C A T, Bassole I H N and Traore A S 2004 Antimicrobial activities of lactic acid bacteria strains isolated from Burkina Faso fermented milk *Pakistan Journal of Nutrition***3** (3) 174.

[32] Sholeva Z, Stefanova S and Chipeva V 1998 Screening of antimicrobial activities among *Bulgarianlactobacilli* strains *Journal of culture collections***2** 15.

[33] Skåra T, Axelsson L, Stefânsson G, Ekstrand B and Hagen H 2015 Fermented and ripened fish products in the northern european countries *Journal of Ethnic Foods***2** 18.

[34] Soetrisno U SS and Apriyantono RRS 2005 Nutrient quality and the safety bekasam as fermented small fish by addition of selected lactic acid bacteria or spontaneous fermentation. PGM**28** (1)38.

[35] Thapa N 2016 Ethnic fermented and preserved fish products of India and Nepal *Journal of Ethnic Foods* **3** 69.

IOP Conf. Series: Materials Science and Engineering **344** (2018) 012006 doi:10.1088/1757-899X/344/1/012006

Isolation and characterization of formacell Lignins from oil empty fruits bunches

S Hidayati[1,*], A S Zuidar[1], W Satyajaya[1], Murhadi[1], and D Retnowati[1]
Department of Agro-industrial Technology, Universitas Lampung
Jl. Prof. Soemantri Brojonegoro Street, No. 1 Bandar Lampung, Lampung, 35145 Indonesia

E-mail: srihidayati.unila@gmail.com

Abstract. Lignin is the largest component in black liquor, it is about 46% of solids total and can be isolated by precipitation using acid and base method. The purpose of this study was to get the best NaOH concentration to produce lignin with yield, solids total content, metoxyle lignins content, weights equivalent of lignin in the black liquor by pulping *formacell* process from oil empty fruits bunches. This study was done with isolation lignin process in black liquor used by NaOH concentration were 5%, 10%, 15%, 20%, 25%, and 30% from volume black liquor and then precipitationed for 10 hours. The result of this research showed the isolation of lignin with NaOH concentration 30% get the pH 5,42%, yield of lignin was 5,67%, solids black liquor total was 65,11%, levels of metoxyle lignin 14,61%, and equivalent weights of lignin was 1787,23. The result of FT-IR identifications of isolates lignin in NaOH concentration 25 and 30% showed a pattern infiltration spektro IR that almost a part that have the same infiltration at the wave numbers that showed lignin had one of the rings lignin was guaiasil, it was building blocks of non wood lignin.

1. Introduction

Black liquor is dark liquid from byproduct of the process that transforms wood into pulp, which is then dried to make paper. Black liquor contains lignin, which is the material in trees that binds wood fibers together and makes them rigid, and which must be removed from wood fibers to create paper. Black liquor is a problem in the pulp and paper industry because it is highly polluting the environment [1], and the rejection of this effluent in nature without any treatment is responsible for serious damage to the environment and constitutes a threat for human health. The black liquor consists of almost all the inorganic chemicals used in pulping and organics in the form of dissolved wood constituents [2]. Lignin in the black liquor, is a mixture of polyphenolic compounds with complex chemical structure that resists to conventional biological treatment processes due to their non-biodegradable nature. The presence of dark colored lignin cause s limiting ligh trasmission in aquatic plants. Lignin in black liquor also contains aliphatic acids, acids, resins and polysaccharides increase the burden of oxygen demand in water. Component of black liquor is lignin which is about 46% of its total solid [3], therefore isolation and separation of lignin is more likely. Black liquor can be a source of lignin feedstock[1][4] [5] [6][7][8][9]. Lignin structure and physico chemical properties depends on isolation method and raw material source [10][11]. Black liquor can be isolated alkaline method or acid method to obtain pure lignin. The research trend is an attempt to harness lignin into more useful chemicals [12]. Commercial use of lignin maybe used carbon fiber, adhesive, polyurethane, polyester, bioplastic, and bio oil for petroleum mixtures of fossils [5,9,13,14,15]. The presence of phenolic ring inside lignin can be utilized for phenolic formaldehyde resins [4,5,6,7,16]. Beside lignin can be used as a filler and reinforcing phases for polymer blends [17,18,19,20,21,22]. The advantage of lignin is to have a high number hidroxyl group can be used for the production of polyol and either through direct utilization or after chemical modification for the production of certain polymers such as polyurethane [23,24,25].

In the world it is attempting to utilize black liquor as a source of lignin feedstock. Some lignin isolation methods include: 1). Klason Method; 2). The Björkman method is also called "Milled Wood Lignin / MWL"; 3). CEL Method, Cellulolythic Enzyme Lignin or "lignin of cellulolytic enzymes"; 4). Technical Lignin Isolation Method, namely the lignin isolation method of residual pulp liquor. The process of lignin isolation from black liquor can use acids such as H_2SO_4, phosphoric acid (H_3PO_4), or HCl [26, 27]. While the basic method can use bases such as NaOH and KOH [28]. In this step about 75% of the lignin is precipitated as the sodium salt. Finally, lignin is separated through a screening process Lignin hydrolysis contains solid lignin residues and large amounts of unhydrolyzed cellulose [27].

The characterization of softwoods and hardwoods black liquor from the pulping of non-wood fibers such as reed canary grass, sugarcane baggasse, wheat straw have been investigated lately. However, none works had been done on the black liquor from black liquor formacell EOFB pulping process. Formacell is one of the organosolved pulping methods that use acetic acid and formic acid as cooking solution [29,30,31,32]. The pressure and temperature can be lower when formic acid is used in pulping compared to those used in alcohol or acetic acid pulping. The advantage of organic acid lignin is an optimal feedstock for many value-added products, due to its lower molecular weight and higher reactivity and organic acid pulping is the retention of silica on the pulp fiber that facilitates the efficient recovery of cooking chemicals.

One research reported that only concentrated on lignin isolation from oil palm black liquor and minor on characterization of lignin [22]. Kraft Black liquor from EOFB resulted optimum lignin precipitation was obtained at pH 2 and comparably result could be obtained at pH 4.5 followed by 1 hour heating by addition of anthraquinone (AQ) in the pulping process showed an improved carbohydrates stabilization and better delignification. NaOH extraction was used in order to obtain lignin for use production of polyurethane [34]. The aim of this research is to know the effect of NaOH concentration in isolation process and lignin characterization on the formacell black liquor from empty palm oil bunches. The use of NaOH in the isolation process can be used to precipitate lignin from black liquor.

2. Methodology

2.1. Raw material

The materials used in this research are black liquor from EOFB formacell pulping, NaCl, NaOH, HCl, water, aquades, ethanol, KBr, filter paper, phenolpthalin indicator, aluminum foil, and clip. The tools used in this research are acid cabinet, acid stove, aluminum cup, digital scales, porcelain cup, oven, desiccator, beaker glass, erlenmeyer, pH meter, dropper drop, volume pipette, stirrer, measuring cup, Funnel, centrifuge, stirrer, titration device and FT-IR (Fourier Transform Infra-Red Spectroscopy) spectrophotometer and SEM (Scanning Electron Microscopy)

2.2. Research methods

This research begins with the process of purifying liquid waste from pulp filtration (black liquor) by using lignin isolation method which refers to isolation method developed by Lubis [35] as shown in Figure 1. A total of 100 ml of filtered black liquor (filtrate) Precipitated lignin by stirring and adding by NaOH with concentrations of 5%, 10%, 15%, 20%, 25%, and 30% (percent v / v) of black liquor volume which then homogenized using a stirrer. Stirring process is done slowly then measured pH of each treatment, then subsequently settled for 10 hours for perfect deposition. The lignin deposit is separated from black liquor using centrifuge (2500 rpm, 20 min). The observations were: pH value, total solids content in black leachate EOFB, lignin yield, lignin methoxyl content, lignin equivalent weight, and lignin analysis with FT-IR spectrophotometer and Scanning electron microscopy (SEM) has been most useful technique for quantification of lignin structural details.

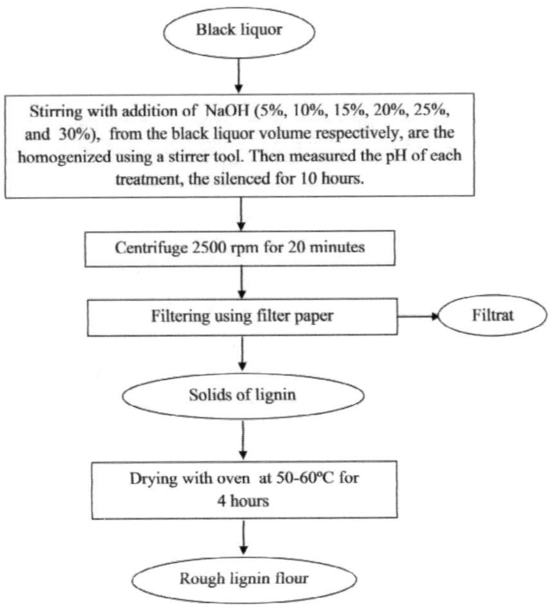

Figure 1. The modified lignin isolation diagram [35]

3. Results and discussion

3.1. Yield of Lignin

The yield was also found to increase for black liquor with a higher total dry solid (TDS) content. The average value of lignin isolate content from various concentrations ranged from 1.48% - 5.67% (Figure 2). The result of variance analysis showed that the NaOH concentration teatment had highly significant effect on yield of lignin.

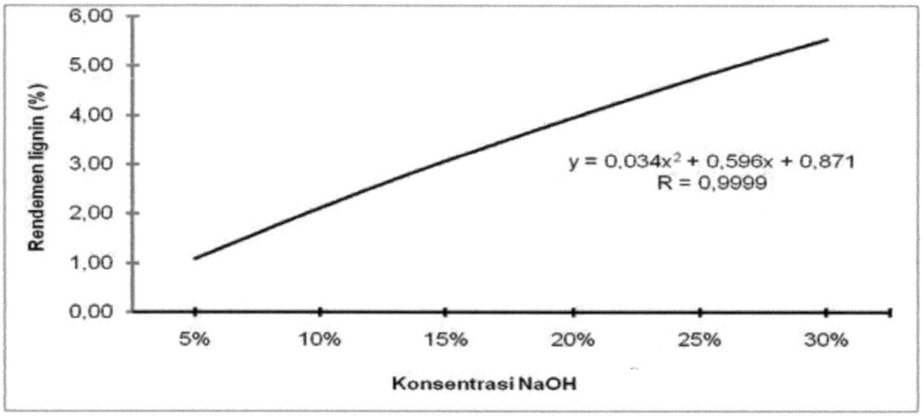

Figure 2. Effect of NaOH concentration on lignin yield

The coagulation of lignin from black liquor occur because the protonation of ionized phenolic group on the lignin molecul. The protonation of phenolic group reduces the electristatic repulsive forces between lignin molecules [36,37] which then become less hydrophylic, leading to precipitation. The highest yield of lignin isolate was found at 30% NaOH concentration with an average of 5.67%. The yield of precipitated activated lignin was influenced by NaOH concentration [38]. The yield of isolate lignin due to the influence of addition factor of NaOH concentration give effect to yield yield of lignin isolate [39]. Yield/equilibrium in the precipitation step of lignoboost process is influenced by diferrent proces condition, i.e, the pH, temperatur and ion strength of black liquor) [41][42]. In this study showed that the yield of lignin isolate tended to increase according to the addition of NaOH concentration as lignin sedimentation solution. The increase of lignin isolate content in the deposition process using NaOH with concentration of 5%, 10%, 15%, 20%, 25%, and 30%, due to the more basic precipitation process where the higher the NaOH concentration the higher the level of base, It is suspected that there is an increasing condensation reaction in the lignin-making units such as para-koumaril alcohol, coniferyl alcohol, and sinapyl alcohol, which will initially undergo repolymerization and form larger molecules of lignin polymer. The higher the concentration, the higher the lignin deposits produced [42]. It is suspected that because of the higher concentration of added NaOH, the OHO ions are consumed by acetyl groups of wood shale during cooking, so the OH^- ion not only dissolves lignin but dissolves other non-lignin components [39]. In this study, the highest lignin precipitate was obtained at an average pH of 5.42 with a 30% NaOH concentration. Increasing the concentration of NaOH results in an increase in pH value due to the addition of base or high alkalinity of the substance increases many OH-ions in water. The greater the pH value of a substance the stronger the degree of base The strength of base can be determined from the scale of basicity which is also expressed by pH value [28].

3.2. Lignin Methoxyl Levels

Methoxyl lignin content in this study has an average value ranges between 14.61 - 20.77% (Figure 3). The result of variance analysis showed that the NaOH concentration treatment had very significant effect on lignin methoxyl content.

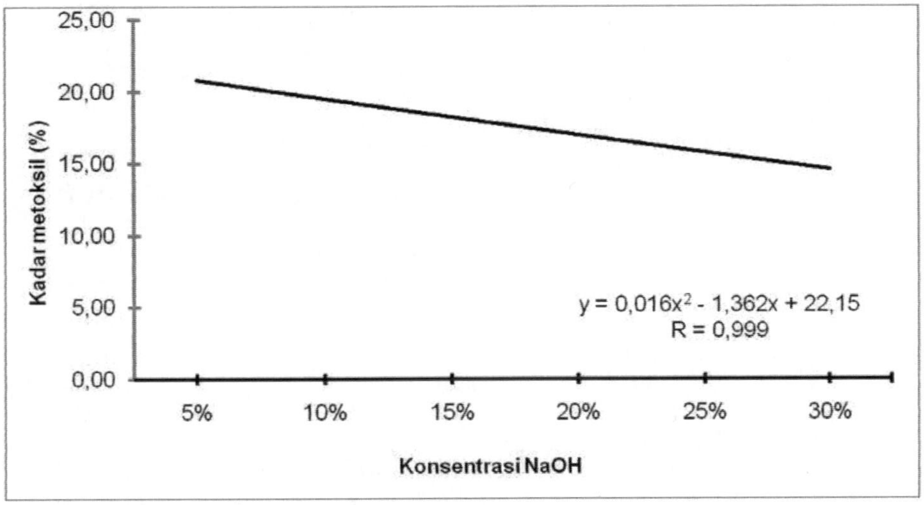

Figure 3. Effect of NaOH concentration on methoxyl lignin levels

The highest levels of methoxyl lignin are present in the addition of 5% NaOH concentration with an average of 20.77%. While the lowest methoxyl lignin levels were found in the addition of 30% NaOH concentration with an average of 14.61%. In general, high methoxyl levels will inhibit lignin reactivity during resin use. Lignin contains phenolic hydroxyl groups which are mostly bonded with adjacent propane phenyl units, allowing the occurrence of lignin bonds with formaldehyde similar to those of a reaction between phenol and formaldehyde [43]. However, in its use as a raw material of lignosulfonate (surfactant), lignin with high methoxyl content is more advantageous because the more -OCH 3 groups contained in lignin, the lignin is increasingly dissolved in water. Such properties are indispensable to lignosulfonate feedstocks.

Meanwhile, low methoxyl levels are suspected because some of the methoxyl groups are degraded and turned into other compounds due to the overuse of acid or base. In alkaline process, phenolic hydroxyl groups are generated by hydrolysis of β-O-4 bond. Low metoxyl levels are caused by changes in methoxyl groups to methyl mercaptan, methyl sulfide, and dimethyl disulfide According

[44]. According to Fengel and Wegener [45], this low metoxyl value is probably caused by the influence of harsh chemicals during isolation, causing the lignin structure to undergo many changes. Low lignin methoxyl levels can be utilized in their use as an adhesive, lignin with a lower methoxyl content is more advantageous than high methoxyl levels, since lignin with low methoxyl content is easier to form gel [46].

3.3. **Lignin's Equivalent Weight**

The average value of the lignin equivalent weight ranges from 1327.64 - 1787.23 (Figure 4). The result of variance analysis showed that the NaOH concentration treatment had very significant effect on the weight of lignin equivalent.

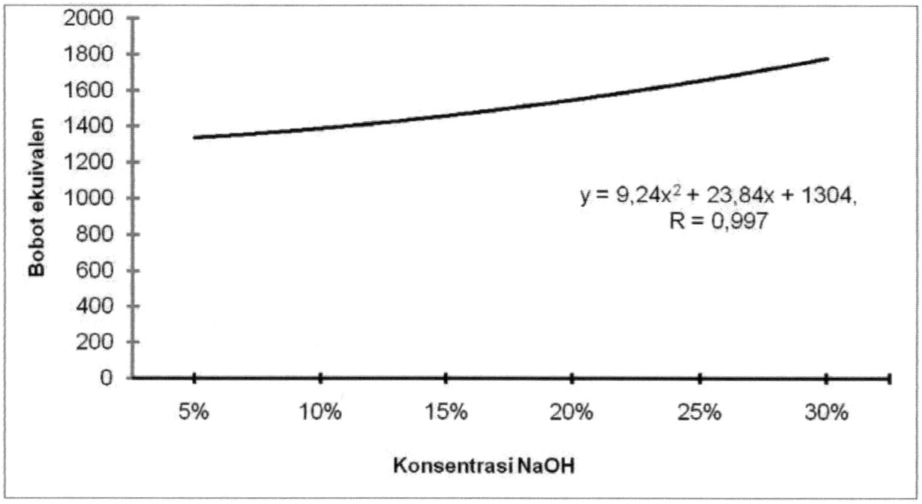

Figure 4. Effect of NaOH concentration on lignin equivalent weight level

At 30% NaOH concentration yields the highest weight of lignin equivalent weight with an average of 1787.23. Meanwhile, the lowest equivalent weight was generated in isolation condition using 5% NaOH concentration with an average of 1327.64. The higher the NaOH concentration as the settling solution, causing the higher lignin equivalent weight. The higher the equivalent weight of a lignin indicates that in lignin insulation, the polymerization proceeds perfectly. In addition, according to Achmadi (1990), the more basic concentrations used at the time of isolation cause lignin to tend to condensate. The condensed lignin composing units form larger molecules so that the weight of the lignin equivalent is increased. The high equivalent weight of EOFB lignin isolates is caused by the lignin structure of EOFB fibers more complex than the lignin structure extracted from the wood. This is due to the complex arrangement of siringil and guaiasil propane units with para-koumaril propane units in EOFB fibers. The standard molecular weight is unknown but is a multiple of 840, the molecular weight of the lignin-making unit [41,47] states the results show that the precipitation yield

IOP Conf. Series: Materials Science and Engineering **344** (2018) 012006 doi:10.1088/1757-899X/344/1/012006

of lignin increases with decreasing pH and temperature and/or with increasing ion strength of kraft black liquor used. The concentration of carbohydrates in lignin decreases with decreasing pH or with increasing temperature, and that an increasing amount of lower molecular weight lignin is precipitated at a higher precipitation yield. This principle is similar to the lignin isolation process using NaOH . According to Santoso [48], the distribution of molecular weight of lignin varies greatly. Lignin is a very complex organic compound, composed of a number of highly variable constituent components, it is difficult to obtain definite molecular weights. By means of chromatographic separation obtained data of lignin molecule weights range of 370-44300 [49]. b. While the results of research [48] states the molecular weight of lignin isolates from black liquor ranged from 304-4010. Based on the equivalent weight obtained in this study, it meets the criteria based on according to Connors *et al.* [49] and Santoso [48].

3.4. **Lignin Analysis with FT-IR Spectrophotometer**

The following is the result of lignin analysis using FT-IR spectrophotometer, using the best sample that is the addition of 25% NaOH concentration is presented in Figure 5 and the addition of 30% NaOH concentration is presented in Figure 6. FTIR was performed to analyze the differentes in the functional groups of the different lignin samples obtained (Namane *et al*, 2016). The FTIR spectrum of commercial kraft lignin (Indulin AT) was utilized as reference.

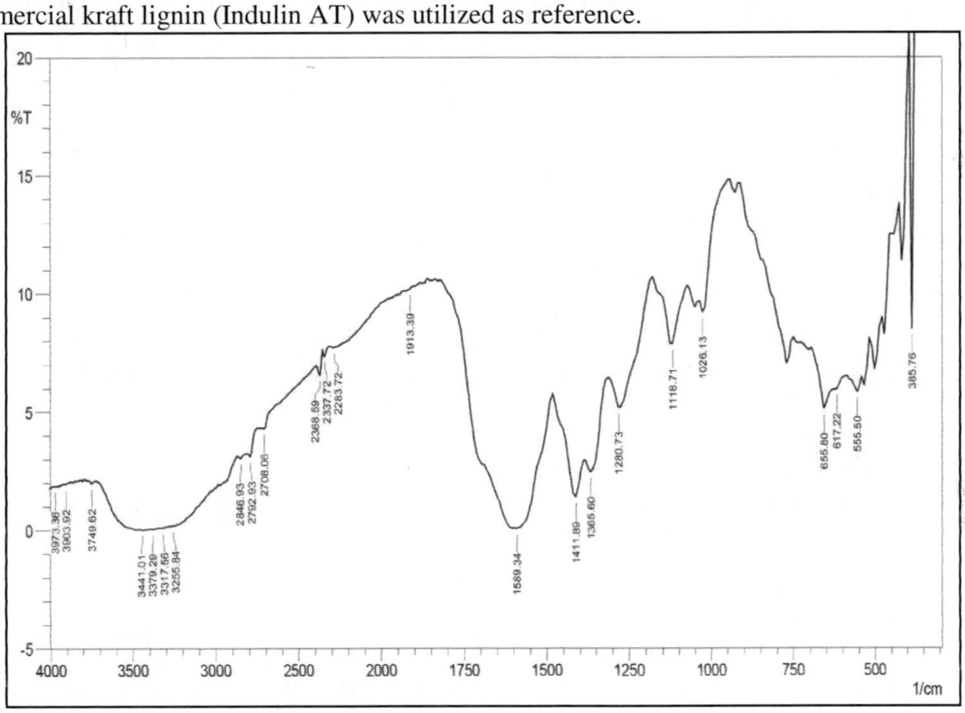

IOP Conf. Series: Materials Science and Engineering **344** (2018) 012006 doi:10.1088/1757-899X/344/1/012006

Figure 5. Results of identification with FT-IR spectrophotometer on lignin isolates from EOFB formacell black liquor results in the addition of NaOH concentration of 25%

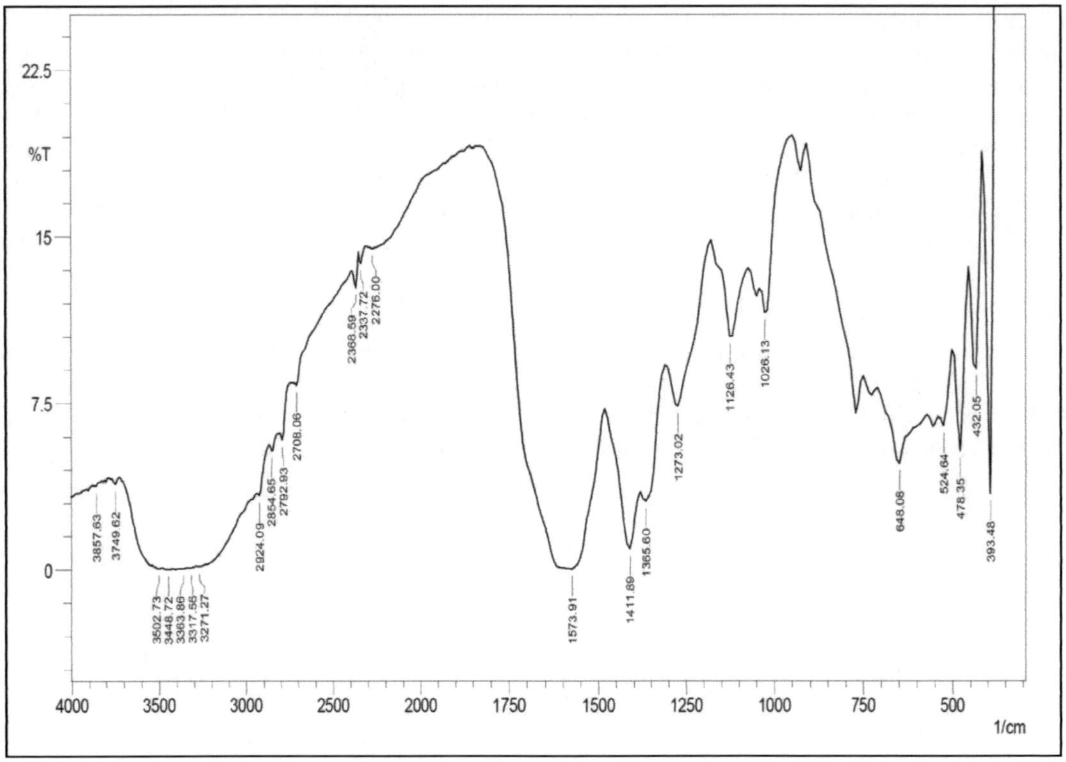

Figure 6. Results of identification with FT-IR spectrophotometer on lignin isolates from EOFB formacell black liquor results in addition of 30% NaOH concentration

Table 1. Fourier transform infrared of two lignin samples [41,50]

No	Lignin isolate NaOH 25%	Lignin isolate NaOH 30%	Indulin-AT	Band Posision (cm-1)*)	Asignment
1	3441.01		3411.36	3450-3400	OH stretch
2	2846.93		2936.36	2940-2820	OH strain on the methyl and methylene groups
3	-		-	1715-1710	The C = O range is unconjugated

					with an aromatic ring
4	-		1668.18	1675-1660	The C = O range is unconjugated with an aromatic ring
5	-		1602.27	1605-1600	Vibration of aromatic rings
6	1589.35	1573.91		1595	Aromatic sceletal vibration, C=O stetching (conjugated)
7	-		1511.36	1515-1505	Vibration of aromatic rings
8	-		1465.91	1470-1460	C-H deformation (asymmetry
9	-		1427.27	1430-1425	Vibration of aromatic rings
10	1365.60		1365.91	1370-1365	In-plane deformation vibration of phenolic OH
11	-		-	1330-1325	Vibration of siringyl ring
12	1273	1273.03	1270.45	1270-1275	Vibration of guaiacyl rings
13	-		1031.82	1085-1030	Deformation of C-H and C-O
14	1026.13	1026.13		1030	C-O of syringyl and guaiacyl ring, C-H bond in guaiacyl ring

The best isolation conditions were lignin isolates with 25% and 30% NaOH concentrations. The lignin isolates were compared with the standard lignin used ie lignin indulin-AT. The purpose of the functional group analysis is to know the functional groups present in lignin from the isolated and standard lignin products used. The two bands at 2900 cm-1 and 2800 cm-1 correspond to methyl (−CH3) and methylene (−CH2) groups. The phenolic OH groups in lignin (band at 1365 cm-1), are produced during chemical process when β-O-4 linkages are cleaved and generate non-etherified hydroxyls. Low intensity of this band in NaOH 30% spectrum shows that small portion of phenolic OH group is generated (due to less β-O-4 linkage cleavage). The intensity of absorption bands at 1268 cm-1 (C-O stretching of guaiacyl ring) spectra is stronger than other lignins because guaiacyl is dominant lignin unit in EOFB. The absorption band range 1030-1025 cm-1 is assigned to deformation vibration of C−H bonds in the guaiacyl ring and also assigned to C-O bonds in both syringyl and guaiacyl [41]. Bands at 1330-1325 cm-1 were attributed to syringil with C-O stretching. Bands at 1217 cm-1 for OPEFB can be attribute to phenolic OH and ether in syringil and guaiacyl [52]. The bands observed at 1030-620 cm-1 were attributed to hemicelluloses and silicates contribution [51].

Lignin is a complex polymer synthesized mainly from three hydroxycinnamyl alcohols differing in their degree of methoxylation: *p*-coumaryl, coniferyl, and sinapyl alcohols [53,54,55]. Each of these monolignols gives rise to a different type of lignin unit called *p*-hydroxyphenyl (H), guaiacyl (G), and syringyl (S) units, respectively, when incorporated into the polymer. Lignin is a polymer of phenolic hydroxyl groups, hydroxyl benzylic and carbonyl groups. The lignin polymer contains characteristic methoxyl groups, hydroxyl phenol groups, and some end-aldehyde groups in the side chain [3]. The presence of absorption bands at wave numbers with a strong intensity of about 1.270-1330 cm-1 in lignin isolates resulting from black liquor of this organosolve delignification process with a 30% NaOH concentration suggests the presence of one of the lignin ligands, ie guaiasil which are the units Lignin constituents in non-wood lignin.

3.5. Assessment of the structure of OPFEB formacell lignin

The Scaniing Electrom Miscoscope (SEM) studies revealed the details on structural and marfologies of lignin. Lignin was analyzed by using SEM that is lignin with treatment of NaOH concentration that is 5% and 30% (Figure 7).

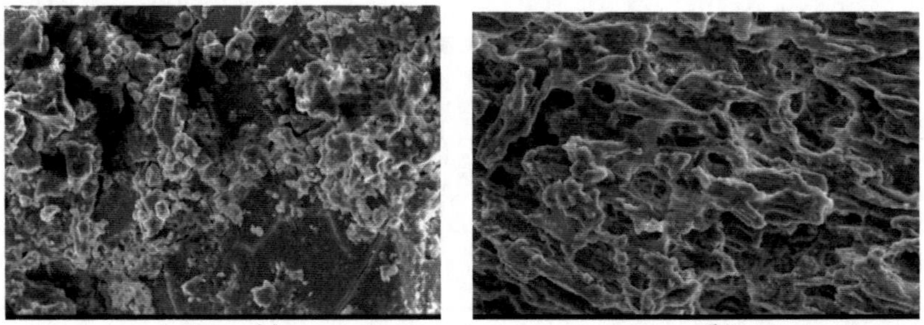

(a) (b)

Figure 7. SEM migrographs of lignin samples (a) isolation by 5% NaOH, (b) isolation by 30% NaOH

The result of SEM analysis showed that lignin isolation treatment using 30% NaOH concentration had a more compact form of agglomeration compared with isolation using 5% NaOH. At the high pH of typical kraft black liquof, the repulsive forces between the ionized hydrophilic groups (mainly phenolic hydroxyl and carboxylate groups) stabilize the colloidal lignin and keep it in solution, thereby preventing lignin agglomeration and precipitation [56]. The results of Namane *et al* [57] showed that the lignin isolation process from liquor derived from formacell produces a constant granular structure.

4. Conclusion

The best lignin isolates were on lignin deposition with 30% NaOH concentration by total solids content in EOFB black liquor, lignin of yield, lignin methoxyl content, lignin equivalent weight, The average lignin yield is 5.67%, the total solid black liquor with an average of 65.11%, the lignin methoxyl content with an average of 14.61% and the weight of the lignin equivalent with an average of 1787.23.. The result of FT-IR identification from lignin isolate at 30% NaOH concentration shows IR spectral absorption pattern which almost most have the same absorption pattern at wave number region. Isolates of lignin at a 30% NaOH concentration showed that lignin has one lignin ring ie guaiasil which is lignin composing units in non-wood lignin.

Acknowledgment

This research work was supported by Fundamental Research Grant 2017, No. 071/SP2LH/LT/DRPM/IV/2017. The authors would also like to acknowledge the contributions and financial support from Ministry of Research, Technology, and Higher Education of the Republic of Indonesia.

References

[1] Zhang AP, Liu CF, Sun RC and Xie J 2013 Extraction, purification and characterization of lignin fractions from sugarcane bagasse *Bioesources*. **8 (2)**: 1604-1641

[2] Louhelainen J, Alen R, Zielinsk I J, and Sagfors P E 2002 Effects of oxidative and non - oxidative thermal treatments on the viscosity and chemical composition of softwood kraft black liquor *J of Pulp and Paper Science* **28(9)**, 285 -291

[3] Sjostrom E 1981 Wood Chemistry, Fundamentals and Applications, Academic Press, New York, 223 p

[4] Abdelwahab N, Nassar M 2011 Preparation, optimisation and characterisation of lignin phenol formaldehyde resin as wood adhesive *Pigment & Resin Technology* **40**, 169-174

[5] Alonso MV, Oliet M, P Rez, Rodriguez F, Echeverra J 2004 Determination of curing kinetic parameters of lignin–phenol–formaldehyde resol resins by several dynamic differential scanning calorimetry methods. *Thermochimica Acta* **419**, 161-167

[6] Khan MA, Ashraf SM, Malhotra VP 2004 Eucalyptus bark lignin substituted phenol formaldehyde adhesives: A study on optimization of reaction parameters and characterization. *J of Applied Polymer Science* **92**, 3514-3523

[7] Mankar S, Chaudhari A, Soni I 2012 Lignin in phenol-formaldehyde adhesives I*nternational J of Knowledge Engineering* **3**, 116-118

[8] Sarkar S, Adhikar B 2001 Jute felt composite from lignin modified phenolic resin. *Polymer Composites* **22**, 518-527.

[9] Tejado A, Pena C, Labidi J, Echeverria J M and I Mondragon 2007 Physico-chemical characterization of lignins from different sources for use in phenol-formaldehyde resin synthesis *Bioresource Technol*. **98**: 1655-1663

[10] Bykov I 2008 Characterization of Natural and Technical Lignins using FTIR Spectroscopy. *PhD thesis*, Lulea University of Technology

[11] Sahoo S, Seydibeyoğlu MÖ, Mohanty AK, and Misra M 2011 Characterization of industrial lignins for their utilization in future value added applications. *Biomass and Bioenergy* **35**, 4230-4237

[12] Anglès MN, Reguant J, Garcia-Valls R and Salvadó J 2003 Characteristics of lignin obtained from steam-exploded softwood with soda/anthraquinone pulping *Wood Science and Technology* **37**, 309-320

[13] Bonini C, M Auria, L Emmanuel, R Ferri, R Pucciarello and AR Sabia 2005 Polyutrethanes and Polyester from Lignin *J Applied Polymer Science*, **98 (3)**: 1451-1456

[14] Kleinert M, and T Barth 2008 Towards in Lignicellulosic Biorefinery: Direct One Step Conversion of Lignin to Hydrogen-Enriched Biofuel *Energy Fuels*. **22 (2)**: 1371-1379

[15] Xu F, Sun J-X, Sun R, Fowler P and Baird M.S 2006 Comparative study of organosolv lignins from wheat straw. *Ind. Crops Prod*. 23, 180–193

[16] Cheng S, Yuan Z, Leitch M, Anderson M, Xu C 2013 Highly efficient de-polymerization of organosolv lignin using a catalytic hydrothermal process and production of phenolic resins/adhesives with the depolymerized lignin as a substitute for phenol at a high substitution ratio *Industrial Crops and Products* **44**, 315-322

[17] Cazacu G, Pascu M.C, Profire L, Kowarski A I, Mihaes M, Vasile C 2004 Lignin role in a complex polyolefin blend. *Industrial Crops and Products*, **20**, 261-273

[18] Gosselink R J A, Abächerli A, Semke H, Malherbe R, Käuper P, Nadif A and Van Dam J E G 2004 Analytical protocols for characterisation of sulphur-free lignin. *Ind. Crop. Prod.* **19**, 271–281

[19] Hatakeyama H, Nakayachi A and Hatakeyama T 2005 Thermal and mechanical properties of polyurethane-based geocomposites derived from lignin and molasses *Composites Part A: Applied Science and Manufacturing* **36**, 698-704

[20] Kadla J F, Kubo S, Venditti R A, and Gilbert R D 2002 Novel hollow core fibers prepared from lignin polypropylene blends. *J. Appl. Polym. Sci.* **85**, 1353–1355

[21] Lora J H, Glasser W G, 2002. Recent industrial applications of lignin: A sustainable alternative to nonrenewable materials. *J. Polym. Environ.* **10**, 39-48

[22] Schorr D, Diouf PN, and Stevanovic T 2014 Evaluation of industrial lignins for biocomposites production. *Industrial Crops and Products* **52**, 65-73

[23] Cateto C A, Barreiro, M F, Rodrigues AE, Brochier-Solan M C, Thielemans W and Belgacem M N 2008 . Lignins as macromonomers for polyurethane synthesis: A comparative study on hydroxyl group determination. *J of Applied Polymer Science* **109**, 3008-3017

[24] Huang J, Zhang L 2002 Effects of NCO/OH molar ratio on structure and properties of graft-interpenetrating polymer networks from polyurethane and nitrolignin. *Polymer* **43**, 2287-2294.

[25] Mahmood N, Yuan Z, Schmidt J, and Xu C 2013 Production of polyols via direct *hydrolysis* of kraft lignin: Effect of process parameters. *Bioresource Technology* **139**, 13-20

[26] Li J 2011 Isolation of lignin from wood, BSc *thesis, Faculty of Technology. SAIMAA University of applied science, USA*

[27] Vishtal A and Kraslawski A 2011 Challenges in industrial applications of technical lignins. *BioRes.* **6**, 3547–3568

[28] Setyawardhani D A 2014 Practical Guidelines for Quantitative Chemistry Analysis. Surakarta

[29] Poppius, Levlin K, Mustonen R, Huovila T, and Sundquist J 1991 Milox pulping with acetic – acid . paperi ja Puu-Paper Timber., 73(2): 154-158

[30] Jimenez L, F Maestre, J L Ferrer and I Pérez 1998 Delignification of wheat straw by use of low-molecularweight organic acids. **Holzforschung 52 (2)**:191–196

[31] Lam H Q, Bigot Y L, Delmas M, and Avigon G 2001 Formic acid pulping of rice straw. *Ind. Crops Prod.*, **14(1)**: 65-71

[32] Hidayati S, Zuidar A S and Satyajaya W 2017 Effect Of Acetic Acid: Formic Acid Ratio On Characteristics Of Pulp From Oil Palm Empty Fruit Bunches (OPEFB). *ARPN J of Engineering and Applied Sciences*, **12(12)**: 3802-3807

[33] Sun R R, Tomkinson J J and Lloyd Jones GG 2000 Fractional characterization of ash-AQ lignin by successive extraction with organic solvents from oil palm EFB fibre. *Polymer Degradation and Stability* **68**, 111-119

[34] Cheng S 2011 Bio-Based Phenolic Resins And Adhesives Derived From Forestry Residues/Wastes And Lignin. *Disertation. Faculty of Natural Resources Management Lakehead University Thunder Bay, Ontario, Canada*

[35] Lubis A A 2007 Lignin Isolation from Black Linde (Black Liquor) Pulp Cooking Process Soda and Pulp Sulphate (Kraft). *(Essay). Bogor Agricultural Institute. Bogor*. 81 p

[36] Sundin J 2000 Precipitation of kraft lignin under alkaline conditions. *Ph.D thesis, Royal Institute of Technology, Stockholm, Sweden*

[37] Vainio U, Maximova N, Hortling B, Laine J , Stenius P, Simola L. K, Gravitis J and Serimaa R 2004 Morphology of Dry Lignins and Size and Shape of Dissolved Kraft Lignin Particles by X-ray Scattering. *Langmui.* **20** 9736-9744

[38] Vasileva T S, Nenkova and K. Stanulov 2007 Obtaining of Phenolic Compounds through Alkaline Depolymer-ization of Technical Hydrolysis Lignin, *Cellulose Chem. Technol* **41**, 7-8:379-384

[39] Heradewi 2007 Lignin Isolation from Black Leafy Process of Organosolv Bunching of Cluster Fibers Empty Palm Oil (EOFB). *Essay Bogor Agricultural Institute. Bogor*. 109 p

[40] Theliander, H 2010 The LignoBoost process: Solubility of lignin. *International Chemical Recovery Conference 29th March - 1st April 2010 Williamsburg, VA, USA*. 33-42.

[41] Zhu W 2013 Equilibrium of Lignin Precipitation The Effects of pH, Temperature, Ion Strength and Wood Origins. *THESIS. Forest Products and Chemical Engineering Department of Chemical and Biological Engineering Chalmers University Of Technology Gothenburg, Swede*

[42] Lin S Y and W D Carlton 1992 Methods in Lignin Chemistry *Berlin Heidelberg: Springer-Verlag*. **69**: 627-642

[43] Syahmani 2000 Isolation, Sulfonation and Lignin Acetylation of Oil Palm Bunches and Studies Its effect on Urea Dissolution Process. *Thesis Faculty of Mathematics and Science, Bandung Institute of Technology. Bandung*

[44] Kirk R E and D F Othmer 1952 *Encyclopedia of Chemical Technology.* ***Vol.3.*** *The Interscience Encyclopedia, Inc., New York* Pp.327-338

[45] Fengel and G Wegener 1984 Wood: Chemistry, Ultrastructure, Reactions de Gruyter, Berlin, Pp. 1-729. X111, 613 S

[46] Damat 1989 Isolation of Lignin from Cooking Solution by Pulp Factory Using H2SO4 and HCl. *Essay). Department of Agricultural Industrial Technology, Faculty of Agricultural Technology, Bogor Agricultural University Bogor* 3 p

[47] Casey P J 1952 Pulp and Paper, Chemistry and Chemical Technology ***Vol. 1****: Pulping and Paper Making. The Wiley Intersince Publisher, Inc., New York*

[48] Santoso A 2003 Synthesis and Establishment of Lignin Resolsinol Formaldehyde Resins for Adhesives Lamina Wood *Dissertation. Post Graduate Program, Bogor Agricultural University, Bogor*

[49] Connors W J, L F Lorenz and T K Kirk 1978 Chromatographic Separation of Lignin Models by Molecular Weight using Sepandhex LH-20. *Holzforschung* **31**
[50] Hergert H L 1971 Infrared Spectra *Willey Interscience, New York* 267-297
[51] Garcia, A. A. Toledano, L. Serrano, I. Egues, M. Gonzalez, F. Marin, and J. Labidi 2009 "Characterization of lignin obatain by selective precipitation", *Separation and Purification Technology*, **68**, pp. 193-198
[52] Lubis M A R, A R Dewi , L Risantoa, LH Zaini, E Hermiatia 2012 . Isolation and Characterization of Lignin from Alkaline Pretreatment Black Liquor of Oil Palm Empty Fruit Bunch and Sugarcane Bagasse. *Mechatronics, Electrical Power, and Vehicular Technology* ***XX . XX-XX* 2**. 1-5 p.
[53] Higuchi T 1997 Biochemistry and Molecular Biology of Wood. *Springer-Verlag, London*
[54] Boerjan W, Ralph J, Baucher M 2003 Lignin biosynthesis. Annu Rev Jornal of Plant Biology **54**: 519–546
[55] Ralph SA, Ralph J, Landucci L 2004 NMR Database of Lignin and Cell-Wall Model Compounds. U.S. Forest Products Laboratory, Madison, WI:http://ars.usda.gov/Services/docs.htm?docid=10491 (July 3, 2006)
[56] Hermans, M. A 1984 High intensity black liquor oxidation, *PhD. Thesis, Law rence University*,
[57] Namane,M; García-Mateos F.J., Sithole, B; Ramjugernath, D; Rodríguez-Mirasol, J And Cordero, T. Characteristics of Lignin Precipitated with Organic Acids as A Source for Valorisation of Carbon Products. *Cellulose Chemistry and Technology* **50 (3-4)**: 355-360

IOP Conf. Series: Materials Science and Engineering **344** (2018) 012007 doi:10.1088/1757-899X/344/1/012007

The synthetic activities of TiO2-moringa oleifera seed powder in the treatment of the wastewater of the coal mining industry

Marhaini, Legiso, Trilestari
Department of Chemical Engineering, Faculty of Engineering, Muhammadiyah University, Palembang, Indonesia

Email: marhainiump@yahoo.co.id

Abstract . To process the coal wastewater, the combination of chemical based technology of Advanced Oxidation Process (AOP) of a strong oxidizer using TiO_2 photocatalyst and biological treatment of moringa seed powder (*Moringa oleifera*) is used in the composite form. AOP can be used as an alternative treatment of coal wastewater which is quite economical and environmentally friendly. The XRD results of TiO_2 powder and the synthesis of TiO_2 - is moringa seed powder in the form of tetragonal crystals. The degradation results of the quality of the coal wastewater using TiO_2 powder reached a decrease of (TSS, Fe, Mn, Zn, Hg, Cu, Co, Cr, Al and Ni) by an average of 70% and the increase of pH value of 7 at 200 minute stirring time. The decrease of the wastewater quality using the synthesis of TiO_2- moringa seed powder by using sunlight and without sunlight is detected negative (-) at 200 minute stirring time.

1. Introduction

The release of heavy metals from the wastewater of coal mining industry has become a threat to the ecosystem [1]. One of the damages that occurs in coal mining activities is the decrease in water pH due to the interaction between the atmosphere, water and rocks. The decrease in the pH of water is not only due to the interaction between the atmosphere, water and rocks, but also due to the coal itself which can lead to acid mine drainage, because generally coal has a humidity range of between 2 - 40%, sulfur content of 0.2 - 8% and ash content of 5 - 40% which can have an effect on the value of coal as an energy source that can cause pollution in its use. Acid mine drainage has a high acidity and is often characterized by a lower pH value below 5 [9]. In addition, acid mine drainage will erode soil and rocks that causes dissolving of various metals such as iron (Fe), cadmium (Cd), manganese (Mn), zinc (Zn), mercury (Hg), nickel (Ni), cobalt (Co) , Cadmium (Cd), lead (Pb) and arsenic, which are heavy metals known for their non-degradable toxicity and are largely non-biodegradable. Therefore, in addition to having a low pH, acid mine drainage also contains high concentrations of metals that can adversely affect both environmental and human health and their presence above specified limits in the body can cause severe damage to the body's vital organs, such as the kidney, the liver and the brain, the reproductive system and the nerves[10].

To overcome this problem, coal wastewater treatment needs to be carried out by combining applied technology, namely chemical-based technology of Advanced Oxidation Process (AOP) using strong oxidizer by using photocatalyst of TiO_2 and biology of Moringa oleifera seed powder in composite form. This advanced oxidation process can be used as an economically viable alternative wastewater treatment[22]. Commercially, TiO_2 powder is also easy to obtain and to produce in large quantities[20]. In addition, TiO_2 is a semiconductor that has a melting point, photoactivity, high thermal and chemical stability, has a non-toxic nature and also one of the best catalysts to be applied in the environment because of its inert biological and chemical

IOP Conf. Series: Materials Science and Engineering **344** (2018) 012007 doi:10.1088/1757-899X/344/1/012007

properties and the price is relatively cheap[12]. Based on its properties, TiO_2 is the most effective photocatalyst to be used, as one of the semiconductor materials, and has been widely studied mainly in the processing of solar energy sources and the processing of hazardous waste[15], and it is also used for hospital waste treatment[7]. TiO_2 photocatalyst is highly efficient in reducing heavy metals of Fe, Cr and Pb in wastewater by 96-98% and TiO_2 photocatalyst through the sunlight shows higher result in degrading dye waste compared to commercial TiO_2[6]. According to previous studies Moringa seeds contain bioactive compounds of rhamnosyloxy-benzyl-isothiocyanate, which is able to adopt and neutralize the particles of sludge and metals contained in the suspension waste with dirt particles floating in water and their effect on the content of bacteria Coli, Moringa oleifera seeds can reduce bacteria Coli by about 28%[14].

2. Research Method

One hundred fifty (150) ml of Moringa seed powder solution is added with 300 gram of sol Tetanium Tetra Isopropoxide (TTIP) which is then stirred with stirrer slowly for 24 hours at room temperature to form a nanoparticle suspension. The suspension is sonicated for 30 minutes. Then, it is evaporated to free water and calcined at a temperature of 500°C for 2 hours. The wastewater of the coal industry of 2 L is added to the photocatalytic synthesis of TiO_2-powder of moringa seed. The activity test is carried out without sunlight and with sunlight for 50 minutes, 100 minutes, 150 minutes, 200 minutes, 250 minutes and is added with 1.0 mL H_2O_2 30% . Then it is put into the reactor, stirred at a speed of 60 rpm. The results of the observation were analyzed using AAS (Atomic Absorption Spectophotometry) to measure the quality of residual wastewater degradation and X-Ray Diffraxion (XRD) for the analysis of the concentration test of the synthesis of TiO_2-Moringa oleifera seed powder.

3. Research Results and Discussion

3.1. Analysis Results of XRD Synthesis of TiO_2- Moringa oleifera Seed Powder

The XRD results of TiO_2 powder and the synthesis of TiO_2- Moringa oleifera Seed Powder in the anatase phase are shown in Figures 1a and 1b.

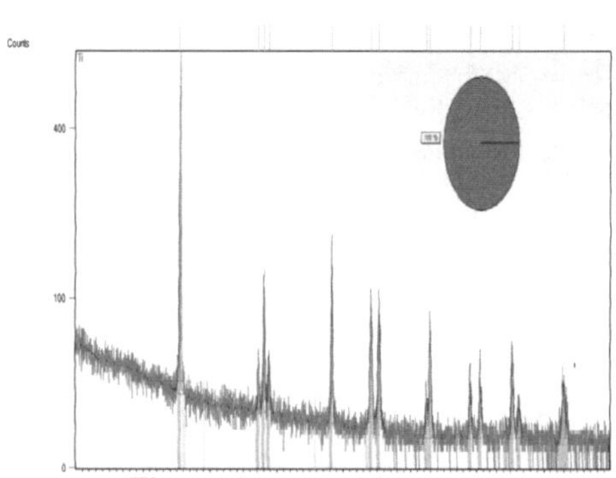

Figure 1a. XRD TiO_2

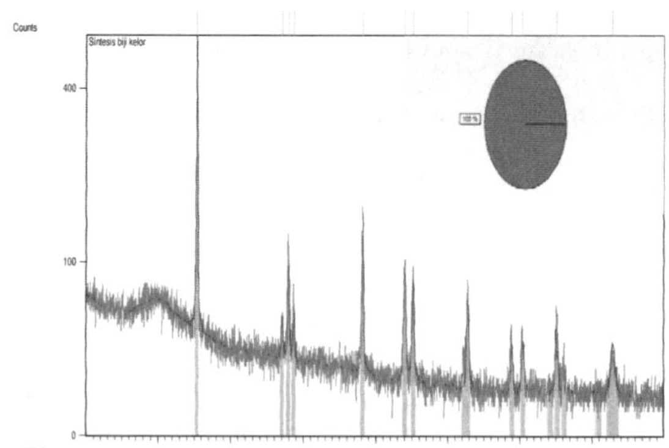

Figure 1b. XRD Synthesis TiO_2-*Moringga oleifera*

The results of the study showed that the powder of TiO_2 and the synthesis of TiO_2-Moringa oleifera seed powder had a tetragonal crystalline form with the space group of 141/amd and the lattice parameter data as shown in Table 1.

IOP Conf. Series: Materials Science and Engineering **344** (2018) 012007 doi:10.1088/1757-899X/344/1/012007

Table 1. Lattice parameters of TiO_2 and Synthesis of TiO_2-Moringa oleifera Seed Powder

Sample	a (Å)	b (Å)	c (Å)	Alpha (°)	Beta (°)	Gamma (°)
TiO_2	3,7760	3,7760	9,4860	90	90	90
Synthesis of TiO_2-Moringa oleifera Seed Powder	3,7760	3,7760	9,4860	90	90	90

3.2. The Effects of the Quality of Coal Wastewater using TiO_2 Crystals

The ability of TiO_2 as a photocatalyst is to provide H^+ with a large oxidation radical. It can provide a continuous hydroxyl radical in accordance with the source of the light. The nature of TiO_2 as a strong oxidizer makes it capable of oxidizing heavy metal compounds contained in coal wastewater. The results of the measurement of the quality of the coal wastewater using TiO_2 with the stirring variables showing a rise in pH up to 7 in 250 minute stirring time and decreasing the quality of wastewater (TSS, Fe, Mn, Zn, Hg, Cu, Co, Cr, Al and Ni) reaching the average of 70% in stirring time of 200 minutes (Table 2). This is because TiO_2 is a strong oxidant that is chemically and biologically inert, producing OH ions and can oxidize and mineralize almost all chemicals that are environmentally friendly [15,23] and can also damage microbial cells by attacking cell walls, cytoplasmic membranes and intracellular structures[18].

Table 2. The Effect of the Quality of Coal Wastewater on TiO_2 Crystals

Parameters	Quality of Wastewater	Maximum Concentration *	Unit	Stirring Time (Minutes)				
				50	100	150	200	250
TSS	70	200	mg/L	60	45	27	12	2
pH	2,68	6 - 9	mg/L	3,5	5	6,3	6,8	7
Fe	15	7	mg/L	14,8	11,5	8,2	5,6	1,9
Mn	10	4	mg/L	9,32	7,35	5,29	2,47	0.88
Zn	5	5	mg/L	4,39	3,67	2,56	0.98	<0,01
Hg	0,0045	0,002	mg/L	0.0040	0,003	0,0021	0,001	<0,001
Cu	2,5	2	mg/L	2,5	2,13	2,01	1,99	1,72
Co	0,5	0,4	mg/L	0,49	0,35	0,31	0,29	0,25
Cr	0,7	0,5	mg/L	0,69	0,62	0,62	0,51	0,45
Al	2,4	0,5	mg/L	1,73	1,23	0,89	0,13	<0,001
Ni	0,25	0,2	mg/L	0,24	0,19	0,001	<0,001	<0.001

Source : Government Regulation No. 82 of Year 2001 (*PP No. 82 Tahun 2001*)
* Ministrial Decree on the Environment No. 51 of Year 2004 (*Kepmen-LH No. 51 Tahun 2004*)

3.3. The Effects of Using Synthesis of TiO_2-Moringa oleifera Powder With Sunlight and Without Sunlight on the Quality of Coal Wastewater

The test of the effects of long illumination shows that the longer the time spent in illumination using sunlight, the more heavy metal concentration (TSS) will decrease and the higher the value of pH will be. TiO_2 photocatalyst material has a very strong oxidizing power when activated by sunlight. The results of the study are shown in Table 3.

IOP Conf. Series: Materials Science and Engineering **344** (2018) 012007 doi:10.1088/1757-899X/344/1/012007

Table 3. Effects of Using Synthesis of TiO_2-Moringa oleifera Seed Powder Without Sunlight on the Quality of Coal Wastewater

Parameter	Quality of the Wastewater (mg/L)	Stirring time (Minutes)									
		Without Sunlight Illumination					With Sunlight Illumination				
		50	100	150	200	250	50	100	150	200	250
TSS	70	55	35	21	7	(-)	56	23	8	(-)	(-)
pH	2.68	4.2	6.5	7	7.2	7,5	4.5	6.4	7	7.2	7.5
Fe	15	13.46	10.21	6.47	1.21	(-)	11.5	5.24	0.001	(-)	(-)
Mn	10	8.32	6.21	3.15	0.001	(-)	10.24	5.98	0.0097	(-)	(-)
Zn	5	4.12	3.07	1.01	<0.001	(-)	3.13	1.45	0.001	(-)	(-)
Hg	0.0045	0.0021	0.0013	<0.001	<0.001	(-)	0.001	0.0001	<0.0001	(-)	(-)
Cu	2.5	2.01	1.001	<0.001	<0.001	(-)	1.99	0.467	0.005	(-)	(-)
Co	0.5	0.32	0.101	<0.001	<0.001	(-)	0.003	0.0001	<0.0001	(-)	(-)
Cr	0.7	0.69	0.037	0.001	<0.001	(-)	0.001	<0.001	<0.001	(-)	(-)
Al	2.4	2.09	1.27	0.001	<0.001	(-)	1.78	0.0497	0.001	(-)	(-)
Ni	0.25	0.199	0.0009	0.001	<0.001	(-)	0.21	0.007	<0.001	(-)	(-)

Source : Government Regulation No. 82 of Year 2001 (*PP No. 82 Tahun 2001*)
* Ministrial Decree on the Environment No. 51 of Year 2004 (*Kepmen-LH No. 51 Tahun 2004*)

The results of the observations with naked eyes showed that the wastewater that was originally yellow looked more clear with the addition of illuminating time, so that the results of the wastewater quality of 200 minutes stirring time showed a negative result, but the results of the analysis without the use of sunlight with 250 minutes stirring time just produced negative water quality. The wastewater treatment conducted through AOP (TiO_2) by sunlight[11-16,4-18] could substantially reduce waste treatment costs and was more favorable from an environmental perspective[17]. The effect of adding 30% H_2O_2 is enlarging the concentration of coal wastewater quality and more ^oOH is produced through reduction reactions. Excessive H_2O_2 can have a negative effect on degrading heavy metals in coal wastewater rather than ^oOH which can partially reshape H_2O_2. HO_2^o can prevent the transfer of photon energy, this is because HO_2^o is a gas molecule that is not dissolved

4. Conclusions

The degradation of coal wastewater quality using TiO_2 shows that the quality of wastewater at 200 minutes of stirring time reaching an average of 70%. The synthesis of TiO_2- Moringa oleifera seed powder with and without illumination of sunlight can be applied as a coal wastewater treatment with a stirring time of 200 minutes resulting in a decrease in the quality of coal wastewater which is detected negative.

References

[1]. Achla Kaushal, S.K., Singh.2017.Removal of Heavy Metals by Nanoadsorbents: A review. Journal of Environment and Biotechnology Research, Vol 6,No.1,96 - 104

[2]. Arung, E.T., 2002. Breakthrough, Seed of Moringa oleifera as River Water Purifier, Suara Merdeka, Jakarta

[3]. Byrne, J.A., Fernandez-Ibanez, P.A., Dunlop, P.S., Alrousan, D.; Hamilton, J.W. 2011. Photocatalytic Enhancement for Solar Disinfection of Water: A Review. Int. J. Photoenergy ,doi:10.1155/2011/798051.

[4]. Blanco, J., Malato, S.,Fernández-Ibañez, P., Alarcón, D., Gernjak, W., Maldonado, M. 2009. Review of feasible solar energy applications to water processes. Renew. Sust. Energ, 13, 1437–1445.

[5]. Byrne, J.A., Dunlop, P.S., Hamilton, J.W., Fernandez-Ibanez, P., Polo-Lopez, I.,Sharma, P.K., Vennard, A.S. 2015. A review of Heterogeneous Photocatalysis for Water and Surface Disinfection. Molecules, 20, 5574–5615.

[6]. Chih-Ho Su.,Chi-Cheng Hu.,Yen-Chun Sun.,Yu-Cheng Hsioo. 2016. Highly active and thermo-stable anatase TiO_2 photocatalysts synthesized by a microwave-assisted hydrothermal method, Journal of the Taiwan Institute of Chemical Engineering

[7]. Chong.M.N.,Cho.Y.J.,Poh.P.E.,Jin.B. 2014. Evaluation of Titanium Dioxide Photocatalytic Technology for the Treatment of Reactive Black 5 dye in Synthetic and Real Grey Water Effluents. Journal of Cleaner Production 89.

[8]. Fatta-Kassinos, D., Meric, S. 2011. Nikolaou, A. Pharmaceutical Residues in Environmental Waters and Wastewater: Current State of Knowledge and Future Research. 399, 251–275.

[9]. Gautama, Rudy Sayoga. 2012. Management of Acid Mine Drainage. Technical Guidance, Reclamation and Post-Mining In Minerals and Coal Mining Activities - Directorate General Of Minerals And Coal, Ministry Of Energy And Mineral Resources.

[10]. Goel,P.K. 2006.Water Pollution Causes Effects and Control (Revised 2nd Edition). New Age International publishers, New Delhi. India

[11]. Gogate, P.R.; Pandit, A.B. 2004. A Review of Imperative Technologies for Wastewater Treatment I: Oxidation Technologies at Ambient Conditions. Adv. Environ. Res. 8, 501–551.

[12]. Hoffmann, M.R., S.T. Martin., W. Choi, dan Bahnemann.D.W 1995. Environmental Applications of Semiconductor Photocatalysis. Chemical Reviews. 95

[13]. Johnson, D. Barrie; Hallberg, Kevin B; Acid mine drainage remediation options: A review; Science of the Total Environment 338 3– 14, 2005

[14]. Juli N., Suriawilis U., Birsyam I., 1986. Exploratory Study of Natural Plant-Based Coagulant Materials and their Effects on Coli Bacterial Content, Ministry of Education and Culture, Institute *of* Technology, *Bandung* (ITB)

[15]. Lu.C.H.,Wu.W.H., Kale.R.B. 2008. Micro Emulsion-Mediated Hydrothermal Synthesis of Photocatalytic TiO_2 Powders.J.Hazard.Mat 154

[16]. Malato, S., Fernández-Ibáñez, P., Maldonado, M.I., Blanco, J., Gernjak, W 2009. Decontamination and Disinfection of Water by Solar Photocatalysis: Recent Overview and Trends. *147*, 1–59.

[17]. Munoz, I., Peral., J.Ayllon, J.A.,Malato, S., Passarinho, P.,Domenech, X. 2006. Life Cycle Assessment of a Coupled Solar Photocatalytic-Biological Process for Wastewater Ttreatment. 40, 3533–3540.

[18]. Oller, I., Malato, S.,Sánchez-Pérez, J. 2011. Combination of Advanced Oxidation Processes and Biological Treatments for Wastewater Decontamination—A review. *409*, 4141–4166.

[19] Robertson, P.K.,Robertson, J.M.,Bahnemann, D.W.2012. Removal of Microorganisms and Their Chemical Metabolites from Water using Semiconductor Photocatalysis. J. hazard. Mater. 211, 161–171

[20]. Slamet., Syakur R., Danumulyo W., 2003. Treatment of Waste of Heavy Metal Chromium (VI) with TiO_2 Photocatalyst. Makara, Technology, 7 (1): 27-32.

[21]. Sutherland, J.P., Folkard, G.K., Mtawali, M.A. and Grant, M.A., Moringa Oleifera as a Natural Coagulant, 1994, Paper 20th WEDC Conference Affordable Water Supply And Sanitation, Colombo, Sri Lanka

[22]. Wang.C.,Liu.H,Qu.Y., 2013.TiO_2-Based Photocatalytic Process for Purification of Polluted Water: Bridging Fundamentals to Applications. Hindawi Publishing Corporation Journal of Nanomaterials Article ID 319637,14

[23]. Widowati, W., Sastiono, A., Jusuf R, R., 2008. Toxic Effect of Metals; Prevention and Overcoming of Pollution. Yogyakarta:

[24]. Wols, B., Hofman-Caris, C. 2012. Review of Photochemical Reaction Constants of Organicmmicropollutants Required for UV Advanced Oxidation Processes in Water. *46*, 2815–2827.

[25]. Zhang, H., Yang, Z., Zhang, X., Maio, N., 2014, Photocatalytic Effect of Wool Fibers Modified with Soley TiO_2 Nanoparticles and N-doped TiO_2 Nanoparticles by Using Hydrothermal, Elsivier,Vol. 254, 106-114

Bamboo reinforced polymer composite - A comprehensive review

S A H Roslan[1], Z A Rasid[1] and M Z Hassan[2]

[1]Malaysia-Japan International Institute of Technology, Universiti Teknologi Malaysia, 54100 Kuala Lumpur, Malaysia
[2]UTM Razak School of Engineering and Advanced Technology, Universiti Teknologi Malaysia, 54100 Kuala Lumpur, Malaysia

E-mail: mzaki.kl@utm.my

Abstract. Bamboo has greatly attention of researchers due to their advantages over synthetic polymers. It is entirely renewable, environmentally-friendly, non-toxic, cheap, non-abrasive and fully biodegradable. This review paper summarized an oveview of the bamboo, fiber extraction and mechanical behavior of bamboo reinforced composites. A number of studies proved that mechanical properties of bamboo fibers reinforced reinforced polymer composites are excellent and competent to be utilized in high-tech applications. The properties of the laminate are influenced by the fiber loading, fibre orientation, physical and interlaminar adhesion between fibre and matrix. In contrast, the presence of chemical constituents such as cellulose, lignin, hemicellulose and wax substances in natural fibres preventing them from firmly binding with polymer resin. Thus, led to poor mechanical properties for composites. Many attempt has been made in order to overcome this issue by using the chemical treatment.

1. Introduction

Bamboo is one of the woody plants that always gain most manufacturers'attention due to several advantages such as ease of growth in diverse climates which make it durability and low cost material when compared to others [1–4]. Eventhough bamboo is one of the old and traditional building materials, it is actively being practiced until today. Bamboo, as shown in **Figure 1.**, are abundantly growing mainly in South America and Asia, including Malaysia. Wong [5] reported that Malaysia has about 70 species of bamboo in which 50% in Peninsular Malaysia, 30% in Sabah and 20% in Sarawak. There are about 10 main genera of bamboo that which are *Bambusa*, *Dinochloa*, *Yushania*, *Chusquea*, *Gigantochloa*, *Phyllostachys*, *Dendrocalamus*, *Racemobambos*, *Thyrsostachys* and *Schizostachyu* [5, 6].

Figure 1. Photo of bamboo plants

Table 1. lists part of bamboo species available in Peninsular Malaysia with their respective uses. Thus, it is no wonder that bamboo is listed as one of the raw material that being harvested the most in Malaysia [7]. Still, Malaysia has not been able to delve it into engineering materials specifically polymer composites. By having excellent properties such as renewable, sustainability and durability, bamboo has potential to be utilized in polymer composite industry [1–4]. Employment of bamboo in polymer composites development has definitely enhanced the socio-economic among the developers that growing and harvesting the bamboo.

Table 1. The available bamboo species in Peninsular Malaysia [6, 8]

Species	Local Name	Uses	State
Bambusa heterostachya	Buluh galah/ tilan/ pering/ pengat	Poles, frames, tooth picks, blinds, skewer sticks	Johor
Bambusa vulgaris	Buluh minyak/ aao/ aro/ gading/ Tamalang/ pa	Ornamental, tooth picks, chopsticks, skewer sticks, shoots as food	Pahang, Perak, Selangor
Dendrocalamus asper	Buluh beting/pering	Shoots as food, higo materials, chopstick	Kedah, Melaka, Pahang, Selangor, Terengganu
Gigantochloa scortechinii	Buluh semantan/ rayah/ gala/ paao/ Seremai/ telur	Handicraft, smallscale industries, incense sticks	Kedah, Kelantan, Perak, Selangor, Terengganu
Gigantochloa wrayi	Buluh beti/ raga	Handicraft, blinds, tooth picks, skewer sticks, shoots as food	Perak
Schizostachyum brachycladum	Buluh nipis/ lemang/ padi/ urat/ rusa/ Pelang	Handicraft, rice vessels (lemant)	Pahang

2. The anatomy of bamboo

As illustrated in **Figure 2.**, the anatomy of a bamboo consists of culm, root, rhizome structure, shoots, branches and leaves. Some species of bamboo produce flowers sproadically, like *Phyllostachys Elegans* in the U.S. and *Phyllostachys edulis 'Moso'* in China [9]. Sporadic flowering means the flowering occur due to environmental factor, rather than genetics.

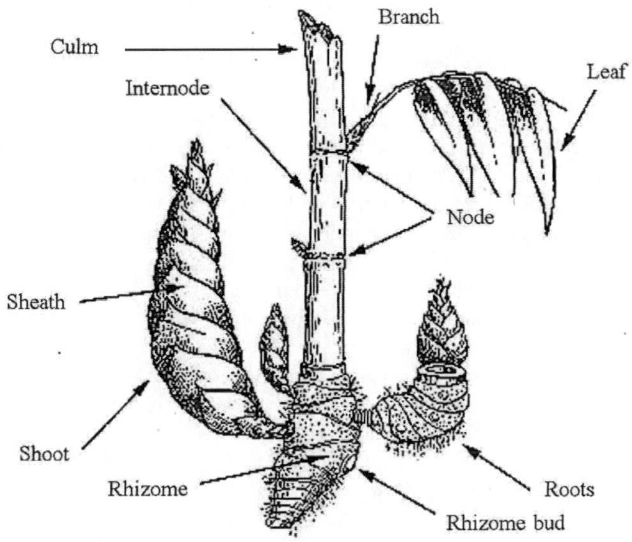

Figure 2. Photo of bamboo's anatomy [10]

Zhang et al. [11] and Jiang et al. [12] stated that the morphology (outward appearance) of bamboo is divided into two major parts; the rhizome and the culm systems. Also, the bamboo plant can be categorized into two parts; the underground part (composed of rhizomes, roots and buds) and above ground part (composed of stems, branches and foliage). **Table 2.** summarizes the functions of existing structure of bamboo. Every functions that carried out by each structure of bamboo are related to each other as to ensure the bamboo plant able to grow steadily.

Table 2. The structure of bamboo [13]

Structure	Description
Rhizome	A horizontal and segmented stem that projecting from the origin plant which travel underground for colonizing new territory purpose. It is covered with a protective sheath. The function of sheath is to give the protection needed by plant in breaching the surface as to form a culm. Rhizome collects, stores and distributes water and nutrients for function of parenchyma and conduction tissue, includes the vegetative production by growing into the new shoots or bamboo culms at their nodes.
Culm	A cylinder-formed woody stem that consists of nodes and internodes. Culms are usually hollow. Still, there are some species of bamboo that have solid culm. A culm is acclaimed by sheath, foliage, leaves, flowers and branching system. It is supported by the root system which allows it to withstand more weight of growing leaves. A culm is fully mature after 3 or 4 years and as it matures, it becomes stronger and harder.
Root	The main function of roots is to support the culms which allows them to stand against exposed damages such as bad weather. Also, allows the culm to hold more weight, giving it the ability to grow more leaves over wider distances. Another function of root is to store nutrients. They form at the base of the culm from the rhizome nodes and generally go no deeper than one foot below the surface. In appearance, the roots are typically symmetrical in size and shape.
Branches	Almost bamboo species grow multiple branches from a single bud that located at the node. However, some genera, like Chusquea, able to grow multiple buds at from each node.
Leaves	Every main parts of bamboo (rhizomes, culm and branches) have leaves. Anatomy of a leaf consists of blade, sheath and ligule. Leaves are first present in the rhizome where they are almost completely comprised of the sheath. At this stage, leaves serve as a protective cover to encase the rhizome as it travels underground. After the rhizome shoots through the soil and becomes a culm, the blade will become the predominant feature. The blade provides the photosynthetic function of the plant by converting sunlight into energy. The appearance of the blade varies among species. The appearance of leaves plays a large role in the identification of bamboo.

3. Bamboo as reinforcing fiber

As a fiber, the overall mechanical properties of bamboo are comparable to or even better than those of wood [14]. Moreover, the specific gravity of bamboo is relatively high when compared to the hardwoods or heavy tropical timber species which are normally used in composites panels manufacturing [13]. Thus, it should be used for the high density composite products such as High Density Fiberboard. These advantages make it highly competitive nature fibre reinforcement in polymer composites.

Table 3. presents the properties of bamboo that are comparable to other natural fibres such as kenaf, jute and sisal [15]. The table also includes the properties of E-Glass, a fibre that commonly used in composite manufacturing and the least one when compared to other types of glass and existing fibre like Kevlar in terms of strength. From the table, the specific young modulus (Specific E) of bamboo is comparable to that of the E-glass fibre. This means the bamboo has a high stiffness with minimum weights. Also, if the string-like bamboo fibre bundles as mentioned by Phong et al. [16] are able to be extracted, those fibres can be used as a replacement for jute and kenaf fibres.

Table 3. The mechanical properties of natural fibres compare with E-glass [15]

Fibres	Density (g/cm^3)	Tensile Strength (MPa)	Young's Modulus, E (GPa)	Specific Young's Modulus, E (E/d)	Elongation at failure (%)	Moisture absorption (%)
E-glass	2.55	2400	73	29	3	-
Bamboo	0.6-1.1	140-230	11-17	10-28	-	8.9
Hemp	1.48	550-900	70	47	1.6	8
Jute	1.46	400-800	10-30	7-21	1.8	12
Kenaf	1.45	930	53	36.5	1.6	-
Ramie	1.5	500	44	29	2	12-17
Sisal	1.33	600-700	38	29	2-3	11

4. The extraction of bamboo fiber

The form of extracted bamboo fiber will determined the final forms of composite whether as laminated fibre reinforced composites, short or randomly oriented fibre or formed into sandwich structure. These forms are essential in order to meet the industry expectation. The common forms of extracted bamboo which are strip, flake, long fibre, short fibre, sliver and powder as shown in **Figure 3.**

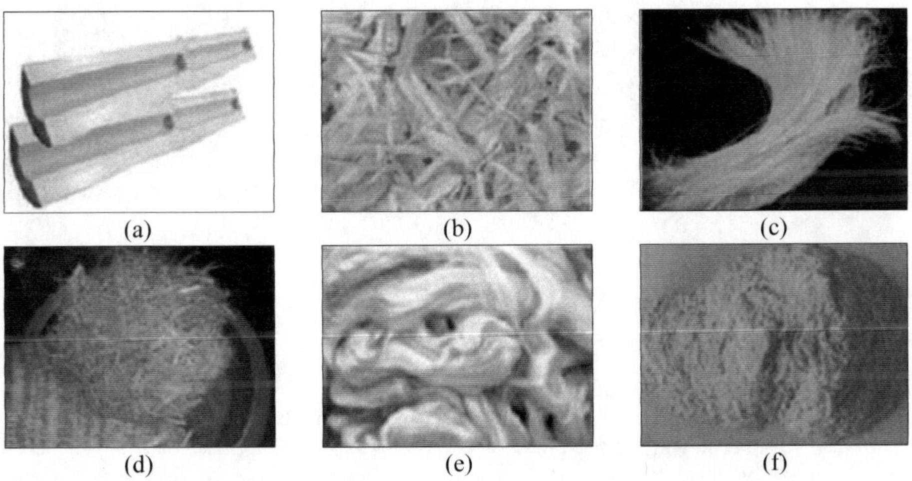

Figure 3. Photos of bamboo that being used as the composite reinforcement in a form of (a) strips (b) flake (c) long fibre (d) short fibre (e) silver and (f) powder [17–20].

In contrast, there is a limited knowledge regarding bamboo fibre extraction, only a few investigations have been done with different processes to define the mechanical properties and the usage of bamboo fibre as reinforced polymer composite. Subsequently, various methods have been used to extract bamboo fibres based on their application in the industries. These processes are classified as chemical, mechanical and combination of both processes.

5. Mechanical properties of bamboo fibre reinforced epoxy composites

Mechanical properties of bamboo fiber reinforced polymer composites (NFPC) are greatly influenced by few factors. The most common factors are listed as follow:

i. Fiber orientation
ii. Physical properties of fiber (short or long fiber)
iii. Interfacial adhesion properties of fibre and matrix

As shown in **Figure 4.**, an example of NFPCs exhibited with different value of tensile strength due to interaction with the different type of matrix resin. Bamboo fibre reinforced epoxy composites also offer different findings under various mechanical testing. Here, the process of bamboo extraction, methods for reinforcing bamboo fibre with epoxy resin and types of mechanical tests conducted severely affect the end results of experiments.

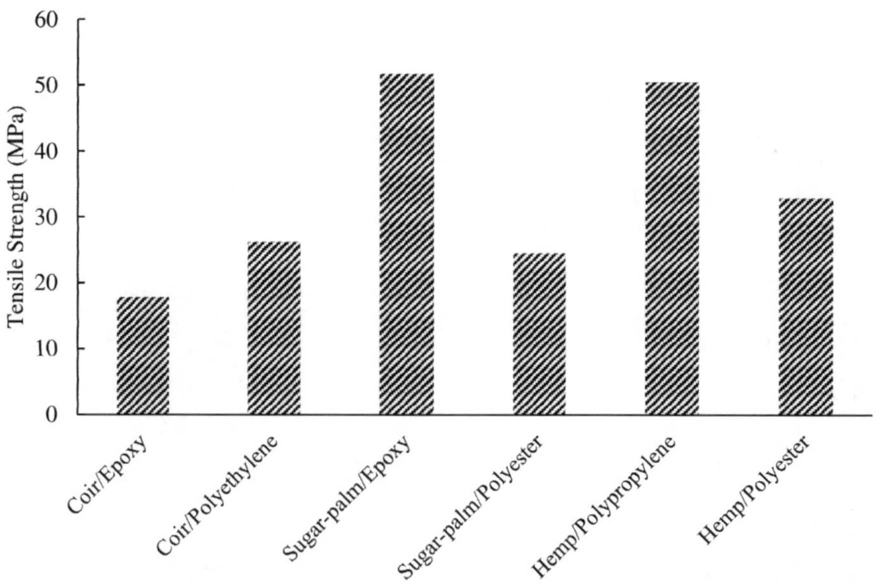

Figure 4. The tensile strength of natural fibres reinforced polymer composites [21–25]

The effect of bamboo fiber orientation

NFPCs are such such composites whose mechanical efficiency is dependent upon the fiber direction. They tend to be stronger in which fibre direction is parallel to axis of applied load. This is because the resistance force exhibited by the composites is equal to the applied force. Moreover, the increasing of fibre degree from the load axis, can cause matrix failure. In this condition, the resistance composite hold against the applied force is not effectively as in the condition where fibres are parallel to axis of applied force.

A study of mechanical behaviors of layered laminate bamboo/epoxy composite was done by Verma and Chariar [26, 27]. In their study, they fabricated three types of LLBC that consist of bamboo silver and epoxy in five layers and each layer have different fibre orientations. The first type of LLBC they made was laminated of (0°/0°/0°/0°/0°) bamboo fibre. The second and third LLBC was composed of (0°/45°/0°/45°/0°) and (0°/90°/0°/90°/0°). Five test specimens were prepared from each type of LLBCs samples respectively using cross cutting and grinding as per ASTM standards D3039M, D3410 and D7264 for tensile, compressive and flexural testing respectively. The experimental data showed that the tensile and compressive properties of LLBCs decreased with increase in lamina angle. **Figure 5.** shows the experimental data obtained from the work conducted by Verma and Chariar [27].

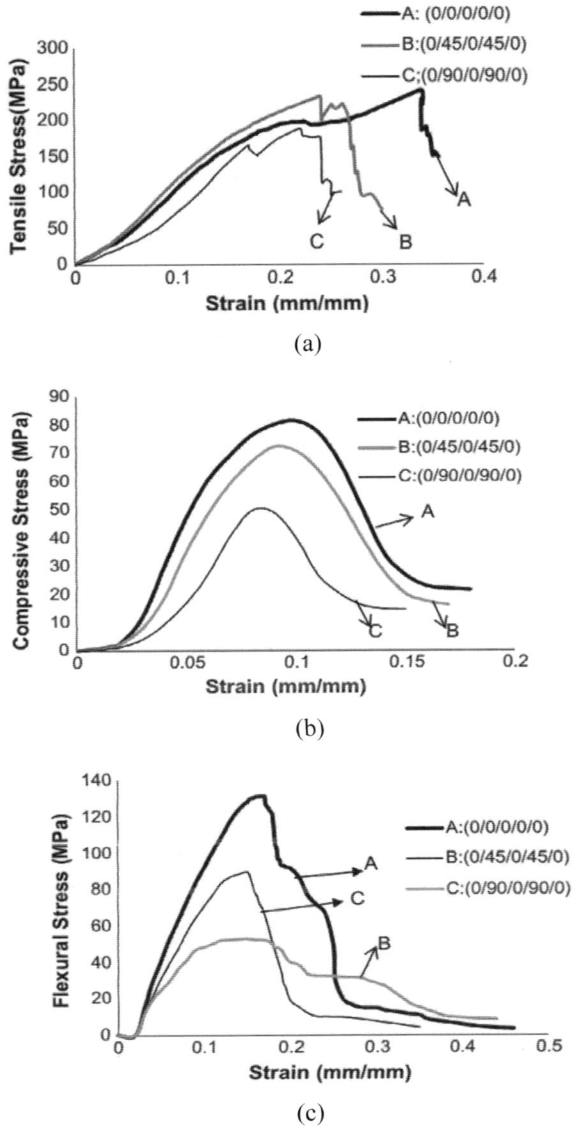

Figure 5. The output data of (a) tensile (b) compression and (c) flexure tests done by Verma and Chariar [27]

Chandrasekhar et al. [28] conducted tensile, flexural and impact tests to investigate the effect fibre orientations in bamboo/epoxy laminates. The orientations were tested at 30°, 45°, 60°, and 90°. The laminated bamboo/epoxy was produced by hand lay-up technique. The investigation was concluded that the maximum flexural and impact strengths were achieved at 60°of bamboo fibre orientation and the tensile strength was maximum in 30° of bamboo fibre orientation. The same investigation was carried out by Roslan et al. [29] where they tested of fibre orientations of laminated bamboo/epoxy composites were at 0°, 45° and 90°. The maximum tensile strength was achieved when the bamboo fibre at 0°, a condition where the fibre is parallel with the load applied.

The effect of physical properties of bamboo fiber

Gupta [30] studied the effect of physical and different of weigth loading of bamboo fibre reinforced epoxy composites. The performance of composites were evaluated by conducting tensile and flexure tests. Before reinforcing the bamboo fibre with epoxy, they cut the fibre into 4, 7 and 10 mm of length. Then mixed with epoxy resin and prepared into 10 wt%, 20 wt% and 30 wt%. **Figure 6.** presents the results obtained from the conducted tests. Based on the data, tensile strength higher when the fibre content at 30 wt% for 4 mm, 7 mm and 10 mm of fibre length respectively. In contrast, the flexure strength showed a fluctuating pattern. They concluded that the excess of bamboo fibers in composite materials deteriorate the mechanical properties of the composite due to the lack of proper bonding between the matrix and fiber around their interface.

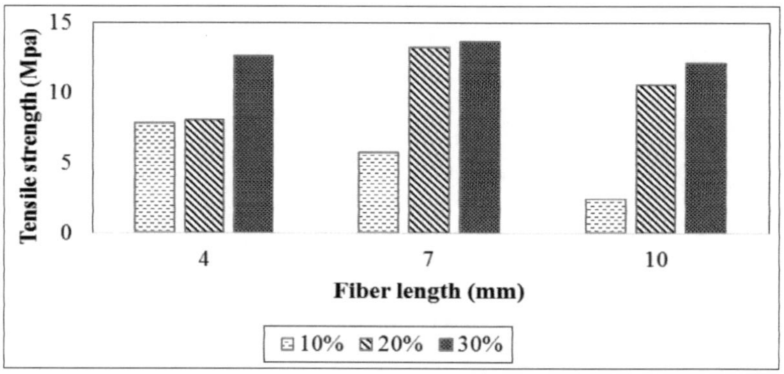

(a)

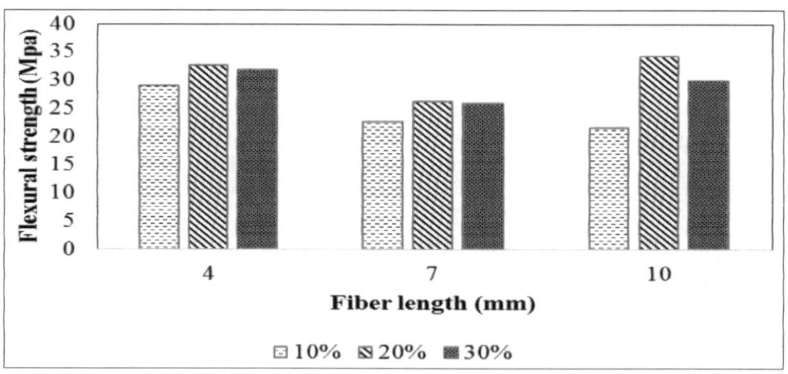

(b)

Figure 6. The data output of (a) tensile and (b) flexure strength by Gupta [30]

On the other hand, Anu Gupta et al. [31] investigated the influence of fibre loading and impingement angle on mechanical and erosion wear behaviour of bamboo/epoxy composites. The results showed the impact strength increased steadily with the increase in fibre loading from 0 to 20 weight%. While the peak erosion rate of the composites with different impingement angle was found to occur at 60° and 75°.

The effect of interfacial adhesion between bamboo fibre and epoxy

The lack of interfacial adhesion between bamboo fibre and epoxy polymer degrades the strength of bamboo/epoxy composite. Thus, chemical treatment was introduced to enhance the bonding between these two interphases. This treatment has been said able to improve the surface of natural fibre by removing the excess moisture, as a result better interlocking with polymer matrix.

Zhang et al. [32] chemically treated bamboo fibre with 4 wt% sodium hydroxide (NaOH) for 1 hour. Based on results from tensile test conducted, the present concentration has significant effect on the fibre tensile strength. Moreover, it was found that the treatment increases the effective surface area for better bonding with matrix by removal of bonding materials such as hemicellulose and lignin. However, according to Mishra et al. [33], higher concentration of NaOH worsened the properties of bamboo fibre. **Figure 7** shows the scanning electron microscopy (SEM) images between the untreated bamboo and alkali treated bamboo fibre [34].

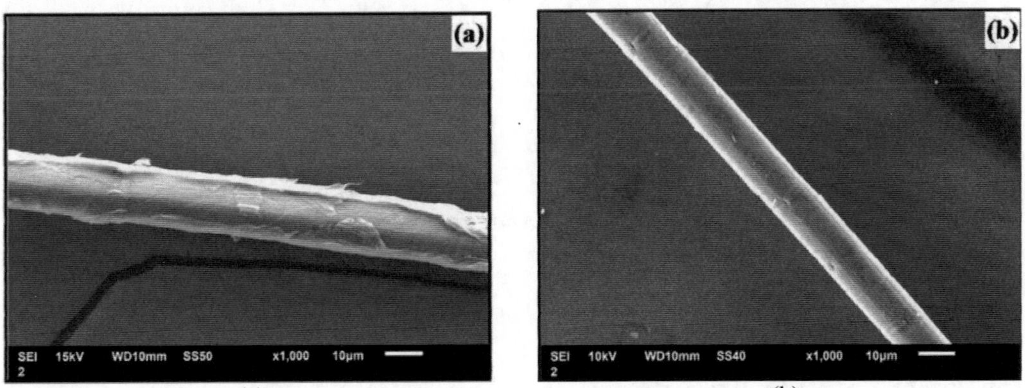

(a) (b)

Figure 7. Images of (a) untreated and (b) alkali treated bamboo fibres

6. Conclusion

Owing to the use of bamboo fibers in a wide range of engineering applications, it can provide a chance for economic development in the rural areas. Moreover, mechanical properties of bamboo reinforced polymer especially bamboo/epoxy composite is comparable to synthetic fibre polymer composites. However, many factors that affect the final mechanical properties of bamboo reinforced polymer including fibre length, fibre orientation, and treatment of the fibre have been studied. In order to improve the adhesion between matrix and bamboo fibre, the chemical treatment is introduced. Research that focusing on optimum the chemical concentration, mainly alkali treatment, should be carried out in order to enhance the interfacial properties of the laminate.

7. Acknowledgement

This study was supported by the Ministry of Higher Education grant No. PY/2016/07135.

References

[1] Holbery J and Houston D 2006 *J. of Maters.* **58** 80-86
[2] Alves C et al. 2005 *J. Clean. Prod.* **18** 313-327
[3] Ghavami K 2005 *Cem. Concr. Compos.* **27** 637-649
[4] Baillie C 2004 *Green Composites: Polymer Composites and the Environment*. CRC Press
[5] Wong K M 1989 *J. Am. Bamboo Soc.* **7** 1-15
[6] Othman A R and Mohamed A 1991 *FRIM Tech. Inf.* **5** 4
[7] Loh Y F 2010 *JEC Compos. Mag.* **55** 29
[8] Razak A O and Azmy H M 1991 *FRIM Tech. Inf.* **4** 6
[9] "Bamboo Flowering," *Bamboo Botanicals*, 2016. [Online]. Available: http://www.bamboobotanicals.ca/html/about-bamboo/bamboo-flowering.html. [Accessed: 05-Aug-2016].
[10] "Understanding the anatomy of bamboo," *Green Pot Enterprise*, 2016. [Online]. Available: http://greenpotenterprises.com/understanding-the-anatomy-of-bamboo/. [Accessed: 05-Aug-2016].
[11] Zhang Q S, Jiang S X and Tang Y Y 2002 Industrial utilization on bamboo: Technical report **26**
[12] Jiang Z H 2007 *People's Republic of China: China forestry publishing house*
[13] Pannipa C 2013 *J. Mater. Sci. Res.* 2 90-102.
[14] Liu H et al. 2008 *Compos. Part A* **39** 1891-1900
[15] Omar M et al. 2012 *Prog. Polym. Sci.* **37** 1552–1596
[16] Phong N T et al. 2012 *J. Mater. Sci. Res.* **1** 144–155
[17] Trujillo E et al. *Compos Part A Appl Sci Manuf* **61** 15–25
[18] Rassiah K 2013 *Aust. J. Basic Appl. Sci.* **7** 247–253
[19] Costa Jr A et al. 2014 *J. Compos. Mater.* 1– 13
[20] "Bamboo Raw materials for bioplastics," *Bamboo Technology Network Europe*, 2016. [Online]. Available: http://www.btn-europe.com/raw-materials-bioplastics.php.
[21] Beckermann G and Pickering K 2008 *Compos. Part A* **39** 979–988
[22] Rouison D, Sain M and Couturier 2006 *Compos. Sci. Technol.* **66** 895–906
[23] Harish S et al. 2009 *Mater. Charact.* **60** 44–49
[24] Brahmakumar M, Pavithrab C and Pillai R 2005 *Compos. Sci. Technol.* **65** 563–569
[25] Mussig J 2008 *J. Polym. Environ.* **16** 94–102
[26] Verma C and Chariar V 2012 *Compos. Part B* **43** 1063–1069
[27] Verma C and Chariar V 2013 *Compos. Part B* **45** 369-376
[28] Chandrasekhar V et al. 2014 *Inter. J. of Eng. Res. and Tech.* **3** 132-134
[29] Roslan S et al. 2015 *Int. J. Automot. Mech. Eng* **12** 2882-2892
[30] Gupta S K "A study on mechanical behavior of bamboo fiber based polymer composites".
[31] G Anu et al. 2011 *Int. J. Polym. Sci.* **2011** 1–10
[32] Zhang X, Wang F and Leon M 2015 *Materials (Basel)* **8** 6597-6608
[33] Mishra S et al. 2008 *Compos Sci Technol* **63** 1377–1385
[34] Bachtiar D, Sapuan S and Hamdan M 2008 *Mater. Des.* **29** 1285–1290

Elastic-Plastic Behaviour of Ultrasonic Assisted Compression of Polyvinyl Chloride (PVC) Foam

N. A. D. Muhalim*, M. Z. Hassan, Y. Daud
UTM Razak School of Engineering and Advanced Technology,
Universiti Teknologi Malaysia, 54100 Kuala Lumpur, Malaysia.
*damiramuhalim@gmail.com

Abstract. The present study aims to investigate the elastic-plastic behaviour of ultrasonic assisted compression of PVC closed-cell foam. A series of static and ultrasonic compression test of PVC closed-cell foam were conducted at a constant cross head speed of 30 mm/min on dry surface condition. For quasi-static test, specimen was compressed between two rigid platens using universal testing machine. In order to evaluate the specimen behavior under ultrasonic condition, specimen was placed between a specifically design double-slotted block horn and rigid platen. The horn was designed and fabricated prior to the test as a medium to transmit the ultrasonic vibration from the ultrasonic transducer to the working specimen. It was tuned to a frequency of 19.89 kHz in longitudinal mode and provided an average oscillation amplitude at 6 μm on the uppermost surface. Following, the characteristics of stress-strain curves for quasi-static and ultrasonic compression tests were analyzed. It was found that the compressive stress was significantly reduced at the onset of superimposed ultrasonic vibration during plastic deformation.

1. Introduction

Multifunctional materials that are lightweight, strong, stiff and tough attracts many engineering disciplines including marine, aircrafts, industrial, wind energy, recreation, road and rail applications. Such material is called polyvinyl chloride (PVC) closed-cell foam which has received attention since 1970s. It provides superior strength to weight ratio which has widely used as core materials in sandwich structures for all composite applications which require damage tolerance and weight saving [1, 2]. Formulations has been refined over the years and the characteristics of PVC closed-cell foams fit the needs of other application as well.

Today, PVC closed-cell foam could be extended to the high frequency cyclic force such as in aircraft structure, high speed machine and gas turbine components. Therefore, this is a requirement to identify the response of this material to this condition. Several studies have been conducted to investigate the properties of PVC closed-cell foam following quasi-static and dynamic conditions in order to determine the material responses [2, 3]. The compressive behavior at low and high strain rate as well as the energy absorption had also been discovered [4, 5]. However, a specific study on the effect of ultrasonic vibration on foam is still limited.

Ultrasonic application is subject to the feasibility of the ultrasonic technology for generation and transmission of acoustic energy at the required intensity and frequency. This interesting phenomena that are associated with intense and inaudible acoustics waves have many promising fields of ultrasonic application which involve scientific, engineering, chemical, industrial, biotechnology and medical [6]. Here, ultrasonic was used in high-intensity applications which refer to power ultrasonic to change the physical and chemical properties of the materials or systems to which it is applied. Common operating frequencies for high-power ultrasonic application is range between 20 and 100 kHz [7].

The versatility of ultrasonic has been exemplified by the wide range of applications that emerge for this technology. Many attempts have been made on the effects of ultrasonic vibration on the mechanical properties of solid metal [8, 9]. It was found that by superimposing an alternating stress on a workpiece, it highly affected the process rate and reduced the force needed for deformation of the workpiece [8]. The effectiveness of superimpose ultrasonic vibration during the deformation process was related to the friction condition, process rate, vibration mode, ultrasonic frequency and amplitude, material properties and type of deformation [9].

This study had triggered the developments of ultrasonic vibration on a material other than the normally metal. Hence, this study aims to investigate the elastic-plastic behaviour of ultrasonic assisted compression on PVC closed-cell foam. It is expected that through in-depth understanding of material response under high cyclic stress provides useful information for PVC closed-cell foam design, fabrication and application.

2. Finite Element Modelling of Ultrasonic Horn

Prior to the ultrasonic compression test, a specifically design of double-slotted block horn was adapted to enable ultrasonic excitation at a frequency of 20 kHz in longitudinal mode. The horn was made of aluminum with dimension 100x100x100 mm and density of 2712 kg/m3. It was fabricated with a double-slotted through both its width and thickness and simulated in a commercial Finite Element (FE) simulation code, ABAQUS 6.14. The slotting configurations were included in this design to maximize vibration amplitude uniformity at the working space. Figure 1 shows the detail of the final profile of this horn.

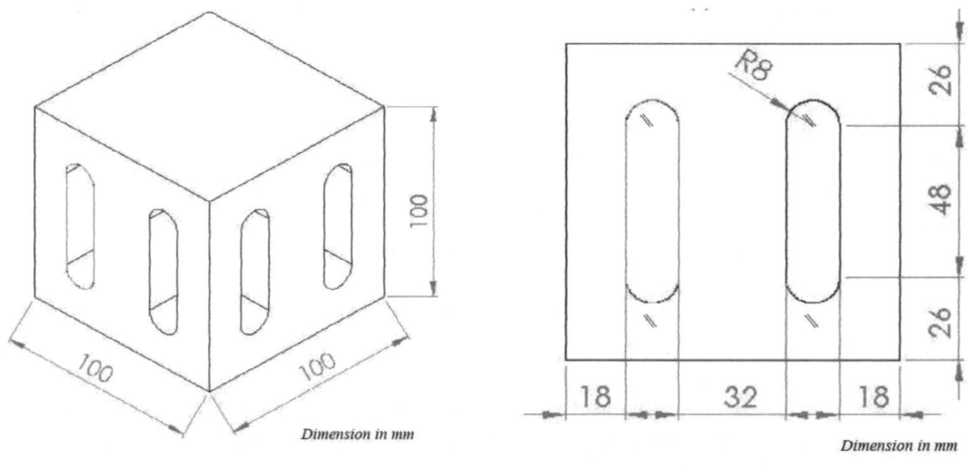

(a) Isometric view (b) Side view

Figure 1. Designed profile horn.

A three dimensional model was developed to calculate the longitudinal natural frequency and mode shape of the horn. A static and dynamic responses of loaded structures is analyzed using a frequency analysis to define the mode shape and natural frequencies with free boundary condition. A vibrational analysis consists of two prescribed step which were frequency step using Lanczos to obtain natural frequencies within the specified frequency range and steady-state dynamics direct step. The frequency step predicts natural frequencies and mode shapes of the system and the steady-state dynamics direct step predicts the stress and displacement amplitude for the applied loading condition. Figure 2 depicts the mode shape of the longitudinal vibrational mode of a double-slotted block horn derived from the FE modelling.

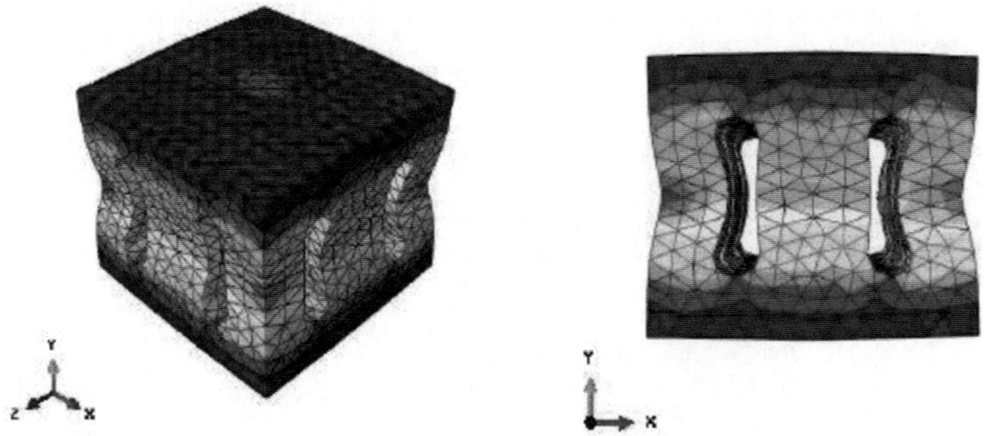

Figure 2. Longitudinal mode shape diagram.

Figure 3 depicts the measured frequency response of a unique nodal on the horn top surface in y-direction which is in longitudinal mode. The extensional length frequency obtained by the simulation was 19.89 kHz. Referring to Figure 3, a clear spectrum at maximum frequency was obtained, as a result from the designed double-slotted block horn at displacement of 0.7 mm.

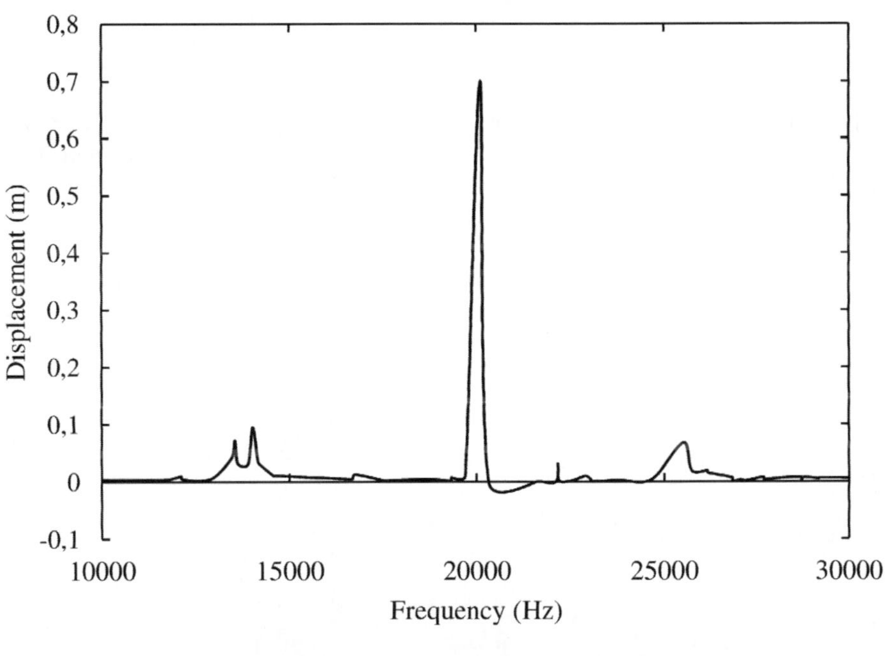

Figure 3. Frequency response in y-direction.

Figure 4 depicts the actual ultrasonic compression test set-up condition. The ultrasonic system consists of several fundamental elements which are electrical power source, ultrasonic generator and ultrasonic transducer. Initially, the ultrasonic generator converts the frequency and power characteristics of the electrical energy received from the power line as required to operate the ultrasonic transducer which attached to a double-slotted block horn. Then, the horn vibrated at 20 kHz frequency in longitudinal mode and transfer the vibrational energy to the working specimen.

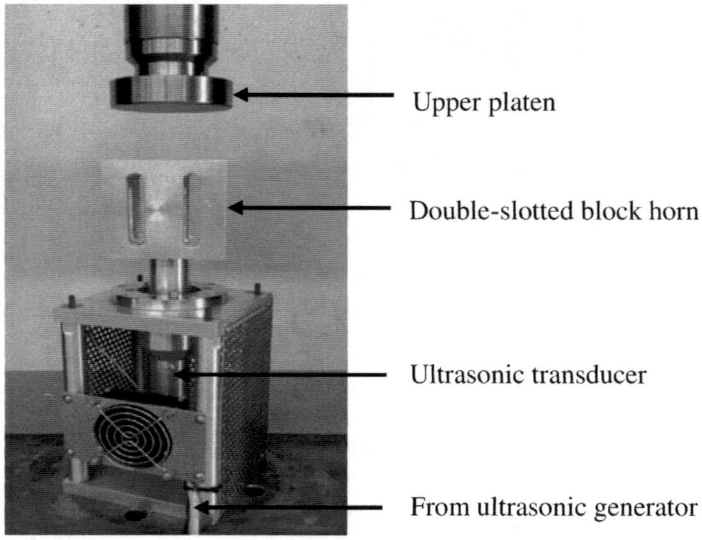

Figure 4. Ultrasonic compression test set-up.

3. Material and Testing

3.1. Specimen Preparation

Cylindrical specimen of PVC closed-cell foam as shown in Figure 5, were used in the static and ultrasonic compression test. Three types of PVC foam which differed by its density were used. Specimens were cut by using laser cutting machine with a laser power of 65 W and cutting speed of 10 mm/s. Details of the foams are given in Table 1.

Table 1. Characterization data for the PVC closed-cell foam samples.

Diameter (mm)	Height (mm)	Density, ρ^* (kg/m^3)	Relative density, ρ^* / ρ_s
20	20	40	0.02857
20	20	60	0.03929
20	20	80	0.05714

ρ_s = 1400 kg/m^3

Figure 5. A cylindrical specimen of PVC closed foam.

3.2. Quasi-Static and Ultrasonic Compression Test

3.2.1. Quasi-static test procedure. A series of static compression test of PVC closed-cell foam was conducted at a constant cross head speed of 30 mm/min on dry surface condition. Normal procedure for the quasi-static compression test was conducted where the specimens was statically compressed between two rigid platens.

3.2.2. Ultrasonic test procedure. The test procedure was then repeated for the ultrasonic compression test, but the lower rigid platen was replaced with a double-slotted block horn. Two sets of ultrasonic compression test were conducted. For the first set, the ultrasonic vibration was applied during post-yield for three short intervals by switching on the ultrasonic generator for duration of 1 second. Successively, when the ultrasonic vibration was discontinued, static compression was continued between these intervals. For the second set, the ultrasonic excitation was applied continuously from the onset of plastic deformation to the completion of the test. Both quasi-static and ultrasonic compression test specimens were stopped until the specimens reduced for more than 50% of its original length. Load-displacement data were recorded through hardware and software of the universal testing machine.

4. Results and Discussion

A typical static compression stress-strain curve is depicted in Figure 6. Foams were initially deformed in elastic behaviour until it strained at approximate 0.04 mm/mm. A steady-state stress was continued after it reach plastic collapse stress and followed by a regime of densification where the stress rises steeply. The densification shown a rapid rise in stress starting from the end of the steady-state region. It was observed that the lowest foam density undergo a larger deformation compared to other foam. It is because the foam had low value of steady-state stress and therefore able to withstand a larger deformation as it comes up to densification. In addition, a comparison of compressive strength with several density of PVC foam showed that for increasing density, the plastic collapse strength increased up to 1.46 MPa. Mechanical properties were evaluated as a function of foam density using three type of compression test frame as shown in Table 2.

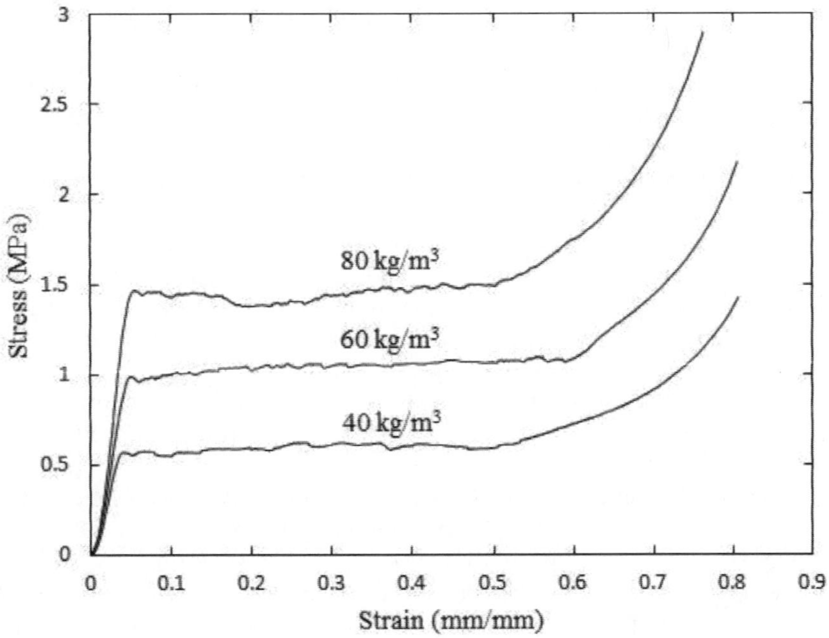

Figure 6. Stress-strain curve for static compression of 40, 60 and 80 kg/m3 of PVC closed-cell foams.

Table 2. Mechanical properties of PVC closed-cell foam.

Type of compression	Density (kg/m^3)	Plastic collapse stress, σ_{pl} (MPa)	Compressive modulus, E_c (MPa)	Steady-state stress, σ_{ss} (MPa)	Stress reduction, σ_r (MPa)	Percentage of stress reduction (%)	Densification strain, ε_D (%)
Static	40	0.57	21.34	0.59	0.00	0.00	52.60
	60	0.99	28.68	1.05	0.00	0.00	59.09
	80	1.46	40.17	1.44	0.00	0.00	55.59
Intermittent ultrasonic	40	0.63	17.90	0.49	0.37	58.96	62.32
	60	1.02	20.11	0.71	0.63	61.95	69.82
	80	1.54	29.25	0.83	0.96	62.71	64.82
Continuous ultrasonic	40	0.61	15.39	0.06	0.35	57.77	79.82
	60	1.02	25.67	0.30	0.63	62.17	79.17
	80	1.54	44.28	0.44	1.05	68.55	79.97

The stress-strain curves for quasi-static and ultrasonic compression are plotted in Figure 7. Foams were initially deformed in elastic behaviour until it strained at a similar value from the previous test. Figure 7 (a) shows the intermittent ultrasonic compression where the stress drop drastically after yield when the ultrasonic vibration was applied and start to regain repetitively at a lower stress value than the yield stress. This process continues for three irregular intervals, corresponding to the three times application of ultrasonic vibration. Then, the stress rises rapidly until it reach more than 50 % of its height reduction. The stress variation in the plastic deformation regime can be explained by the repetitive ultrasonic vibration application. It was observed that the stress reduced when ultrasonic vibration applied.

Significant stress reduction has been recorded for the continuous ultrasonic compression as shown in Figure 7 (b) when the ultrasonic vibration applied after it reached plastic collapse stress. It can be observed that the stress reduced up to 50 % when the ultrasonic excitation applied. In addition, the densification regime was also eliminated since the stress continue to drop until the completion of the test. It suggested that the excitation of high frequency at lower platen has remarkably reduced the stress.

Details of the mechanical properties for intermittent and continuous ultrasonic compression test are summarize in Table 2. Besides the remarkable stress reduction given by the intermittent and continuous ultrasonic compression test, densification strain value for continuous ultrasonic compression test were the highest compared to other test. It suggested that the application of ultrasonic excitation at the onset of plastic deformation until completion of the test has restrict the stress from rising steeply for densification process to take place. This is occurred because the foams has the lowest value of steady-state stress and therefore able to withstand a larger deformation such as it eliminate the densification regime.

It was clearly observed that ultrasonic compression has remarkably lowered the stress distribution compared to static compression. It has also been reported by the previous researchers [10-12] who showed that the stress flow of compressive deformation reduced by superimposing ultrasonic vibration on the static load in compression test. This phenomena was defined as Blaha effect or known as acoustoplastic effect. It was described as a decrease in the flow stress during deformation at an increase strain rate during plastic deformation under a constant stress [13-15]. Another researchers [10-12] who claimed that the stress reduction was due to the change of elastic-plastic properties and material softening that occurred under the influence of ultrasonic vibration.

It is allegedly reported that the exact time when the ultrasonic excitation is superimposed to the static compression could also influence the amount of stress reduction. This is because if the ultrasonic vibration applied during the weakest point of the compressed foam, definitely further stress drop is obtained due to these combined effects. Besides, another researcher showed that the material flow stress in ultrasonic compression test was under the influence of friction [16]. It occurred when the oscillatory velocity of the lower platen exceeds the specimen velocity. Friction vector is a result from the relative motion between the lower platen and specimen which normally acts in perpendicular direction to the specimen motion. But the friction vector is reversed when the lower platen is vibrated. This is because the motion of the lower platen is parallel to the specimen motion, such that it assists the motion to the specimen in the working direction and therefore reduced the friction force then stress.

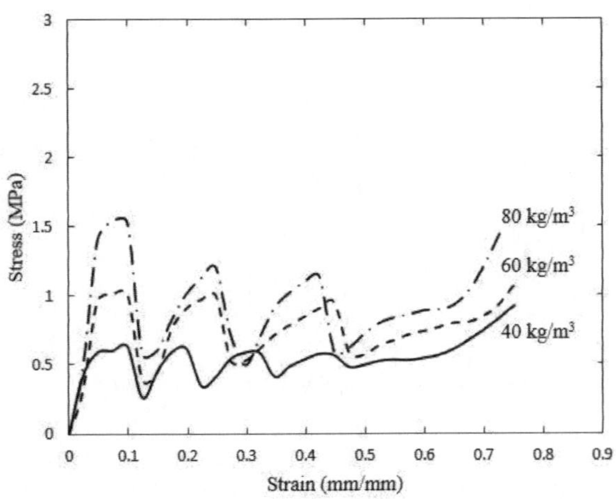

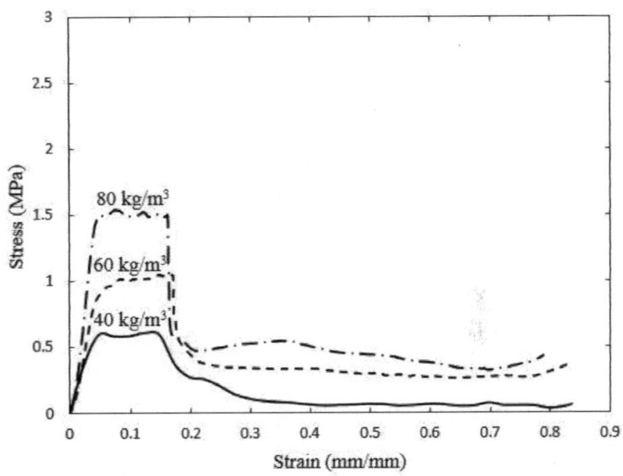

(a) (b)

Figure 7. Comparison of the three different density of PVC closed-cell foam under (a) intermittent ultrasonic and (b) continuous ultrasonic compression test results.

The influence of superimposing ultrasonic vibration during compression test on the energy absorption capability was studied. Energy absorption subjected to foam density were calculated. This can be understood by observing the bar chart in Figure 8. To ensure the value of energy absorption are comparable within tests and specimens, the calculation was limited to 50 % of strain and below. The highest energy absorption was recorded by the highest density of foam. This is because if the density is too low, the densification zone is reached and a very high force is obtain before all the energy has been dissipated. On the contrary, if the density is too high, the force exceeds the critical value before enough energy has been absorbed, while the material compressive strain remain only partially utilized [17]. Another substantial observation from the figure is that the energy absorption of foam was reduced when ultrasonic vibration applied in the test. This means that the ability of the foams to absorb energy was lowered by the application of ultrasonic vibration in the compression test. The material softening effect described earlier was proven by this lowered energy absorption as it reduced the material toughness. It is therefore, with the presence of ultrasonic vibration during compression test, foams can withstand low resistance to deformation when stressed.

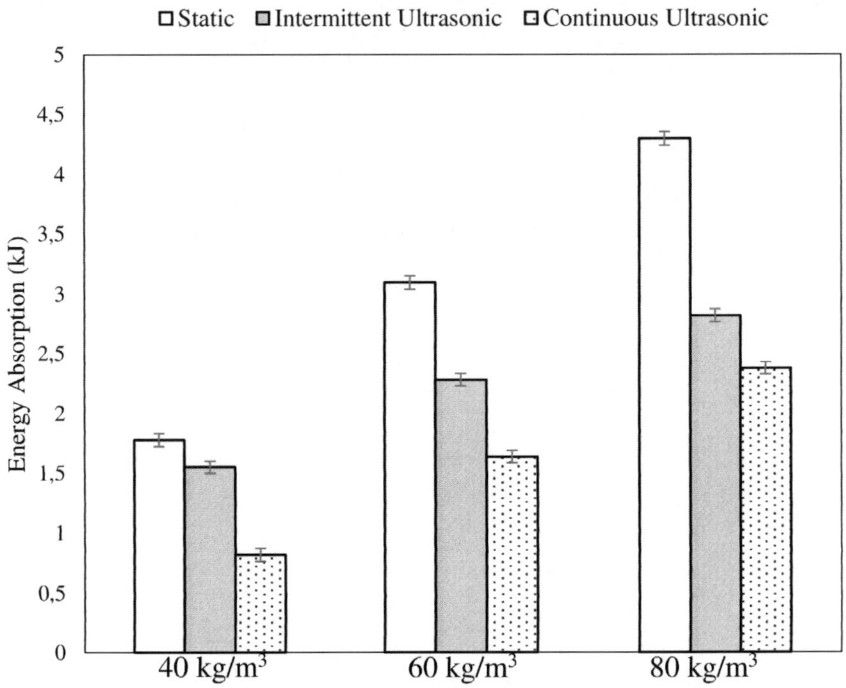

Figure 8. Energy absorption for quasi-static and ultrasonic compression of PVC closed-cell foam.

5. Conclusion

The application of ultrasonic vibration during the plastic deformation process of compression test on foam has remarkably reduced compressive stress and lowered the energy absorption. This effects was related to the material softening such as it exhibited a special temporary softening material properties during the application of ultrasonic vibration. The elastic-plastic behaviour of compressed foam under the ultrasonic compression test is similar to the normally compressed metal in the literatures. It demonstrated that superimposed ultrasonic excitation on the compression test worked effectively in other materials, particularly foam apart of metal. Further investigation on the underlying mechanisms of the stress reduction can be carried out and it is expected that through in-depth understanding of material response under high cyclic stress provides useful information for PVC closed-cell foam design, fabrication and application.

Acknowledgement

This study was supported by the Ministry of Higher Education grant No. PY/2016/07135.

References

[1] Stewart, R., At the core of lightweight composites. Reinforced Plastics, 2009. 53(3): p. 30-35.
[2] Atas, C. and C. Sevim, On the impact response of sandwich composites with cores of balsa wood and PVC foam. Composite Structures, 2010. 93(1): p. 40-48.
[3] Assarar, M., A. El Mahi, and J.-M. Berthelot, Evaluation of the dynamic properties of PVC foams under flexural vibrations. Composite Structures, 2012. 94(6): p. 1919-1931.
[4] Luong, D.D., D. Pinisetty, and N. Gupta, Compressive properties of closed-cell polyvinyl chloride foams at low and high strain rates: Experimental investigation and critical review of state of the art. Composites Part B: Engineering, 2013. 44(1): p. 403-416.
[5] Gibson, L.J. and M.F. Ashby, Cellular solids: structure and properties. 1999: Cambridge university press.

[6] Abramov, O.V., High-intensity ultrasonics: theory and industrial applications. Vol. 10. 1999: CRC Press.
[7] Gallego-Juárez, J.A. and K.F. Graff, Power Ultrasonics: Applications of High-intensity Ultrasound. 2014: Elsevier.
[8] Blaha, F. and B. Langenecker, Ultrasonic investigation of the plasticity of metal crystals. Acta Metallurgica, 1959. 7(2): p. 93-100.
[9] Astashev, V.K. and V.I. Babitsky, Ultrasonic processes and machines: dynamics, control and applications. 2007: Springer Science & Business Media.
[10] Abdul Aziz, S. and M. Lucas. The effect of ultrasonic excitation in metal forming tests. in Applied Mechanics and Materials. 2010. Trans Tech Publ.
[11] Hung, J.-C. and C. Hung, The influence of ultrasonic-vibration on hot upsetting of aluminum alloy. Ultrasonics, 2005. 43(8): p. 692-698.
[12] Daud, Y., M. Lucas, and Z. Huang, Modelling the effects of superimposed ultrasonic vibrations on tension and compression tests of aluminium. Journal of Materials Processing Technology, 2007. 186(1–3): p. 179-190.
[13] Blaha, F. and B. Langenecker, Tensile deformation of zinc crystal under ultrasonic vibration. Naturwissenschaften, 1955. 42(556): p. 0.
[14] Lucas, M., A. Gachagan, and A. Cardoni, Research applications and opportunities in power ultrasonics. Proceedings of the Institution of Mechanical Engineers, Part C: Journal of Mechanical Engineering Science, 2009. 223(12): p. 2949-2965.
[15] Malygin, G., Acoustoplastic effect and the stress superimposition mechanism. Physics of the Solid State, 2000. 42(1): p. 72-78.
[16] Ibrahim, I.N., The mechanics of ultrasonic tube bending. 1983, Aston University.
[17] Avalle, M., G. Belingardi, and R. Montanini, Characterization of polymeric structural foams under compressive impact loading by means of energy-absorption diagram. International Journal of Impact Engineering, 2001. 25(5): p. 455-472.

IOP Conf. Series: Materials Science and Engineering **344** (2018) 012010 doi:10.1088/1757-899X/344/1/012010

Controlling corrosion rate of Magnesium alloy using powder mixed electrical discharge machining

M A Razak[1, 2], A M A Rani[1, 3], N M Saad[3], G Littlefair[4] and A A Aliyu[1, 5]

[1]Mechanical Engineering Department, Universiti Teknologi PETRONAS, Bandar Seri Iskandar, 32610, Perak, Malaysia
[2]Manufacturing Section, Universiti Kuala Lumpur Malaysian Spanish Institute, Kulim Hi-Tech Park, 09000, Kedah, Malaysia
[3]Biomedical Technology, Universiti Teknologi PETRONAS, Bandar Seri Iskandar, 32610, Perak, Malaysia
[4]School of Engineering, Faculty of Science, Engineering and Built Environment, Deakin University, Geelong, VIC 3220, Australia
[5]Department of Mechanical Engineering, Faculty of Engineering, Gwarzo Road, Bayero University Kano, Nigeria

Email: alhapis@unikl.edu.my

Abstract. Biomedical implant can be divided into permanent and temporary employment. The duration of a temporary implant applied to children and adult is different due to different bone healing rate among the children and adult. Magnesium and its alloys are compatible for the biodegradable implanting application. Nevertheless, it is difficult to control the degradation rate of magnesium alloy to suit the application on both the children and adult. Powder mixed electrical discharge machining (PM-EDM) method, a modified EDM process, has high capability to improve the EDM process efficiency and machined surface quality. The objective of this paper is to establish a formula to control the degradation rate of magnesium alloy using the PM-EDM method. The different corrosion rate of machined surface is hypothesized to be obtained by having different combinations of PM-EDM operation inputs. PM-EDM experiments are conducted using an opened-loop PM-EDM system and the in-vitro corrosion tests are carried out on the machined surface of each specimen. There are four operation inputs investigated in this study which are zinc powder concentration, peak current, pulse on-time and pulse off-time. The results indicate that zinc powder concentration is significantly affecting the response with 2 g/l of zinc powder concentration obtaining the lowest corrosion rate. The high localized temperature at the cutting zone in spark erosion process causes some of the zinc particles get deposited on the machined surface, hence improving the surface characteristics. The suspended zinc particles in the dielectric fluid have also improve the sparking efficiency and the uniformity of sparks distribution. From the statistical analysis, a formula was developed to control the corrosion rate of magnesium alloy within the range from 0.000183 mm/year to 0.001528 mm/year.

1. Introduction

The quality of an implant is important to determine the success of implantation [1]. Biomedical implant can be divided into permanent and temporary use. A temporary implant should allow sufficient duration for newly generated tissue to replace the bone defect [2-5]. A second surgery to dissemble the temporary implant can be avoided by applying biodegradable implant [6]. Magnesium and its alloys have been established as being compatible for biodegradable implant [7-10]. Currently, the same type of implant is applied on both children and adult with only considering the difference in size. The real situation in the fractured bone healing process is that the different healing rate among the children and adult requires different corrosion rate of biodegradable implant. A temporary implant is normally applied to the children aged under six years for up to six months and adult for up to ten months [11-13]. However, it is difficult to control the degradation rate of magnesium alloy to be applied for children or adult biodegradable implant application. A method is required to control the degradation rate of magnesium alloy during the fabrication process. The objective of this paper is establishing a formula to control the degradation rate of magnesium alloy using the PM-EDM method. The different corrosion rate of machined surface was hypothesized to be obtained by having different combinations of PM-EDM operation inputs.

Machining magnesium alloy using conventional methods, such as milling, turning, and drilling, causes a built-up edge and chatter. The most important precaution needed to bear in mind whilst machining magnesium alloy is that the formation of fine chips and dust are highly flammable. The melting point of magnesium is 650°C, and this metal is only stable below its melting point. Non-traditional machining, such as the EDM process is preferable to machine magnesium based material, especially to machine the high geometrical accuracy required by bio-medical implants [14, 15].

Recently, new exploratory research works have been initiated to improve the efficiency of EDM process using PM-EDM method, a modified EDM process by adding powder in the dielectric fluid [16]. In PM-EDM, the added conductive particles help to improve the sparking efficiency during the ignition process. It was found that the addition of conductive particles as shown in Figure 1 led to increase the gap size between the tool electrode and the workpiece which subsequently ensuing in a reduction of the electrical discharge power density [17]. The added conductive particles have also lowered the dielectric strength, creating early and uniform electric discharges at low energy to improve the machined surface quality. In PM-EDM process, the added particles can be suspended in the dielectric fluid in the same machining tank or a separate tank. The use of a stirrer and circulating pump are to ensure the uniform distribution of the conductive particles [18-23].

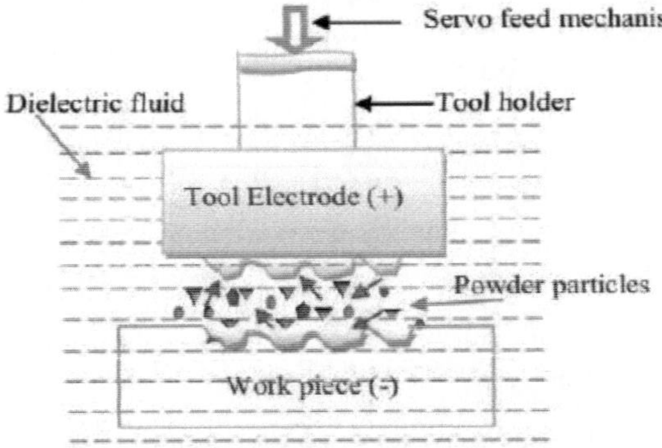

Figure 1. Electrical discharge during PM-EDM process [17].

Most of the researchers reported the advantages of PM-EDM in reducing the surface roughness, material removal rate, tool wear rate, and the modification of the machined surface [24-26]. For instance, it was found that the added titanium powder in the PM-EDM process on SKD61, SKD11, and SKT4 die steels improved the material removal rate by 42.1% as compared to conventional EDM (C-EDM) method [24]. On the other hand, it was suggested that some of the removed materials got deposited on the machined surface [27]. It was also reported that a significant decreased in surface crack density on the machined surface with silicon powder PM-EDM [28]. Figure 2 indicates a better surface quality obtained by the PM-EDM method compared to C-EDM method which possesses several micro cracks. This phenomenon will result in reducing the metals' corrosion rate. Previously, an investigation indicates that the surface roughness of magnesium alloy processed using the PM-EDM method has been reduced by 44% [29]. However, there was no study have been done to investigate the effect of PM-EDM method on the corrosion rate of magnesium alloy. The advantages of the PM-EDM method deserve deeper clarification in order to meet the needs of high precision industries, such as the bio-medical and aerospace industries.

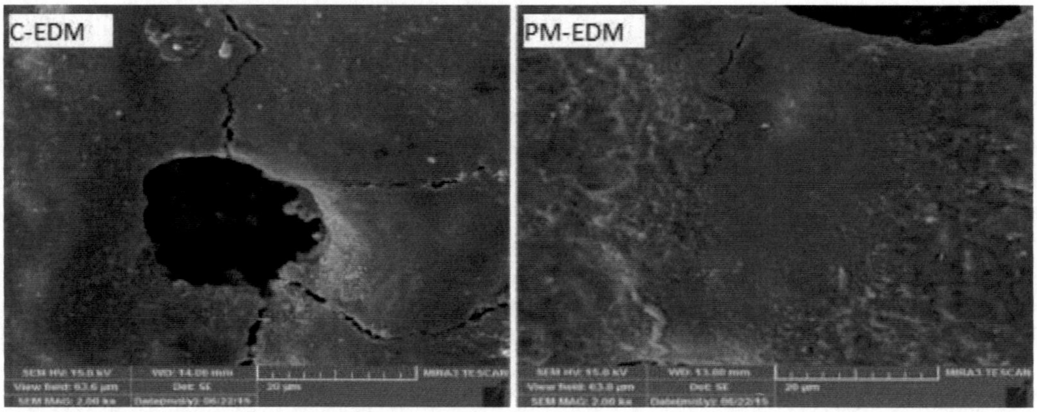

Figure 2. SEM micrographs of the machined surface by C-EDM and PM-EDM [30].

2. Materials and methods

PM-EDM experiments were conducted using Mitsubishi EA8 EDM die sinker with attached to opened-loop PM-EDM dielectric circulation system. The workpiece material used in this research was an AZ31 magnesium alloy and the tool electrode was copper. A constant cutting depth of 2 mm was maintained throughout the experiments. A preliminary experiment was conducted to determine the value range of the operation inputs. This research was conducted with four operation inputs as shown in Table 1. Each experiment was repeated diligently three times to ensure data accuracy. Zinc nanoparticles with an average size of 80 nm with three different concentrations were mixed in the dielectric fluid. Zinc was selected in this investigation due to the biocompatibility and it was one of the major components of the AZ31 magnesium alloy. Zinc is an effective alloying element for magnesium and can assist to improve its corrosion resistance [31]. By the deposition of zinc particles on the machined surface, the corrosion rate of magnesium alloy is hypothesized to be reduced significantly because the electrode potential of zinc (-0.76 V) is higher and more protected than magnesium (-2.37 V) [32].

Table 1. Experimental operation inputs.

Operation Input	Value
Zinc powder concentration	1, 2 and 3 g/l
Peak current	38, 47 and 55 A
Pulse on-time	16, 32 and 64 μs
Pulse off-time	128, 256 and 512 μs

The in-vitro corrosion tests were conducted with complying the ASTM G59-97 [33] and the simulated body fluid with components as shown in Table 2 were used. The corrosion rate of the machined surface was analyzed using the CH1608E electrochemical analyzer. Results from the corrosion test were analyzed using Design Expert software. Responses from the analysis of variance (ANOVA) were fitted into quadratic mathematical model.

Table 2. Hank's solution chemical composition [34].

Component	g/l
Sodium chloride, NaCl	8.00
Potassium chloride, KCl	0.40
Calcium chloride, $CaCl_2$	0.14
Sodium bicarbonate, $NaHCO_3$	0.35
Magnesium chloride hexahydrate, $MgCl_2.6H_2O$	0.60
Magnesium sulfate heptahydrate, $MgSO_4.7H_2O$	0.06
Sodium phosphate dibasic, Na_2HPO_4	0.06
Potassium phosphate monobasic, KH_2PO_4	0.60
Glucose.$2H_2O$	1.0
pH	6.8

3. Results and discussion

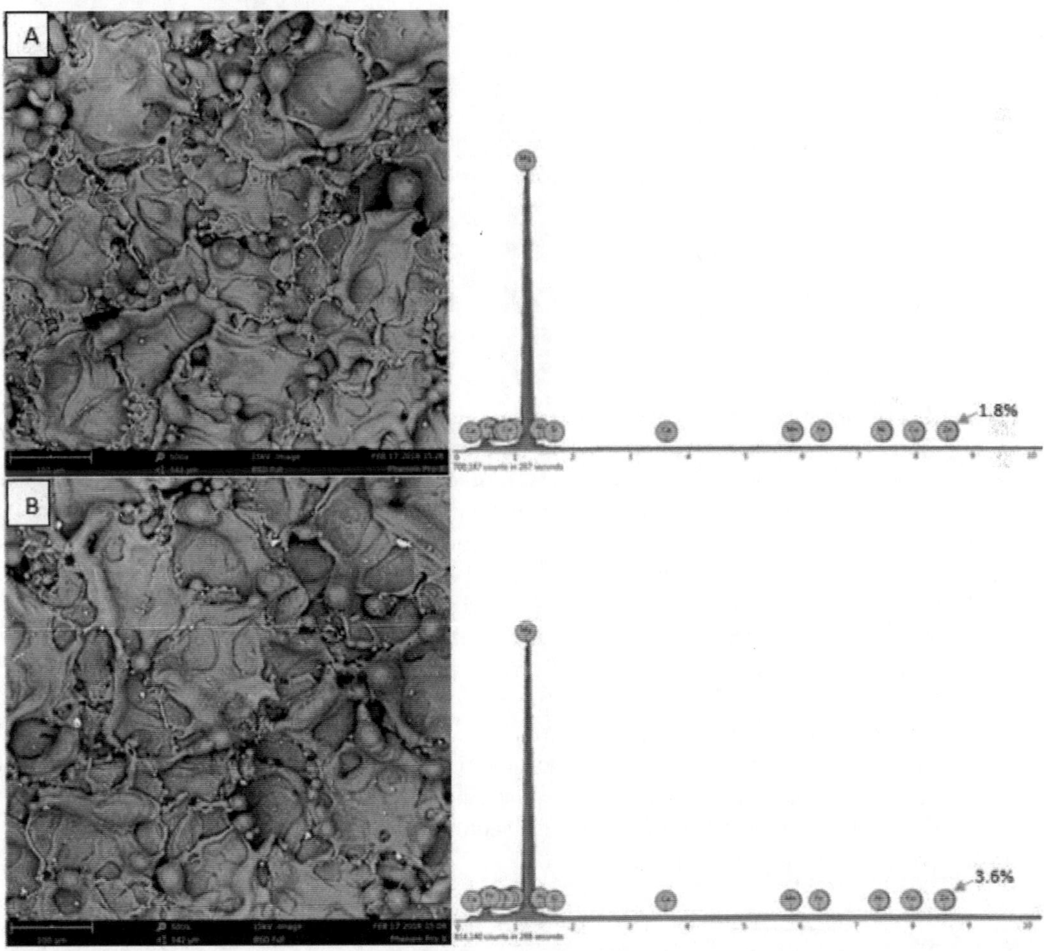

Figure 3. SEM images and EDX spectrum of magnesium alloy processed by; A) C-EDM method and B) PM-EDM method.

Initially, an experiment has been conducted using both C-EDM and PM-EDM method on the magnesium alloy specimen by applying the same operation inputs (47 A peak current, 80 V gap voltage, 16 μs pulse on-time and 512 μs pulse off-time). The objective was to evaluate the feasibility of the PM-EDM method in improving machined surface quality. The zinc particles with 1 g/l powder concentration were used in PM-EDM experiment. The surface roughness obtained from PM-EDM experiment was 4.971 μm. The result shows a reduction of 10.61% surface roughness compared to C-EDM method (5.561 μm). The element composition analysis indicates that a higher weight percentage of zinc was found from the specimen processed by the PM-EDM method which was 1.8% higher than that from C-EDM method as presented in Figure 3. Some of the zinc particles mixed in the dielectric fluid get deposited on the machined surface during the experiment due to high localized temperature at the cutting zone. The suspended zinc particles in the dielectric fluid have also improve the sparking efficiency and the uniformity of sparks distribution [30].

The morphology disclosed from both machined surfaces was not much different. There were micro cracks, craters, globules and voids formed on both machined surfaces. From the corrosion test results as presented in Figure 4, it was found that the specimen processed by PM-EDM method obtained 24.15% lower corrosion rate than the specimen processed by C-EDM method. From this fact, the PM-EDM method seems promising to improve the machined surface quality and reduce the corrosion rate of magnesium alloy.

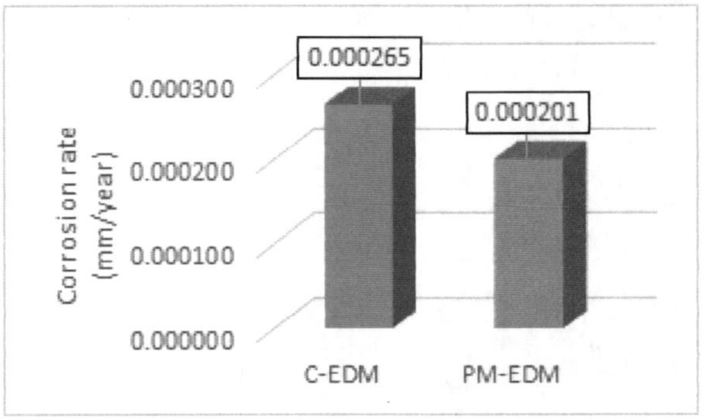

Figure 4. Corrosion rate of magnesium alloy processed by C-EDM and PM-EDM method.

Figure 5 presents the PM-EDM main effects plot of means with data response of means for corrosion rate. From the response table of means, pulse off-time with the delta value 0.000804 mm/year was identified as the most significant operation input affecting the changes of magnesium alloy corrosion rate. The results confirmed that the pulse off-time was inversely proportional to the response. The lower pulse off-time results in higher corrosion rate. The higher pulse off-time results in lower corrosion rate. This was due to the smoother machined surface obtained by the higher pulse off-time, hence obtaining the lower corrosion rate and vice versa. The zinc powder concentration with the delta value 0.000212 mm/year was also significantly affect the corrosion rate, which was more significant compared to peak current with the delta value 0.000172 mm/year and pulse on-time with the delta value 0.000157 mm/year. The presence of zinc particles in the dielectric fluid and high localized temperature at the cutting zone causes some zinc particles to get melted and deposited on the machined surface, hence improving the surface structure compared to the processed without zinc particles suspended in the dielectric fluid.

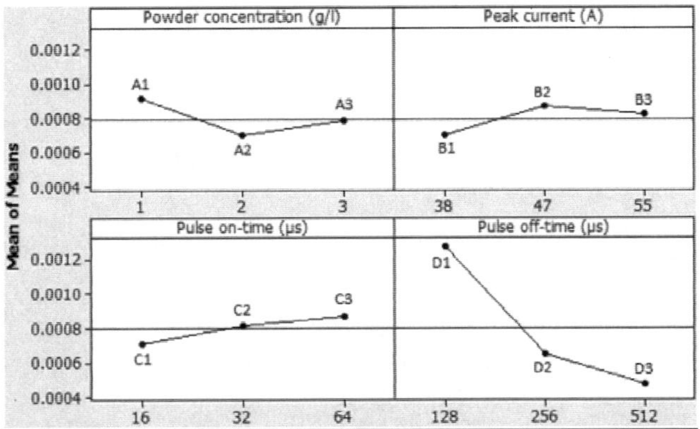

Level	Powder concentration (g/l)	Peak current (A)	Pulse on-time (µs)	Pulse off-time (µs)
1	0.000910	0.000698	0.000710	0.001275
2	0.000698	0.000870	0.000817	0.000647
3	0.000786	0.000825	0.000866	0.000471
Delta	0.000212	0.000172	0.000157	0.000804
Rank	2	3	4	1

Figure 5. Main effects plot of means with data response of means for corrosion rate (mm/year).

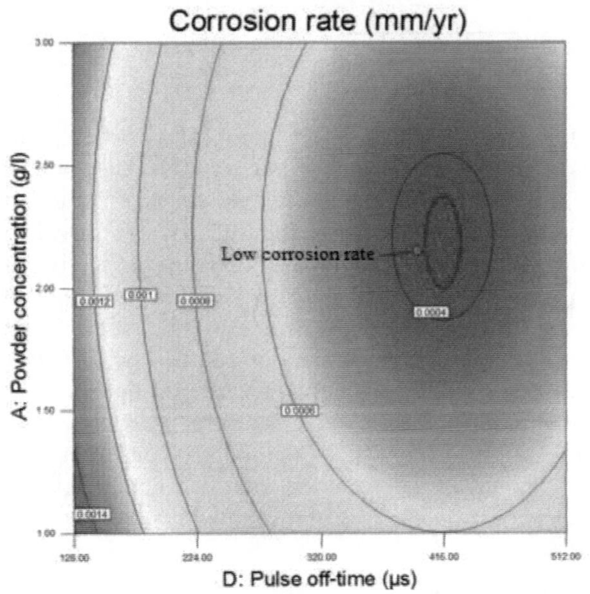

Figure 6. Interaction between zinc powder concentration and pulse off-time affects the corrosion rate of magnesium alloy.

Figure 6 presents the interaction between zinc powder concentration and pulse off-time affecting the corrosion rate of magnesium alloy machined surface processed by PM-EDM method. Within the range from 128 µs to 512 µs, the lowest corrosion rate has been obtained at 416 µs. Meanwhile, the highest corrosion rate was obtained at lowest pulse off-time, 128 µs. The higher pulse off-time applied in the EDM process contributing in better flushes of debris from the cutting area. A longer resting time allows the reduction of heat on the machined surface and the electrode. This phenomenon results in smoother machined surface with less formation of the cracks, craters and voids.

IOP Conf. Series: Materials Science and Engineering **344** (2018) 012010 doi:10.1088/1757-899X/344/1/012010

Table 3. ANOVA for response surface quadratic model.

Source	Sum of squares	df	Mean square	p-value
Model	3.51E-06	8	4.39E-07	< 0.0001
A-Zinc powder concentration	6.96E-08	1	6.96E-08	< 0.0001
B-Peak current	7.22E-08	1	7.22E-08	< 0.0001
C-Pulse on-time	1.1E-07	1	1.1E-07	< 0.0001
D-Pulse off-time	2.91E-06	1	2.91E-06	< 0.0001
A^2	5.74E-08	1	5.74E-08	< 0.0001
B^2	2.78E-08	1	2.78E-08	< 0.0001
C^2	7.7E-09	1	7.7E-09	< 0.0001
D^2	3.25E-07	1	3.25E-07	< 0.0001
Residual	2.09E-12	10	1.31E-13	
Cor Total	3.51E-06	24		

The interaction analysis reveals that there were interactions between all operation inputs during PM-EDM experiments. All four inputs were depending on each other in influencing the response. An interaction occurred when the effect of one input changes depending on the level of another input. It happened because the operation inputs determining the size of discharge energy. However, the ANOVA explicates that the interactions between two operation inputs were not significant affecting the machined surface corrosion rate since all the p-values for the interactions were more than 0.1. They were AB (zinc powder concentration with peak current), AC (zinc powder concentration with pulse on-time), AD (zinc powder concentration with pulse off-time), BC (peak current with pulse on-time), BD (peak current with pulse off-time) and CD (pulse on-time with pulse off-time). Therefore, these measurements were withdrawn from the ANOVA reduced quadratic analysis as presented in Table 3 to obtain the formula on PM-EDM operation inputs to control the corrosion rate of magnesium alloy. The model has eight degrees of freedom and total correlations of 24 degrees of freedom. The determined constants for final equation in terms of actual factors are:

Corrosion rate = -7.81276×10^{-4}
Zinc powder concentration = -6.62901×10^{-4}
Peak current = 1.42548×10^{-4}
Pulse on-time = 1.19244×10^{-5}
Pulse off-time = -9.13054×10^{-6}
Squared zinc powder concentration = 1.50184×10^{-4}
Squared peak current = -1.45265×10^{-6}
Squared pulse on-time = -1.08257×10^{-7}
Squared pulse off-time = 1.09936×10^{-8}

The quadratic mathematical model is given as equation (1). The formula has been elaborated and the final formula is shown in equation (2) where CR is the corrosion rate, W is the zinc powder concentration, X is the peak current, Y is the pulse on-time and Z is the pulse off-time.

$$CR = \beta_0 + \beta_W W + \beta_X X + \beta_Y Y + \beta_Z Z + \beta_{WW} W^2 + \beta_{XX} X^2 + \beta_{YY} Y^2 + \beta_{ZZ} Z^2 \quad (1)$$

$$CR = (-7.81276 \times 10^{-4}) + (-6.62901 \times 10^{-4})W + (1.42548 \times 10^{-4})X + (1.19244 \times 10^{-5})Y + (-9.13054 \times 10^{-6})Z + (1.50184 \times 10^{-4})W^2 + (-1.45265 \times 10^{-6})X^2 + (-1.08257 \times 10^{-7})Y^2 + (1.09936 \times 10^{-8})Z^2 \quad (2)$$

With the established formula, the corrosion rate of magnesium alloy can be forecasted by changing the values of PM-EDM operation inputs as long as the values are within the limits. The range of values for each operation inputs are:

- Zinc powder concentration: 1 g/l to 3 g/l
- Peak current: 38 A to 55 A
- Pulse on-time: 16 μs to 64 μs
- Pulse off-time: 128 μs to 512 μs

The lowest corrosion rate computed by this formula is 0.000183 mm/year from the combination of 2 g/l zinc powder concentration, 38 A peak current, 16 μs pulse on-time and 512 μs pulse off-time. Meanwhile, the highest corrosion rate computed by this formula is 0.001528 mm/year from the combination of 1 g/l zinc powder concentration, 47 A peak current, 64 μs pulse on-time and 128 μs pulse off-time. To verify the formula, a confirmation test was conducted by applying the combination of 1 g/l zinc powder concentration, 38 A peak current, 16 μs pulse on-time and 128 μs pulse off-time. The predicted value was 0.0012 mm/year and the actual value was 0.0013 mm/year. It brought to 91.67% similarity between the predicted and actual values.

4. Conclusions

It can be concluded that the pulse off-time was the most significant operation input affecting the corrosion rate of magnesium alloy with the higher pulse off-time ensuing in lower corrosion rate. The zinc particles mixed in the dielectric fluid was also found significantly affecting the changes of the corrosion rate of magnesium alloy. Among the three levels of zinc powder concentration examined, the 2 g/l zinc powder concentration obtained the lowest corrosion rate. The different combination of PM-EDM operation inputs was found results in different surface characteristics on the machined surfaces hence resulting in different corrosion rate. The presence of zinc particles at the cutting zone contribute into improving sparking efficiency and uniformity of sparks distribution. The deposition of zinc particles on the machined surface results in lower smoother machined surface with less formation of micro cracks. These factors influencing the changes of corrosion rate on the machined surface. With the developed formula, the corrosion rate of magnesium alloy can be controlled by controlling PM-EDM operation inputs. For future investigation, it is recommended to conduct in-vivo corrosion test using an implant processed by PM-EDM method.

References

[1] M. Genisa, Z. A. Rajion, D. Mohamad, A. Pohchi, M. R. A. Kadir, and S. Shuib, "Effect of Different Angle Scanning on Density Estimation Based on Hounsfield Unit on CT and CBCT," *Sains Malaysiana,* vol. 44, pp. 1331-1337, 2015.

[2] H. Park, J. S. Temenoff, and A. G. Mikos, "Biodegradable orthopedic implants," in *Engineering of Functional Skeletal Tissues*, ed: Springer, 2007, pp. 55-68.

[3] L. Nassif and M. El Sabban, "Mesenchymal stem cells in combination with scaffolds for bone tissue engineering," *Materials,* vol. 4, pp. 1793-1804, 2011.

[4] J. L. Olson, A. Atala, and J. J. Yoo, "Tissue engineering: current strategies and future directions," *Chonnam Medical Journal,* vol. 47, pp. 1-13, 2011.

[5] J. S. Temenoff and A. G. Mikos, "Injectable biodegradable materials for orthopedic tissue engineering," *Biomaterials,* vol. 21, pp. 2405-2412, 2000.

[6] P. H. Long, "Medical devices in orthopedic applications," *Toxicologic Pathology,* vol. 36, pp. 85-91, 2008.

[7] M. T. Andani, N. S. Moghaddam, C. Haberland, D. Dean, M. J. Miller, and M. Elahinia, "Metals for bone implants. Part 1. Powder metallurgy and implant rendering," *Acta Biomaterialia,* vol. 10, pp. 4058-4070, 2014.

[8] P. R. Cha, H. S. Han, G. F. Yang, Y. C. Kim, K. H. Hong, S. C. Lee, *et al.*, "Biodegradability engineering of biodegradable Mg alloys: Tailoring the electrochemical properties and microstructure of constituent phases," *Scientific Reports,* vol. 3, pp. 1-6, 2013.

[9] F. Witte, "The history of biodegradable magnesium implants: a review," *Acta Biomaterialia,* vol. 6, pp. 1680-1692, 2010.

[10] S. Zhang, X. Zhang, C. Zhao, J. Li, Y. Song, C. Xie, *et al.*, "Research on an Mg–Zn alloy as a degradable biomaterial," *Acta Biomaterialia,* vol. 6, pp. 626-640, 2010.

[11] O. Islam, D. Soboleski, S. Symons, L. Davidson, M. Ashworth, and P. Babyn, "Development and duration of radiographic signs of bone healing in children," *American Journal of Roentgenology,* vol. 175, pp. 75-78, 2000.

[12] W. Jackson, T. Theologis, C. Gibbons, S. Mathews, and G. Kambouroglou, "Early management of pathological fractures in children," *Injury,* vol. 38, pp. 194-200, 2007.

[13] P. V. Giannoudis, E. Jones, and T. A. Einhorn, "Fracture healing and bone repair," *Injury,* vol. 42, pp. 549-550, 2011.

[14] F. Klocke, M. Schwade, A. Klink, and A. Kopp, "EDM machining capabilities of Magnesium (Mg) alloy WE43 for medical applications," *Procedia Engineering,* vol. 19, pp. 190-195, 2011.

[15] M. A. Razak, A. M. Abdul-Rani, T. V. V. L. N. Rao, S. R. Pedapati, and S. Kamal, "Electrical discharge machining on biodegradable AZ31 magnesium alloy using Taguchi method," *Procedia Engineering,* vol. 148, pp. 916-922, 2016.

[16] A. Singh and R. Singh, "Effect of powder mixed electric discharge machining (PMEDM) on various materials with different powders: A review," *International Journal for Innovative Research in Science and Technology,* vol. 2, pp. 164-169, 2015.

[17] M. A. Tawfiq and A. S. Hameed, "Effect of powder concentration in PMEDM on surface roughness for different die steel types," *International Journal of Current Engineering and Technology,* vol. 5, pp. 3323-3329, 2015.

[18] H. Kansal, S. Singh, and P. Kumar, "Technology and research developments in powder mixed electric discharge machining (PMEDM)," *Journal of Materials Processing Technology,* vol. 184, pp. 32-41, 2007.

[19] S. Mohal, H. Kumar, and S. Kansal, "Nano-finishing of materials by powder mixed electric discharge machining (PMEDM): A review," *Science of Advanced Materials,* vol. 7, pp. 2234-2255, 2015.

[20] A. M. Nanimina, A. M. A. Rani, and T. L. Ginta, "Assessment of powder mixed EDM: A review," in *MATEC Web of Conferences*, 2014, p. 04018.

[21] S. Kumar and U. Batra, "Surface modification of die steel materials by EDM method using tungsten powder-mixed dielectric," *Journal of Manufacturing Processes,* vol. 14, pp. 35-40, 2012.

[22] M. A. Razak, A. M. Abdul-Rani, and A. M. Nanimina, "Improving EDM efficiency with silicon carbide powder-mixed dielectric fluid," *International Journal of Materials, Mechanics and Manufacturing,* vol. 3, pp. 40-43, 2015.

[23] Y. F. Tzeng and C. Y. Lee, "Effects of powder characteristics on electrodischarge machining efficiency," *The International Journal of Advanced Manufacturing Technology,* vol. 17, pp. 586-592, 2001.

[24] B. T. Long, N. H. Phan, N. Cuong, and V. S. Jatti, "Optimization of PMEDM process parameter for maximizing material removal rate by Taguchi's method," *The International Journal of Advanced Manufacturing Technology,* pp. 1-11, 2016.

[25] A. Bhattacharya, A. Batish, and N. Kumar, "Surface characterization and material migration during surface modification of die steels with silicon, graphite and tungsten powder in EDM process," *Journal of Mechanical Science and Technology,* vol. 27, pp. 133-140, 2013.
[26] P. Janmanee, "Performance of Cu-Cr-Zr electrodes in electrical discharge machining of tungsten carbide composite material base using Taguchi method," *International Journal of Machining and Machinability of Materials,* vol. 18, pp. 412-425, 2016.
[27] A. Batish and A. Bhattacharya, "Mechanism of material deposition from powder, electrode and dielectric for surface modification of H11 and H13 die steels in EDM process," *Materials Science Forum,* vol. 701, pp. 61-75, 2012.
[28] C. Prakash, H. Kansal, B. Pabla, and S. Puri, "Experimental investigations in powder mixed electric discharge machining of Ti–35Nb–7Ta–5Zrβ-titanium alloy," *Materials and Manufacturing Processes,* pp. 1-13, 2016.
[29] A. Abdul-Rani, M. Razak, G. Littlefair, I. Gibson, and A. Nanimina, "Improving EDM Process on AZ31 Magnesium Alloy towards Sustainable Biodegradable Implant Manufacturing," *Procedia Manufacturing,* vol. 7, pp. 504-509, 2016.
[30] M. Shabgard and B. Khosrozadeh, "Investigation of carbon nanotube added dielectric on the surface characteristics and machining performance of Ti–6Al–4V alloy in EDM process," *Journal of Manufacturing Processes,* vol. 25, pp. 212-219, 2017.
[31] M. M. Avedesian and H. Baker, *ASM specialty handbook: magnesium and magnesium alloys*: ASM international, 1999.
[32] R. K. Rajput, *Material Science and Engineering*, 3rd ed. New Delhi: S.K. Kataria & Sons, 2012.
[33] A. Standard, "G59–97: Standard Test Method for Conducting Potentiodynamic Polarization Resistance Measurements," *West Conshohocken, PA: ASTM International,* 2009.
[34] A. R. H. Bidhendi and M. Pouranvari, "Corrosion study of metallic biomaterials in simulated body fluid," *Metalurgija,* vol. 17, pp. 13-22, 2011.

IOP Conf. Series: Materials Science and Engineering **344** (2018) 012011 doi:10.1088/1757-899X/344/1/012011

The effect of environmental factors on selected mechanical properties of zirconium dioxide

W Wirwicki [1], A Andrzejewska[1], M Andryszczyk[1] and P Siemianowski[1]

[1] UTP University of Science and Technology in Bydgoszcz, Faculty of Mechanical Engineering, Department of Biomedical Engineering, Al. Prof. S. Kaliskiego 7, 85-796 Bydgoszcz, Poland

Email: mateusz.wirwicki@utp.edu.pl

Abstract. In many centers around the world, research studies are carried out on the mechanical strength of dental materials and glued joints. A literature review shows the variety of testing techniques related to analyzing the strength and durability of the material itself and the glued joints. In dental ceramics, zirconium dioxide is most often used as a base material, and chemically it consists of 97% ZrO2 and 3% Y2O3. This study was to determine the mechanical properties of zirconium dioxide under different environmental conditions. The material is used for the production of dental crowns and tooth bridges in the CAD/CAM technology. This medium is currently one of the most advanced-generation materials used for prosthetic and implant restorations. They were then subjected to a three-point bending test on the Instron ElektroPlus E3000 durability machine. Storage conditions and time have a positive influence on reducing variation in zirconium resistance for active forces and destructive stresses.

1. Introduction

In many centers around the world, research studies are carried out on the mechanical strength of dental materials and glued joints. A literature review shows the variety of testing techniques related to analyzing the strength and durability of the material itself and the glued joints. The most extensive research development was observed for dental adhesives and materials used for the production of dental crowns. Research results related to the analysis of mechanical properties are used in dental practice to improve the quality of service. In dental ceramics, zirconium dioxide is most often used as a base material, and chemically it consists of 97% ZrO2 and 3% Y2O3. Another issue with modern prosthetics is the aging of dental materials in the oral cavity. Teeth are mechanically exploited on a daily basis under changing physical and chemical conditions (including temperature and pH changes). Under such conditions, the teeth can lose their natural colour, damage the enamel, and weaken their structure and age. In order to minimize these destructive factors, the authors of the paper [2] present the processes of regeneration of the natural tooth. A review of the scientific literature allows to list four basic research methods for zirconium dioxide with other dental materials or natural teeth. These research methods are durability testing for: bending, stretching, shearing, compression. The bending tests are carried out on all dental materials in order to find out and analyze their basic mechanical properties [1].

This study was to determine the mechanical properties of zirconium dioxide under different environmental conditions.

IOP Conf. Series: Materials Science and Engineering **344** (2018) 012011 doi:10.1088/1757-899X/344/1/012011

2. Methodology

Zirconium dioxide called 3M ESPE was used for static testing. The material is used for the production of dental crowns and tooth bridges in the CAD/CAM technology. This medium is currently one of the most advanced-generation materials used for prosthetic and implant restorations. All samples were subjected to an the authors' own machining technique, which consisted of trimming with a Buehler ISOMET 5000 saw blade and grinding with a 1200 grit sandpaper. The final machining was performed using the Mazak Vertical Center Smart 430A milling machine. Ceramic materials used in dentistry are characterized by shrinkage during the process of curing and sintering. According to the manufacturer of the Cyrkon Lava material, technological shrinkage is at 20%. Prior to static testing, samples were measured with a Mitutoyo caliper. Two samples with dimensions of 1.5 mm x 1.5 mm (A and A') and large samples with dimensions of 2 mm x 2.5 mm (B and B') were used for the analysis. Results of the geometry analysis are presented in Table 1.

Table 1. Analysis of the geometry of the samples tested from zirconium dioxide

Dimensions	Mean values of width and height in mm	Standard deviation in mm	Relative standard deviation in%
1.5 mm x 1.5 mm	1.46	0.039	2.68
	1.44	0.036	2.51
2 mm x 2.5 mm	2.03	0.09	4.42
	2.43	0.07	3.01

The samples were divided into four groups (A, A ', B, B') of 30 samples per group. The samples of 1.5 mm x 1.5 mm (A') and 2 mm x 2.5 mm (B') in the number of 30 were immersed in a NaCl 0.9% saline solution. The samples were kept in that solution for 3 years. They were then subjected to a three-point bending test on the Instron ElektroPlus E3000 durability machine. The upper arm movement was at 0.5mm/min. A strain gauge of ± 5 kN was used to measure the force. The static three-point bending test was performed in accordance with the PN-EN 843-1 standard.

3. Results and Discussion

Statistical analysis was performed using the Statisticia 13 software. All descriptive data are shown in tables as mean, standard deviations and variation coefficient. The Schapiro-Wilk test was used to determine compliance with the normal distribution. The statistical comparison between the sample types was made for the mean using Student's t-test, and the Levene's test for the standard deviation. Specimens marked with A were non-normalized samples. Samples marked letter B is a normalized sample. Absence of apostrophe meant testing the samples after complete processing and synthesis. Samples marked with apostrophe in addition to complex. The treatments were stored in an environment that imitates the oral environment. The results of the descriptive statistics on the durability of the samples are presented in Table 2.

The statistical analysis does not indicate significant differences in the force required to destroy the samples due to storage conditions for both small samples (A) and large samples (B). It can only be noted that the destructive force has increased by approximately 18.37 N for large samples (B) stored under conditions that mimic the oral cavity environment.

Table 2. Results of descriptive durability statistics for three-point bending of different types of samples and comparison to normal distribution

Sample type	Mean	SD	Relative SD	Min	Max	Shapiro-Wilk test critical value	Shapiro-Wilk test p value
Fmax [N]							
A	185.07	33.69	18%	97.00	248.00	0.958	0.278
A`	189.17	9.64	5%	166.69	211.55	0.975	0.672
B	367.90	42.80	12%	283.00	461.00	0.973	0.628
B`	386.27	28.55	7%	315.71	418.96	0.919	0.090
Σmax							
A	947.56	182.35	19%	494.63	1299.40	0.951	0.175
A`	933.53	51.60	6%	827.90	1047.42	0.976	0.706
B	1151.46	115.88	10%	879.58	1380.80	0.981	0.840
B`	1109.54	75.92	7%	924.98	1259.60	0.988	0.980

An additional analysis of the results relying on Levene's test, in both cases, indicates the presence of statistically significant differences in standard deviation. In both cases, the spread of results in standard deviation was reduced. In the case of small samples (A), the deviation decreased by 24.05[N] and for large samples - by 14.24 [N]. The results obtained in this way point to zirconium stabilization that occurs in the aquatic environment.

An analogical analysis for the σmax results also fails to indicate any significant difference between the mean values. Damage stress results slightly decrease in the oral cavity imaging medium, namely by 14.04 MPa for the small samples (A) and 41.91MPa for the large samples (B), respectively. There was a statistically significant decrease in the result variability for the small samples (A) and the tendency to reduce the variability of the large samples (B). Also, these parameters can indicate the beneficial effects of oral conditions on zirconium stabilization.

Table 3. Statistical analysis of the results of comparison of sample results with respect to storage conditions by sample size

Sample	Mean		Mean difference	t Student test critical value	t Student test p Value	SD		SD difference	Levene's test critical value	Levene's test p value
	Dry	Wet				Dry	Wet			
Fmax [N]										
AvsA`	185.07	189.17	4.10	-0.64	0.526	33.69	9.64	-24.05	16.73	<0.001
BvsB`	367.90	386.27	18.37	-1.96	0.056	42.80	28.55	-14.24	3.99	0.042
σ_{max}										
AvsA`	947.56	933.53	-14.04	0.41	0.688	182.35	51.60	-130.75	16.68	<0.001
BvsB`	1151.46	1109.54	-41.91	1.66	0.104	115.88	75.92	-39.96	3.35	0.072

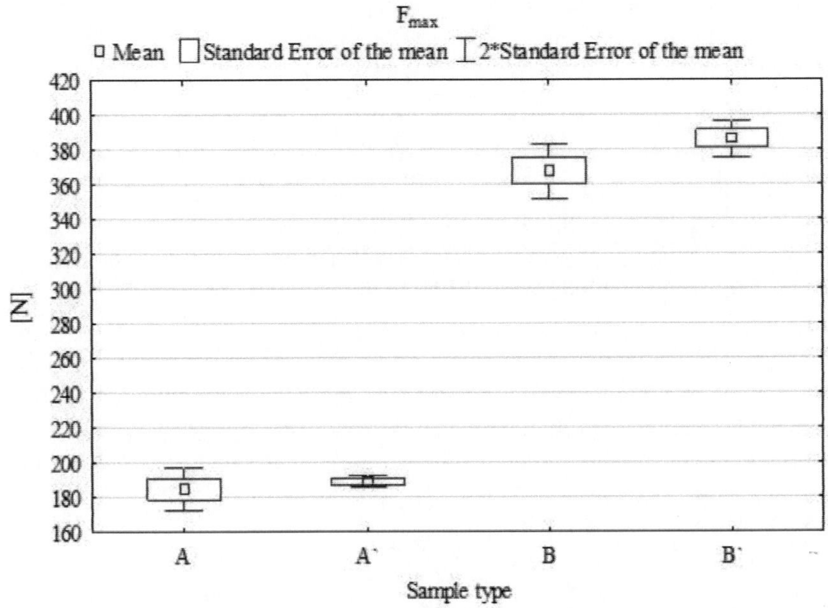

Figure 1. Breakdown of the results in terms of the force needed to destroy the sample by sample size and storage conditions

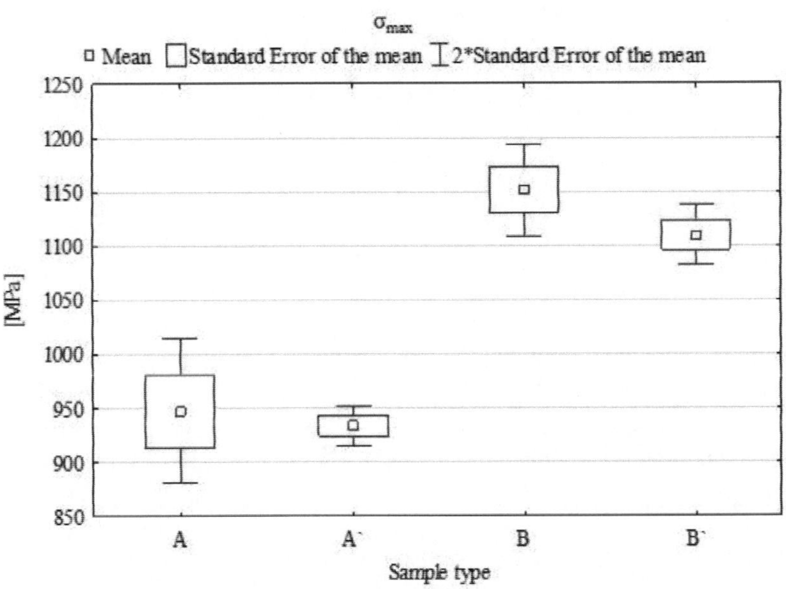

Figure 2. Breakdown of the results in terms of the force needed to destroy the sample by sample size and storage conditions

4. Conclusions

The following conclusions were drawn based on the conducted research: The study did not show any significant change in zirconium stored in oral-like conditions within the range of occurring forces and destructive stresses. Storage conditions and time have a positive influence on reducing variation in zirconium resistance for active forces and destructive stresses.

References

[1] S. S. Scherrer, M. Cattani-Lorente, E. Vittecoq, F. de Mestral, J. a Griggs, and H. W. A. Wiskott, "Fatigue behavior in water of Y-TZP zirconia ceramics after abrasion with 30 μm silica-coated alumina particles.," Dent. Mater., vol. 27, no. 2, pp. e28–42, Feb. 2011

[2] E. d'Incau, C. Couture, and B. Maureille, "Human tooth wear in the past and the present: tribological mechanisms, scoring systems, dental and skeletal compensations.," Arch. Oral Biol., vol. 57, no. 3, pp. 214–29, Mar. 2012

IOP Conf. Series: Materials Science and Engineering **344** (2018) 012012 doi:10.1088/1757-899X/344/1/012012

Uncanny valley: A preliminary study on the acceptance of Malaysian urban and rural population toward different types of robotic faces

T T Tay[1,*], Raymond Low[1], H J Loke[1], Y L Chua[2] and **Y H Goh[1]**

[1]Faculty of Engineering, Tunku Abdul Rahman University College, Jalan Genting Klang 53300 Setapak, Kuala Lumpur, Malaysia

[2]Department of Mechanical Engineering, College of Engineering, Universiti Tenaga Nasional, Jalan Ikram-Uniten, 43000, Kajang, Selangor, Malaysia

Email: taytt@acd.tarc.edu.my

Abstract. The proliferation of robotic technologies in recent years brings robots closer to humanities. There are many researches on going at various stages of development to bring robots into our homes, schools, nurseries, elderly care centres, offices, hospitals and factories. With recently developed robots having tendency to have appearance which increasingly displaying similarities to household animals and humans, there is a need to study the existence of uncanny valley phenomenon. Generally, the acceptance of people toward robots increases as the robots acquire increasing similarities to human features until a stage where people feel very uncomfortable, eerie, fear and disgust when the robot appearance become almost human like but not yet human. This phenomenon called uncanny valley was first reported by Masahiro Mori. There are numerous researches conducted to measure the existence of uncanny valley in Japan and European countries. However, there is limited research reported on uncanny valley phenomenon in Malaysia so far. In view of the different cultural background and exposure of Malaysian population to robotics technology compared to European or East Asian populations, it is worth to study this phenomenon in Malaysian context. The main aim of this work is to conduct a preliminary study to determine the existence of uncanny valley phenomenon in Malaysian urban and rural populations. It is interesting to find if there are any differences in the acceptance of the two set of populations despite of their differences. Among others the urban and rural populations differ in term of the rate of urbanization and exposure to latest technologies. A set of four interactive robotic faces and an ideal human model representing the fifth robot are used in this study. The robots have features resembling a cute animal, cartoon character, typical robot and human-like. Questionnaire surveys are conducted on respondents from urban and rural populations. Survey data collected are analysed to determine the preferred features in a humanoid robot, the acceptance of respondents toward the robotic faces and the existence of uncanny valley phenomenon. Based on the limited study, it is found that the uncanny valley phenomenon existed in both the Malaysian urban and rural population.

IOP Conf. Series: Materials Science and Engineering **344** (2018) 012012 doi:10.1088/1757-899X/344/1/012012

1. Introduction

The term "uncanny valley" refers to a sense of unease and discomfort when people look at increasingly realistic human. This concept was identified by Masahiro Mori, professor of engineering at Tokyo Institute of Technology as Bukimi no Tani Genshō in 1970 [1]. Prof Mori first reported on this phenomenon in a Japanese journal named Energy. In the subsequent years after its first publication, the paper received almost no attention. His work is only properly translated to English in 2012 [2]. He wrote a paper [1] on how he envisioned people's reactions to robots that looked and acted almost like a human. In particular, he hypothesized that a person's response to a human-like robot would abruptly shift from empathy to revulsion as it approached, but failed to attain, a life-like appearance. This descent into eeriness is known as the uncanny valley. Mori suggested that robot builders should not try to design and build human-like robots, either in motion or appearance. According to Mori's hypothesis, designers must seek a moderate level of realism for the physical appearance of robots and virtual reality agents in order to avoid falling into the uncanny valley [3].

However, with recent advancement in social robots where researchers start building robots with appearance and movements resembling human [4], the concept of uncanny valley has attracted attentions among researches. In recent years, many robots have and will be developed to work alongside humans at homes, offices, hospitals, hotels, etc. Whether the robots' appearance should look human or not is still being debated. If the robot should look human, then should the robot be a male or female? Has short or long hair? Has straight or curly hair? Fair or tanned? Which human features to be incorporated and which features that can be ignored? These are some of the many questions that will be in the mind of roboticists. Different individuals will invoke different feelings when encounter different types of robotic faces. They may be influenced by factors such as personal experiences, upbringing, education level, community they are in and so on. Some people might accept and like a human-like robotic faces while some people will express a certain degree of fear, insecurity and loathe toward the human-like robot.

There are numerous studies conducted in other parts of the world identifying the existence of uncanny valley phenomenon [3],[5],[6],[7]. However, most Malaysians have not heard or aware of uncanny valley phenomenon. There is no known work done yet as far as the authors' knowledge on determining whether this phenomenon exists among Malaysian population. In view of the different cultural background and exposure of Malaysian population to robotics technology compared to European or East Asian populations, it is worth to study this phenomenon in Malaysian context. Malaysian populations may be influenced by local factors such as local taboos, exposure to technology, cultural and social norm. This study to measure the existence of uncanny valley phenomenon among Malaysian urban and rural population is timely to gather preliminary data in order for roboticists to research and design inclusive robots that suit Malaysian populations' psyche.

2. Uncanny valley

Figure 1 below illustrated the concept of uncanny valley [2]. Industrial robots commonly used in factories, do not have a face or legs. Industrial robots can only rotate, extend and retract their articulated arm. Industrial robot's appearance does not look human-like. Industrial robots are designed based purely on their functionalities. Therefore, industrial robots must have functionalities that equal or exceed human while it does not matter whether they look human-like or not. People have less affinity to industrial robot because of its lack of similarities to human being.

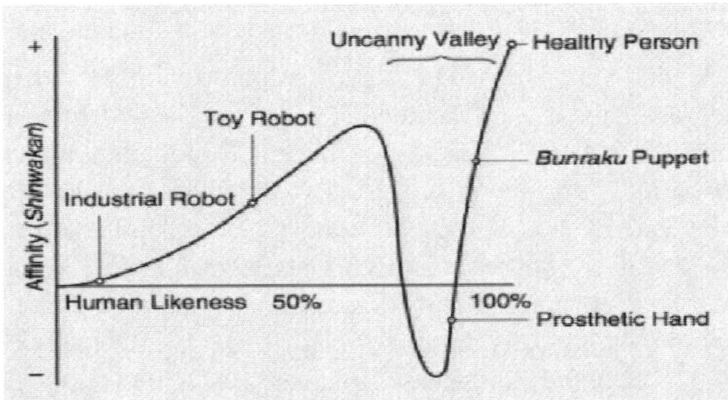

Figure 1. Graph affinity vs human likeness indicating the uncanny valley phenomenon [2].

On the other hand, in designing toy robots, robot designers will focus more on the robot's appearance rather than its functionalities. Toy robots typically have human-like features such as a face with a pair of eyes, body, two legs and two arms. There are many toy robots available in the market such as Paro and RoboHon as illustrated in Figure 2. Adult and children alike will be happy and enjoy interacting with these toy robots. Due to the advancement of silicon technology, people have difficulties in differentiating between a prosthetic hand and a real hand. Some of these prosthetic hands even have realistic skin colour, fingernails, veins and wrinkles [2]. These prosthetic hands are built very similar to human form. However, user will feel the stiffness and coldness during a handshake with its limp boneless grip and lack of tissue. Due to these feelings, people will lose affinity toward it and the hand will become uncanny although it looks real. This example describes the uncanny valley phenomenon. Mori doesn't think, on close inspection, a Bunraku puppet appears very similar to a human being. Its realism in terms of size, skin texture, and so on, does not even reach that of a realistic prosthetic hand. But when audiences enjoy a puppet show in the theatre, they are seated at a certain distance from the stage. The puppet's absolute size is ignored, and its total appearance, including hand and eye movements, is close to that of a human being. So, given human tendency as an audience to become absorbed in this form of art, audiences might feel a high level of affinity for the puppet [2]. Human beings themselves are the final goal of robotics designer, with 100% likeness and maximum affinity.

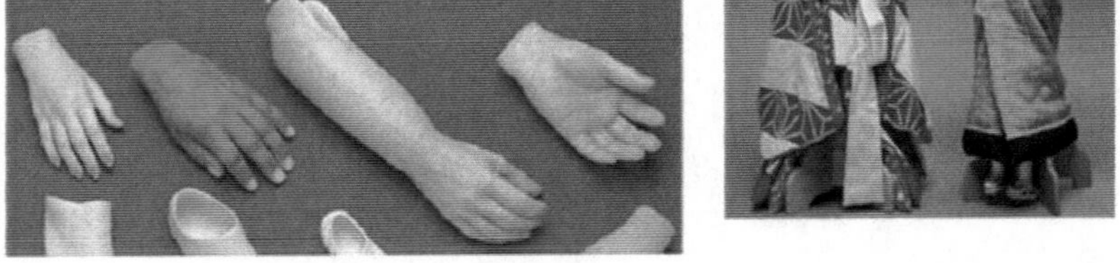

Figure 2. Paro, RoBoHon, Bunraku puppets and prosthetic hands (clockwise from top left).

3. Methodology

As illustrated in Figure 3, this work is divided into three phases. In the first phase, four suitable robotic faces are designed and fabricated. The robotic faces are designed to have separate distinct looks: Robot 1, cute animal; Robot 2, cartoon character; Robot 3, a typical robot look; and Robot 4, human-like. Robot 1, Robot 2 and Robot 3's characteristics are benchmarked to numerous popular cartoon and movie characters. Robot 4 is benchmarked to characteristics of a human being. The design of the robotic faces consists of mechanical design, prototyping, and electronics design. The robots' structures are designed and simulated in 3D CAD environment. The structures, levers, and motor brackets are 3D printed using plastic material. Miniature geared servomotors are used to actuate the eyebrows, lips, ears, face muscles, and head movement. Motion detection sensors are used to detect the presence of human in order for the robot to interact with human. Coloured light emitting diodes are used to represent the robots' eyes. The interactions are controlled by an embedded microcontroller. The controller is programmed to incorporate friendly and unfriendly behaviours on the robotic faces as illustrated in Figure 4. Phase two consists of designing questionnaire surveys and conducting surveys at selected urban and rural populations. The surveys are conducted at Setapak neighbourhood, Kuala Lumpur (urban) and Kelapa Sawit village, Johor (rural). A total of 50 respondents, 25 at urban area and 25 at rural area, volunteered to take part in the surveys. The respondents are selected randomly from the identified neighbourhood and village. The respondents' genders are spread almost equally with 52% male and 48% female. Figure 5 shows the experimental setup. In the surveys, besides the four robots, a pleasant looking human model is included as Robot 5 to represent an ideal robot as illustrated in Figure 6. Third phase involves analysis of the data collected. It is to determine the existence of uncanny valley among urban and rural population in Malaysia and to determine the preferred features in a humanoid robot.

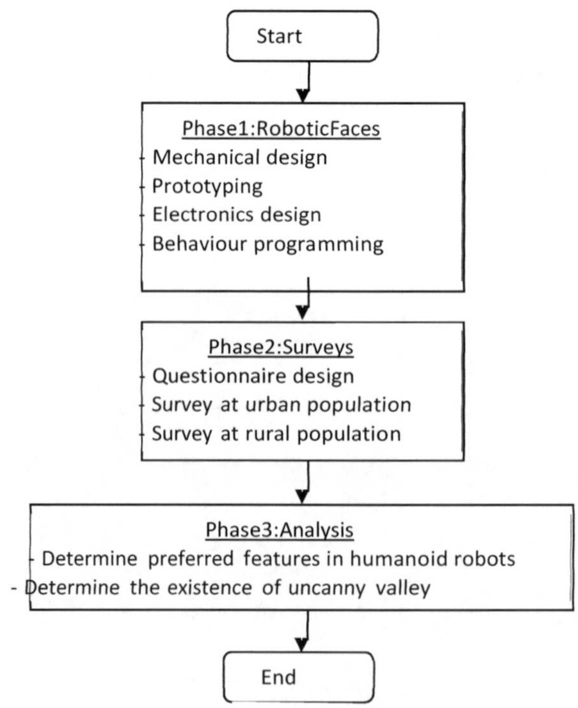

Figure 3. Research methodology flowchart.

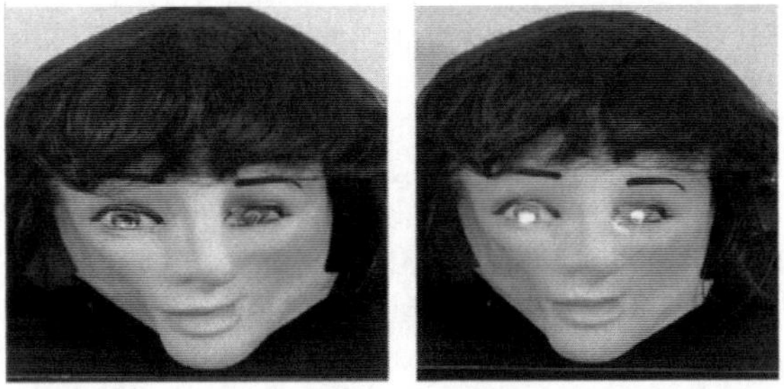

Figure 4. Friendly (left) and unfriendly (right) faces.

Figure 5. Experiment setup.

Figure 6. Robot 1, cute animal; Robot 2, cartoon character; Robot 3, a typical robot look; Robot 4, human-like; and Robot 5, actual pleasant looking model (clockwise from top left).

4. Results and discussions

The data gathered from the respondents are analysed to determine the preferred features in a humanoid robot. Perceived likeness of the robots to human and human acceptance to the robots are also ranked. Based on the rank of perceived likeness to human and the rank of human acceptance to the robots, the uncanny valley graphs are plotted. The results and discussions are presented in the following sub sections.

4.1. Preferred features in a humanoid robot

In this experiment, the respondents both in the urban and rural areas are requested to indicate their preferred features in a humanoid robot. Before the experiment started, the respondents are briefed on the concept of humanoid robot, a robot that resembles human in appearance and with human capabilities. The respondents are requested to select from a list of five features namely, limbs, face, eyes, ears and nose. The respondents can pick more than one features. The respondents' preferred features are summarized in Figure 7 below. There is not much variation observed in the result, with limbs the most preferred feature at 21.5% and nose the least preferred feature at 18%. The features ranked from the most preferred to the least preferred are robot's limbs, face, eyes, ears and nose. According to the respondents, face acts as a front of a robot, akin to human face where all the other features are attached to. Hence, face is essential in every humanoid robot. Eyes allow robots to see and limbs to allow robots to work and interact with other humans and the surrounding. Limbs allow robots to be useful, which is the actual motivation of having robot in a household in the first place. The respondents felt that ears and nose are of relatively less importance to a humanoid robot. In robotics, a microphone will allow a robot to hear without the need for elaborate ears. Up to now, it is difficult for layman to imagine a robot that can smell, more so using nose. This insight will be useful for robot designers to identify features to be included or to emphasize in a robot design.

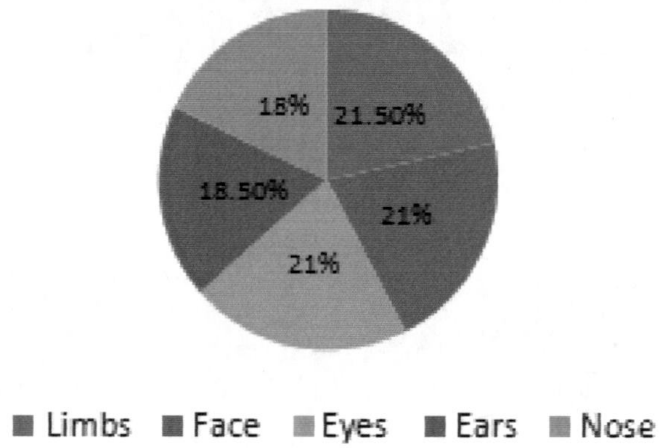

Figure 7. Preferred features in a humanoid robot.

4.2. Likeness to human

In this experiment, the respondents rank the robots' likeness to human being. Table 1 and Table 2 summarize the results obtained from the surveys conducted on urban and rural populations respectively. Both the surveys returned similar results. Robot 5, the human model is ranked most likeness to human at 94%. This is expected as the model is indeed a real and pleasant looking human being. Understandably, Robot 4 the human-like robot, with wigs and silicon skin, is ranked second most likeness to human being. This robot is intentionally built to closely resembles human being. Robot 3, iron man inspired robot is ranked third likeness to human being. It has a head with eyes. Respondents associated the robot with the ironman character in the movie. Robot 2 and Robot 1, cartoon and cute animal inspired robots are ranked by the participants as not having likeness to human. Robot 1 is designed to closely resembles a small animal, in this case a bird while Robot 2 resembles a cartoon character. Robot 2 is ranked higher than Robot 1. This ranking is used to plot the x-axis of the uncanny valley graph where Robot 1 the least likeness to human being to Robot 5 the most likeness to human being.

IOP Conf. Series: Materials Science and Engineering **344** (2018) 012012 doi:10.1088/1757-899X/344/1/012012

Table 1. Likeness to human ranking for various robots (urban population)

Robot	Likeness to Human (Urban)					Total Respondents	Likeness Normalised to %
	Very not likeness	Not likeness	Neutral	Likeness	Very likeness		
	1	2	3	4	5		
1	22	1	0	1	1	25	26
2	1	19	3	1	1	25	46
3	0	4	21	0	0	25	57
4	0	1	1	23	0	25	78
5	2	0	0	0	23	25	94

Table 2. Likeness to human ranking for various robots (rural population)

Robot	LikenesstoHuman(Rural)					Total Respondents	Likeness Normalised to %
	Very not likeness	Not likeness	Neutral	Likeness	Very likeness		
	1	2	3	4	5		
1	20	1	1	2	1	25	30
2	0	14	9	2	0	25	50
3	1	7	13	2	2	25	58
4	3	2	2	18	0	25	68
5	1	1	0	1	22	25	94

4.3. Human acceptance

In this experiment, the respondents are required to rank the five robots from 1 to 5 according to their preferred choice to bring home. The results are summarized in Table 3 and Table 4 below. Rank 1 is the first choice followed by Rank 2 for the second choice and so on. This ranking can be implied as the respondents' acceptance level of the robots. Usually a person will invite another person to his/her home only when he/she accepts the other person as a close friend or a family member. Based on the survey data computed from urban area, Robot 1, Robot 2 and Robot 3 acceptance rate are at 70% or more. It can be implied that, human beings can accept Robot 1, Robot 2 and Robot 3 into their home. Many respondents cannot accept Robot 4 that looks like human being. Most respondents felt that the robot is creepy because of the robot's close resemblance to human being. Surprisingly for Robot 5, the ideal human model, the urban respondents' acceptance rate is at 43% only which is below expectation. Urbanites are usually cautious of being close to a stranger, let alone to bring the stranger home. In current urban scenario, sadly friends usually have to make appointment before visiting one another. Most of the time, the meeting will be held in a neutral venue such as coffee shops, restaurants and malls as opposed to at home as in the olden days. Urbanites are more comfortable having a clear-cut machine (robot) home compared to a stranger or a robot resembling a stranger. The acceptance rate for the ideal robot might be higher had the respondents have the chance to get to know the robot better. However, this is still not possible with the current technologies available.

The results are slightly different for rural populations. The acceptance rate for Robot 1, Robot 2 and Robot 3 are at 67%, 68% and 64% respectively. Although indicating acceptance, it is slightly lower than in the urban area. The acceptance for Robot 4 is low at 34% as expected. The acceptance rate for Robot 4 is similar to the urban respondents. The acceptance rate for Robot 5, the ideal human model is much higher at 67%. This acceptance rate is high and similar to the acceptance rate for Robot 1 and Robot 2. The rural respondents are more welcoming to strangers into their home. The community spirit in rural setting is still high where usually anyone can freely walk into a neighbour's house for a short chat. The houses in a village are usually situated near one another with the villagers sharing the same common amenities. This encourages acceptance and trust among the villagers. Furthermore, rural respondents are less weary toward strangers due to the relatively low crime rate environment.

Table 3. Human acceptance ranking for various robots (urban population).

Robot	HumanAcceptance (Urban)					Total Respondents	Acceptance Normalised to %
	Rank #5	Rank #4	Rank #3	Rank #2	Rank #1		
1	1	4	3	10	7	25	74
2	0	2	13	5	5	25	70
3	1	2	7	6	9	25	76
4	13	7	2	3	0	25	36
5	10	10	0	1	4	25	43

Table 4 . Human acceptance ranking for various robots (rural population).

Robot	Human Acceptance (Rural)					Total Respondents	Acceptance Normalised to %
	Rank #5	Rank #4	Rank #3	Rank #2	Rank #1		
1	3	3	8	4	7	25	67
2	1	3	9	9	3	25	68
3	2	6	6	7	4	25	64
4	17	4	1	1	2	25	34
5	2	9	1	4	9	25	67

4.4. Uncanny valley

The relationship between the respondents' acceptance toward the five robots against the measure of robots' likeness to human beings is plotted. The relationship graphs are illustrated in Fig. 8 for urban and rural populations. As shown in Figure 8, there is a pronounce dip in acceptance level for Robot 4, indicating the presence of uncanny valley phenomenon in both the urban and rural Malaysian populations. As stated by Mori in [1], a person's response to a humanlike robot would abruptly shift from empathy to revulsion as it approached, but failed to attain, a lifelike appearance. This descent into eeriness is known as the uncanny valley. This is demonstrated in the Robot 4 design that closely resembles a human being with wig as its hair and silicon skin. However, the design is still not realistic enough to persuade the respondents to accept the robot as part of human. Hence, the respondents felt eeriness, uncomfortable and are scared to interact with the robot. With the shift toward individualism, especially in the urban area, it is unsure whether the acceptance level will increase for a robot that almost 100% resembles a human being.

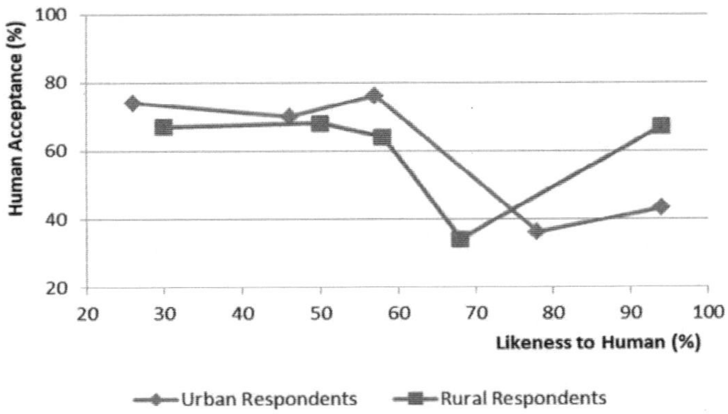

Figure 8. Uncanny Valley graph for urban and rural respondents.

5. Conclusion

In this study, the existence of uncanny valley phenomenon among Malaysian urban and rural population are measured. From the limited study, it is found that the uncanny valley phenomenon is indeed present in Malaysian population. This outcome is in line to many similar studies in other countries. It is interesting to note that for urban population, the acceptance level for a robot that almost 100% resembles a human is still low. Besides uncanny valley, the key robotic features preferred by Malaysians are also measured. This research is important that it provides a local perspective to a global phenomenon. The data will come in handy for social and educational robot designers targeting Malaysian or Asian population. However, further improvement can be made on the robots' physical capabilities to enable meaningful social interaction with human respondents. There is also a plan to extend this survey to include a larger number of respondents from different educational background, race and geographical locations.

Acknowledgments

The authors would like to thank the respondents from Setapak township and Kelapa Sawit village for their participation in this work.

References

[1] Mori M Bukimi 1970 no tani [The uncanny valley] J. Energy 7(4) 33-35
[2] Mori M, MacDorman, K F and Kageki, N 2012 The Uncanny Valley [From the Field] IEEE Robotics and Automation Magazine 19(2) 98-100
[3] Seyama J and Nagayama R S 2007 The uncanny valley: Effect of realism on the impression of artificial human faces J. Presence: Teleoperators and Virtual Environments 16(4) 337-351
[4] Kobayashi H, Ichikawa Y, Senda M and Shiba T 2003 Proc. IEEE Int. Conf. on Intelligent Robots and Systems (Las Vegas: IEEE) pp. 1123-1128
[5] Woods S N, Dautenhahn K and Schulz J 2006 Exploring the design space of robots: Children's perspectives J. Interacting with Computers 18(6) 1390-1418
[6] Appel M, Weber S, Krause S and Mara M 2016 Proc. 11th ACM/IEEE Int. Conf. on Human Robot Interaction (Christchurch: IEEE) pp. 411-412
[7] Burleigh T J, Schoenherr J R and Lacroix G L 2013 Does the uncanny valley exist? An empirical test of the relationship between eeriness and the human likeness of digitally created Faces J. Computers in Human Behaviour 29(3) 759-771

Experimental verification of numerical calculations of railway passenger seats

B Ligaj[1,*], M Wirwicki, K Karolewska and A Jasińska

UTP University of Science and Technology in Bydgoszcz, Faculty of Mechanical Engineering, Department of Biomedical Engineering, Al. Prof. S. Kaliskiego 7, 85-796 Bydgoszcz, Poland

Email: bogdan.ligaj@utp.edu.pl

Abstract. The construction of railway seats is based on industry regulations and the requirements of end users, i.e. passengers. The two main documents in this context are the UIC 566 (3rd Edition, dated 7 January 1994) and the EN 12663-1: 2010+A1:2014. The study was to carry out static load tests of passenger seat frames. The paper presents the construction of the test bench and the results of experimental and numerical studies of passenger seat rail frames. The test bench consists of a frame, a transverse beam, two electric cylinders with a force value of 6 kN, and a strain gauge amplifier. It has a modular structure that allows for its expansion depending on the structure of the seats. Comparing experimental results with numerical results for points A and B allowed to determine the existing differences. It follows from it that higher stress values are obtained by numerical calculations in the range of 0.2 MPa to 35.9 MPa.

1. Introduction

The construction of railway seats is based on industry regulations and the requirements of end users, i.e. passengers. The two main documents in this context are [1], [2] and [3]. The structure of seats, as in sitting spaces in the vehicle, concerns two particular components: the seat itself and the seat frame for fixing it to the rail vehicle structure. Due to the requirements of the regulations, the system should be considered holistically. The main design criterion is rigidity, with further criteria being specified as mass, durability, strength, etc.

Due to the functional and normative criteria for the seats and their attachments to the rail vehicle structure, three types of solutions have been developed: fixing the seats to the floor on two legs (Figure 1a), attaching the seats to the side wall and the floor on one leg (Figure 1b)) and fixing the seats to the side wall only (Figure 1c), with the lattermost being the most challenging solution

Experimental studies of rail passenger seats are based on the guidelines of the UIC 566 [2]. This document standard formulates the following conditions for carrying out tests under static load conditions (Figure 2):

a) armrest load:

- force F4 = F5 = 0.75 kN applied to the free end of the armrest, load acting upwards (force direction: vertical)
- force F6 = F7 = 0.75 kN applied to the free end of the armrest, load acting on the side (force direction: horizontal),

b) seat load:

- force F1 = 1.0 kN applied to the seat centrally along the width, 50 mm from the edge (force direction: vertical)

- force F2 = 1.2 kN applied to the seat centrally along the width of the seat, 50 mm from the edge (force direction: vertical)

c) backrest load:

- force F3 = 1.5 kN applied centrally to the upper part of the backrest (force direction: horizontal)

d) hinged table load:

- force F8 = 0.75 kN applied centrally to the table (force direction: vertical).

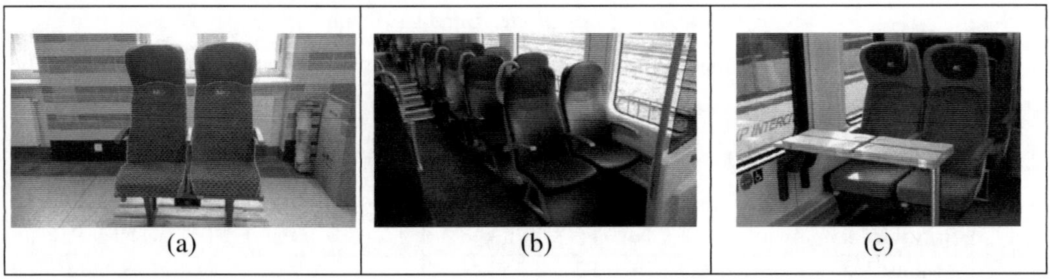

Figure 1. Methods of fixing the seats to the frame of a rail vehicle: a - to the floor on two legs [2], b - to the side wall and to the floor on one leg [3], c – outboard system, to the side wall only [3]

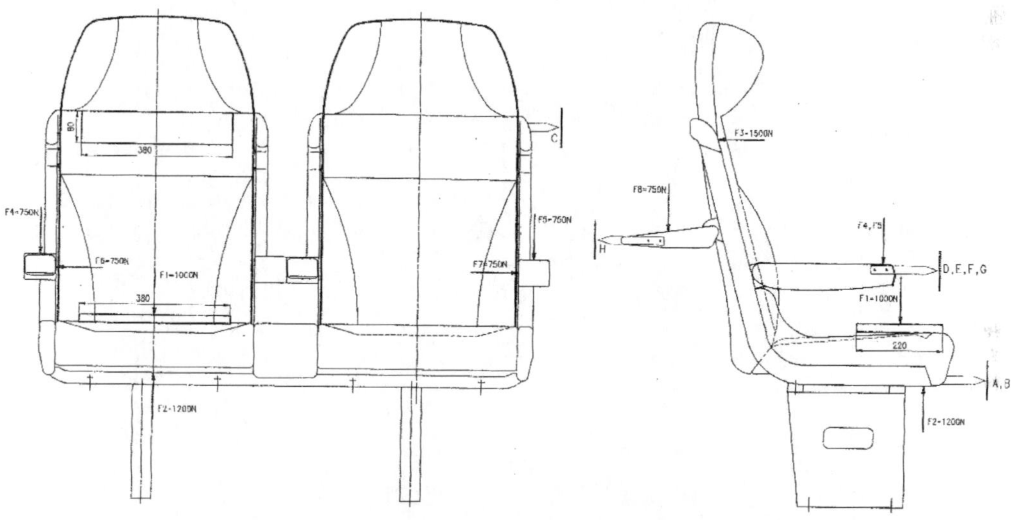

Figure 2. Loading the seats during static load tests [1]

The study was to carry out static load tests of passenger seat frames. The paper presents the construction of the test bench and the results of experimental and numerical studies of passenger seat rail frames.

2. Methods

2.1 Bench for experimental studies involving seats

The test bench is shown in Figure 3. It consists of: a frame, a transverse beam, two electric cylinders with a force value of 6 kN, a strain gauge amplifier. It has a modular structure that allows for its expansion depending on the structure of the seats. The bench is designed so as to make it possible to

IOP Conf. Series: Materials Science and Engineering **344** (2018) 012013 doi:10.1088/1757-899X/344/1/012013

examine the seats of any mounting system to the frame of the rail vehicle, meeting the requirements as set out in the UIC 566 standard [2]

2.2 Test results

Experimental studies of the passenger seat frame were carried out in accordance with the UIC 566 [2] standard. The test program included separate tests for loads applied to the seat, backrest and armrest. A detailed study program is provided in Table 1.

Table 1. Study program and test result

Lp	Area of load application	Load value	Stress value [MPa]	
			Point A	Point B
1	Seat load	F1 = 1.0 kN	75 MPa	70 MPa
2	Seat load	F2 = 1.2 kN	86 MPa	65 MPa
3	Backrest load	F3= 1.5 kN	125 MPa	138 MPa
4	Backrest load	F4 = F5 = 0.75 kN	109 MPa	118 MPa
5	Armrest load	F6 = F7 = 0.75 kN	39 MPa	48 MPa

Figure 3. Test bench

2.3 Results of numerical calculations

Numerical analysis of strains and stresses in the test objects was performed using the finite element method using the ABAQUS 6.6-4 software. Three-dimensional finite elements C3D4H (Fig. 4) were used for discretization of the above 3D models. A material model was used to describe material properties. For this purpose, the material model was described as a Young's modulus of E = 2.1×10^5 MPa and a Poisson's ratio of v = 0.3. The results of numerical calculations are presented in Table 2.

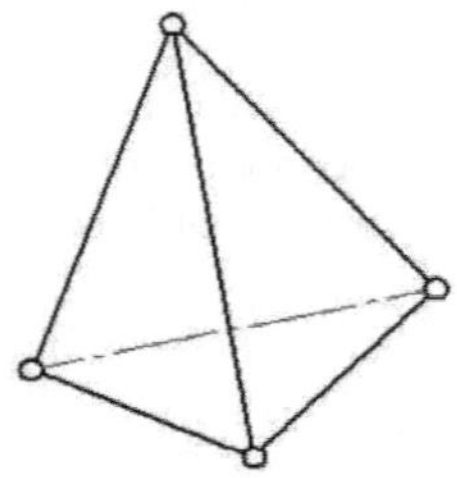

Figure 4. Four-mode 3D C3D4H element used in numerical calculations

Table 2. Results of numerical calculations

Lp	Area of load application	Load value	Stress value [MPa]	
			Point A	Point B
1	Seat load	F1 = 1.0 kN	94.5 MPa	93.5 MPa
2	Seat load	F2 = 1.2 kN	100.7 MPa	84.5 MPa
3	Backrest load	F3= 1.5 kN	158.2 MPa	173.9 MPa
4	Backrest load	F4 = F5 = 0.75 kN	142.5 MPa	135.4 MPa
5	Armrest load	F6 = F7 = 0.75 kN	45.7 MPa	48.2 MPa

3. Results and Discussion

Comparing experimental results with numerical results (Table 3) for points A and B allowed to determine the existing differences. It follows from it that higher stress values are obtained by numerical calculations in the range of 0.2 MPa to 35.9 MPa.

Table 3. Comparison of experimental results and numerical results

Lp.	Area of load application	Point A			Point B		
		Test (Table 1)	Calculation (Table 2)	Difference	Test (Table 1)	Calculation (Table 2)	Difference
1	Seat load	75 MPa	94.5 MPa	19.5 MPa	70 MPa	93.5 MPa	23.5 MPa
2	Seat load	86 MPa	100.7 MPa	14.7 MPa	65 MPa	84.5 MPa	19.5 MPa
3	Backrest load	125MPa	158.2 MPa	33.2 MPa	138 MPa	173.9 MPa	35.9 MPa
4	Backrest load	109 MPa	142.5 MPa	30.4 MPa	118 MPa	135.4 MPa	17.4 MPa
5	Armrest load	39 MPa	45.7 MPa	6.7 MPa	48 MPa	48.2 MPa	0.2 MPa

The reason for the differences should be seen in the manner of loading the seats during experimental research. As a result of the load applied to the seat structure, the deflection is deformed, which can affect the direction of operation and the value of the load. In case of numerical calculations it is difficult to accurately model the way of loading the seats and their susceptibility resulting from the applied materials for the seat and the back

4. Conclusions

This study performed static load tests of passenger seat frames according to UIC 566 (3rd Edition, dated 7 January 1994) and the EN 12663-1: 2010+A1:2014. Construction of the test bench and the results of experimental and numerical studies of passenger seat rail frames are presented. The test bench which consists of a frame, a transverse beam, two electric cylinders with a force value of 6 kN, and a strain gauge amplifier has a modular structure that allows expansion depending on the structure of the seats. Experimental results were compared with numerical results for points A and B. The numerical analysis was able to calculate the stresses at points A and B, with a difference in the range of 0.2 MPa to 35.9 MPa compared to the experimental results.

References

[1] European standard EN 12663-1:2010+A1:2014: Railway applications. Structural requirements of railway vehicle bodies. Locomotives and passenger rolling stock (and alternative method for freight wagons), 2010
[2] International Union of Railways UIC 566: Loadings of coach bodies and their components, 3rd Edition, January 7, 1994.
[3] International Union of Railways UIC 567 (E): General provisions for coaches, 2nd Edition, November 2004
[4] Website: http://www.intap.com.pl/intap-fotele
[5] Website: http://swr.podkarpackakolej.net/archiwum/2015-06/4

Experimental studies of glued Aluminum-glass joints

B Ligaj [1], M Wirwicki[1], K Karolewska[1], A Jasińska[1]

[1] *Department of Biomedical Engineering, Faculty of Mechanical Engineering, University of Science and Technology in Bydgoszcz (UTP), Poland*

Email: Bogdan.ligaj@utp.edu.pl

Abstract. Glued steel-glass or aluminum-glass joints are to be found, among other things, in vehicles (cars, buses, trains, trams) as windscreen assembly pieces for the supporting structure. For the purposes of the experiments, samples were made in which the top beam was made of the AW-2017A aluminum alloy and the bottom beam was made of thermally reinforced soda-lime glass whereas the glued joints were made of one-component polyurethane glue Körapur 175. The tests were performed under four-point bending conditions at monotonic incremental bending moment values on the Instron 5965 durability machine. The experimental study of the durability of glued joints under four-point bending conditions with the monotonic incremental bending moment allows to determine the values of stresses, whose value is related to initiation of damage of the tested joint.

1. Introduction

At the initial stages of constructing the construction of the machine components are carried out on samples. Their shape and dimensions are based on norms, examples of which are develop [2, 5, 6]. The results of the tests performed on standard samples are presented in [11]. There are also many examples of research conducted using non-standard samples [3, 14] from a wide range of scientific disciplines. Glued steel-glass or aluminum-glass joints are to be found, among other things, in vehicles (cars, buses, trains, trams) as windscreen assembly pieces for the supporting structure. This application requires the combination of high durability properties to be maintained for the intended service life so as to ensure a high level of safety for vehicle users. Conducting experiments of this type of connections, in terms of their sturdiness and durability, allows to select their geometrical features so that they can meet the normative and utility requirements. In the paper [9] the results of investigations of glued aluminum-glass joints with static and creep loading were presented. Obtained results allow determining the minimum life of the connectors. The tests were performed on samples of the original shape corresponding to the construction of the structural element.

The paper [1] describes the study of glue joints in which nanoparticle glue was used. The use of nanoparticles has been associated with the ability to diagnose damaged joints. In the study, overlapping samples were used. In the work [4] the defectability and static strength of adhesive joints, produced using five different types of dispersion methods, were carried out on the basis of examinations of overlapping samples. In the article [8], adhesion tests in glass supporting structures are presented in relation to the effect of different substrates (glass, structural steel and stainless steel, aluminum alloy) and surface preparation (sanding of the glass surface). The original test samples consisted of a glass rectangle adhered to a cuboid with glass, steel and aluminum alloy.

This paper is to present a method of assessing the damage of the glued aluminum-glass joint under conditions of a monotonically increasing bending moment [3].

2. Method

A new type of specimens, being the subject of the patent application in Poland fig 1. [10], was used in the studies. The sample consists of two beams: top one (longer) made of structural material such as steel, aluminum alloy, and bottom one (shorter) that is part of the glass, which be tempered, layered or traditional glass. The top beam may have any cross-section, e.g. rectangular, square, c-shaped, etc., which is characterized by different values of the moment of inertia. The bottom beam, meanwhile, is made of rectangular glass. Tensometric sensors are mounted to the both beams in order to determine the deformations of the individual beams under load.

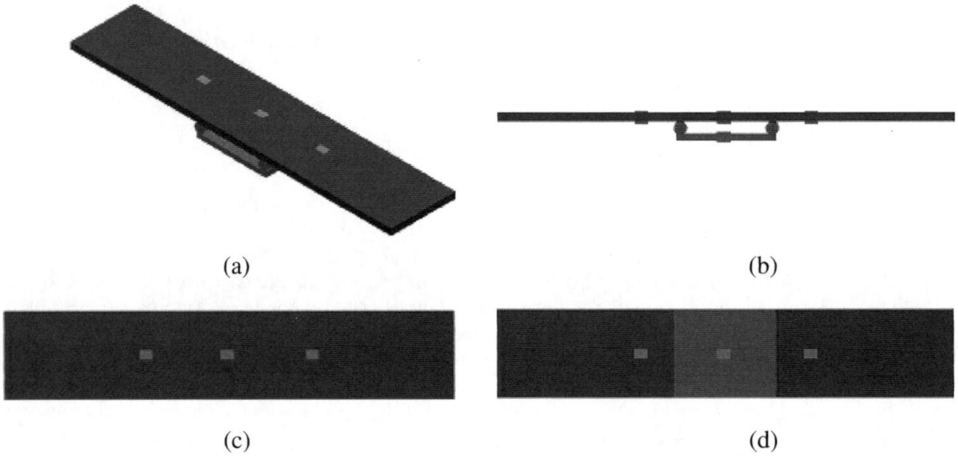

Figure 1. General construction of an innovative test specimen for glued joints: a - axonometric view, b - side view, c - top view, d - bottom view

For the purposes of the experiments, samples were made in which the top beam was made of the AW-2017A aluminum alloy (AlCu4MgSi alloy mark according to the PN-EN 573-3: 2010), and the bottom beam was made of thermally reinforced [12], whereas the glued joints were made of one-component polyurethane glue Körapur 175 [13]. The geometrical dimensions of the test sample are shown in Figure 2 and the physical form of the sample object is shown in Figure 3. The selected mechanical properties of the AW-2017A aluminum alloy are shown in Table 1, while the properties of sodium glass are presented in Table 2.

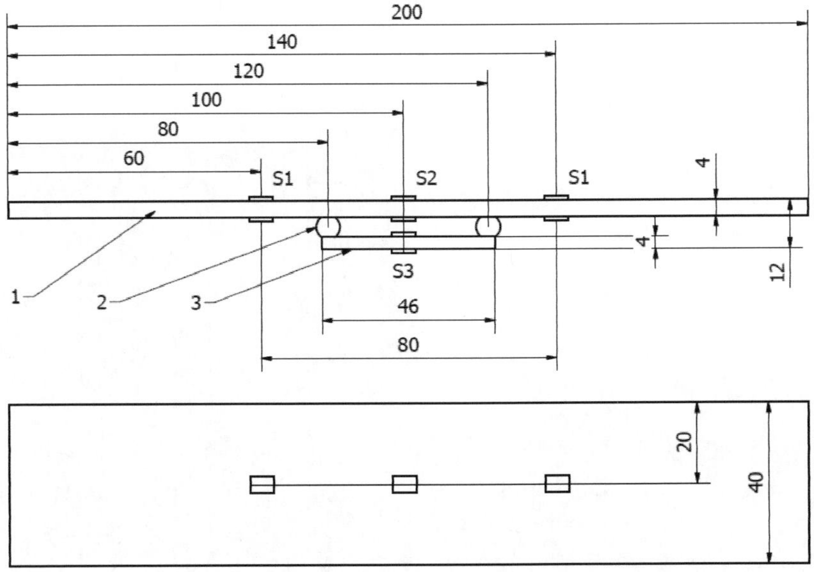

Figure 2. Geometric dimensions of the sample used in strength tests

(a) (b)

(c)

Figure 3. Physical form of research object

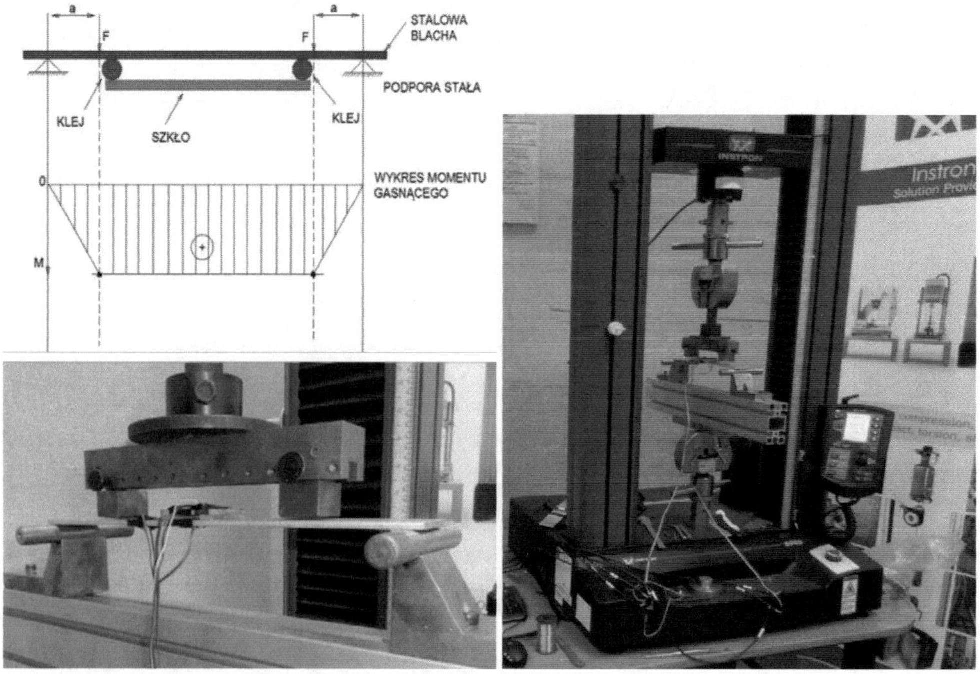

(a) (b)

Figure 4. Method of loading sample (a) and construction of test stand (b)

Table 1. Mechanical properties of the AW-2017A aluminium alloy [7]

	$R_{p0,2}$	R_m	E	A_5	Z
	MPa	MPa	MPa	%	%
Mean value	320	470	73077	16,1	23,0
Standard deviation	1,3	4,5	1304	0,8	0,8

Table 2. Mechanical properties of thermally reinforced soda-lime glass [12]

	R_c	R_g	E	G	
	MPa	MPa	MPa	MPa	-
Data by manufacturer	39	83	72000	30000	0,23

The tests were performed under four-point bending conditions at monotonic incremental bending moment values on the Instron 5965 durability machine. The sample allows bending tests to be performed at bending loads while maintaining a constant bending moment in the glued joint area. Figure 4a shows the method of loading the sample during testing, while Figure 4b presents the test bench.

3. Result and Discussion

The experimental studies of glued joints under monotonically increasing bending moment values allowed to determine the stress values in the top beam made of aluminum alloy and the bottom beam made of thermally reinforced soda lime glass. Figure 5 shows the course of stress changes in the individual components of the sample: S1 - Stress in the top beam beyond the place of the glued joints, S2 - Stress in the upper bar between the glued joints, S3 - Stress in the bottom beam between the glued joints. The intersection of the stress lines in the top beam marked S1 and S2 was assumed to be the criterion for identifying the initiation of damage to the glued joint. Combining the top beam with the bottom beam using two glued joints causes the top beam to become more rigid in the middle section.

Then the stress value (S1) in the top beam beyond the bottom beam fixing point is higher than the stress value (S2) between the glued joints. This is the case until the joints carry the load from the top beam to the bottom beam. In the event of initiation of damage to the glued joint, the load value transmitted by the bottom beam decreases. This results in increased stress (S2) in the top beam between the glued joints. Therefore, the stress value corresponding to the point of intersection of the stress waveform curve for S1 and S2 can be considered the limit stress value that the glued joint is able to transfer. For samples where the thickness of the adhesive layer was 3 mm, the glued joint damage occurred at a stress range of 20 MPa to 38 MPa.

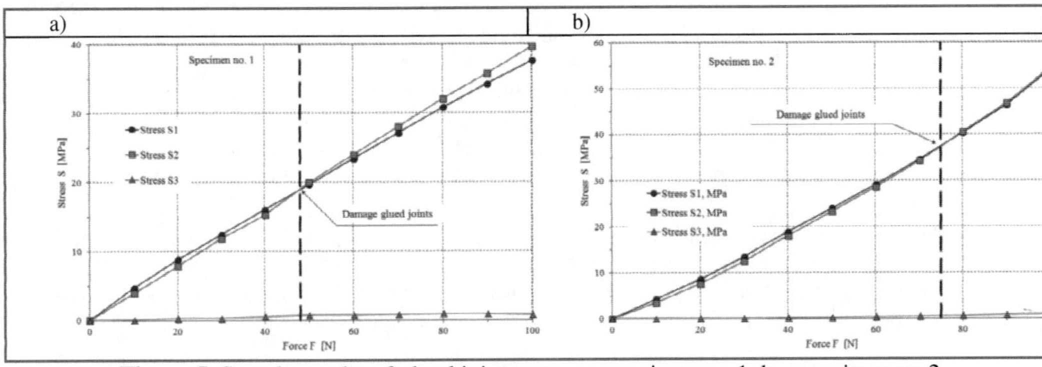

Figure 5. Sample results of glued joint tests: a – specimen no.1, b – specimen no.2

4. Conclusion

The new type of specimen has been developed to allow for the testing of glued joints of different structural rigidity, with special emphasis on the steel-glass, aluminum-glass, steel-aluminum and similar connections. The experimental study of the durability of glued joints under four-point bending conditions with the monotonic incremental bending moment allows to determine the values of stresses, whose value is related to initiation of damage of the tested joint. Obtained results are difficult to relate to the results of other authors because of many factors affecting the properties of the joints and the way the samples are prepared. In the next work of our team will be presented results of studies of adhesive joints implemented with the use of normative samples and new samples. Preliminary investigations indicate the suitability of a new type of specimen to evaluate the durability of the glued joints of machine components. In order to confirm their usefulness, a broader program of research will be carried out including: various combinations of materials, thickness of adhesive layer.

REFERENCES

[1] Akpinar I. A.; Gultekin K.; Akpinar S., Research on strength of nanocomposite adhesively bonded composite joints, COMPOSITES PART B-ENGINEERING, vol. 126, 2017, pp.143-152.

[2] European standard PN-EN 1465:2009: Adhesives. Determination of tensile lap-shear strength of bonded assemblies, 2009

[3] Gaurav Jain, Aditi Narad, Lalit C. Boruah, Balakrishnan Rajkumar, Comparative evaluation of shear bond strength of three resin based dual-cure core build-up materials: An In-vitro study, J. Conserv. Dent. 2015 Jul-Aug; 18(4), pp. 337–341,doi: 10.4103/0972-0707.159754.

[4] Kim Cheol-Hwan; Choi Jin-Ho, Effects of dispersion methods and surface treatment of carbon nano-tubes on defect detectability and static strengths of adhesive joints, COMPOSITE STRUCTURES, vol. 176, 2017, pp. 684-691.

[5] International standard ISO 5173: Destructive tests on welds in metallic materials. Bend tests, 2009.

[6] International standard ISO 6892-1: Metallic materials. Tensile testing. Part 1: Method of test at room temperature, 2016.

[7] Ligaj B., Szala G.: Hybrid Calculation Method of Fatigue Life (in polish), Monograph, Publishing Institute for Sustainable Technologies – National Research Institute, Bydgoszcz-Radom, 2013.

[8] Machalicka K., Eliasova M., Adhesive joints in glass structures: effects of various materials in the connection, thickness of the adhesive layer, and ageing, INTERNATIONAL JOURNAL OF ADHESION AND ADHESIVES, vol. 72, 2017, pp. 10-22.

[9] Marquesa E.A.S., Carbasa R.J.C., Silvab Fábio, da Silvac Lucas F.M., de Paivad D.P.S., Magalhãesd Fernão D.: Use of master curves based on time-temperature superposition to predict creep failure of aluminium-glass adhesive joints, INTERNATIONAL JOURNAL OF ADHESION AND ADHESIVES, vol. 74, 2017, pp. 144-154.

[10] Patent application no. P.422884, Ligaj B., Jasińska A., Karolewska K., Test specimen for bonding materials with significantly different physical properties, especially steel-glass, Polish Patent Office, registered on september 18, 2017.

[11] Szala G., Ligaj B.: Application of hybrid method in calculation of fatigue life for C45 steel (1045 steel) structural components, International Journal of Fatigue 91 (2016), pp. 39–49.

[12] Website: https://www.continentaltrade.com.pl/szklo-sodowo-wapniowe.

[13] Website: http://www.kommerling.com.pl/rodzaje/kleje-poliuretanowe/item/7-koerapur-175.

[14] Wirwicki M., Topolinski T., Determining the S-N Fatigue Curve for Lava Zirconium Dioxide, MATERIALS, INDUSTRIAL, AND MANUFACTURING ENGINEERING RESEARCH ADVANCES, Book Series: Advanced Materials Research, vol. 845, 2014, pp. 153-157, DOI: 10.4028/www.scientific.net/AMR.845.153.

A review of mechanical and tribological behaviour of polymer composite materials

K Prabhakar[1], S Debnath[1], R Ganesan[1] and K Palanikumar [2]

[1] Department of Mechanocal Engineering, Curtin University, Miri, Malaysia.
[2] Department of Mechanical Engineering, Sri Sai Ram Institute of Technology, Chennai, Tamilnadu, India.

Abstract. Composite materials are finding increased applications in many industrial applications. A nano-composite is a matrix to which nanosized particles have been incorporated to drastically improve the mechanical performance of the original material. The structural components produced using nano-composites will exhibit a high strength-to-weight ratio. The properties of nano-composites have caused researchers and industries to consider using this material in several fields. Polymer nanocomposites consists of a polymer material having nano-particles or nano-fillers dispersed in the polymer matrix which may be of different shapes with at least one of the dimensions less than 100nm. In this paper, comprehensive review of polymer nanocomposites was done majorly in three different areas. First, mechanical behaviour of polymer nanocomposites which focuses on the mechanical property evaluation such as tensile strength, impact strength and modulus of elasticity based on the different combination of filler materials and nanoparticle inclusion. Second, wear behavior of Polymer composite materials with respect to different impingement angles and variation of filler composition using different processing techniques. Third, tribological (Friction and Wear) behaviour of nanocomposites using various combination of nanoparticle inclusion and time. Finally, it summarized the challenges and prospects of polymer nanocomposites.

1. Introduction

Composite materials are macroscopic combination of two or more distinct materials having a recognizable interface between each other. Composites are not only preferred for their structural properties, but also used for their tribological, thermal and electrical properties to achieve desired properties for a given range of applications. Different types of polymers such as thermoplastics, thermosetting plastics and elastomers have been utilized for manufacturing polymer nanocomposites. Polymer nano-composites have superior properties if constructed properly.

Polymer matrix based nano-composites have generated a significant amount of attention in current research in the area of nanotechnology. The structural components produced using nano-composites will exhibit a high strength-to-weight ratio. For instance, an epoxy containing carbon nanotubes can be utilized to produce wind turbine cutting edges which results in a strong but light weight material, which in turn allows for longer windmill blades. Practically, these longer blades increase the amount of electricity generated by each wind turbine. A typical polymer nano-system is represented in Figure 1 where phase changes occurs from a natural nanoparticle which is added to polymer resin and then is further reinforced with fibers such as Glass, Carbon or Aramid fibers. Finally, it is cured using resin transfer molding (RTM) to a Polymer nanocomposite.

IOP Conf. Series: Materials Science and Engineering **344** (2018) 012015 doi:10.1088/1757-899X/344/1/012015

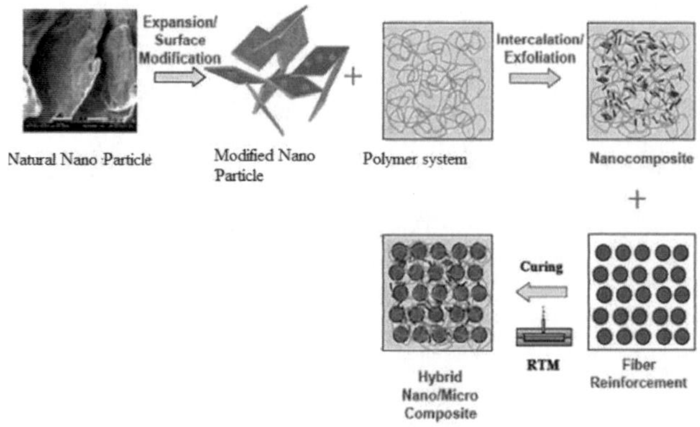

Figure 1. Typical Polymer nano system [36]

Nano-composite is fundamentally a multiphased solid material where one of the manufacturing phases has no less than one dimension under 100 nm. In recent years nano-reinforcements have recieved the greatest attention as a result of intense research and development. The main advantages of nanocomposites is its ability to achieve a desirable mechanical strength, toughness, increased wear resistance properties, as well as improved electrical and thermal conductivity by the addition of nano-particles in as small a volume as 2-5 % of weight. Based on the matrix used, material composites are classified as Metal matrix composites, Polymer matrix composites and Ceramic matrix composites.

Polymer nano-composites are being increasingly used as engineering materials in which tribological properties are of considerable importance. Fillers, for example, glass, carbon and textile fibers are incorporated within numerous polymers to enhance their tribological properties. Nano-particles such as carbon-nanotube, nano-clay, titanium oxide (TiO_2) and silicon carbide (SiC) are added in volumes of 2-5 % by weight which, in turn enhances the mechanical properties dramatically. The reduction in wear is due mainly to preferential load support or the load bearing capacity of the reinforcement components, by which the action of abrasive mechanisms to the wear of the materials is highly suppressed. Figure 2 shows microscopic images of nanotubes, nanographite, nanoclay and nanosilica for information and visualization of these microstructures.

When nano particles are embedded in the polymer, the resulting composite material is known as a polymer nano-composite. Polymer nano-composites consist of a polymeric material (e.g. Thermoplastics, thermosets or elastomers) having nano-particles or nano-fillers dispersed in the polymer matrix which may be of different shapes (e.g. fibers, spheroids or platelets) with at least one of the dimensions less than 100 nm. Figure 3 shows sample images for nanostructures of a DNA and carbon nanotube for reference.

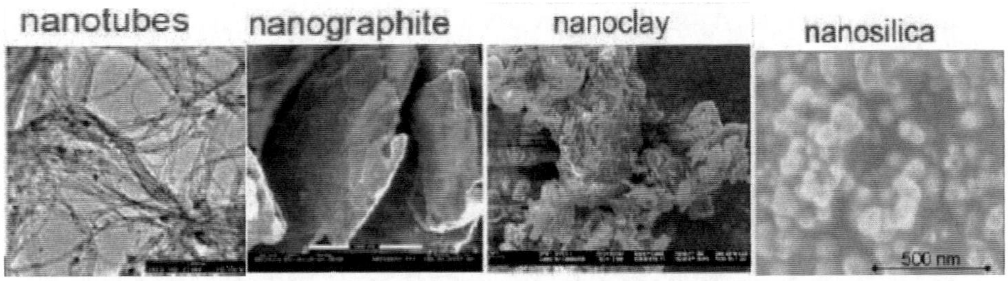

Figure 2: Typical nano particles used for polymer nano composites [35]

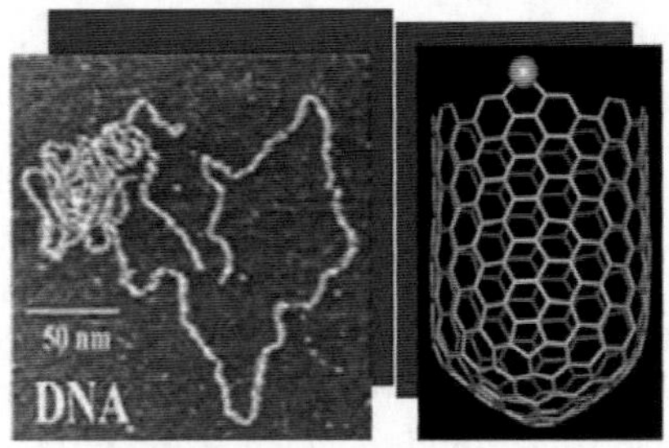

Figure 3: Examples of a nanostructure AFM image of a DNA and a carbon nano tube [34]

2. Review of mechanical and wear behaviour

2.1 Mechanical behaviour of polymer composite materials

Various polymer materials, inclusive of fillers depending on the necessity of the applications are reviewed in this section. Polyester resin with glass fiber reinforced fillers have been employed to analyse the mechanical properties such as tensile strength, flexural strength and Young's Modulus for single and multiple fibers by Ya-Jung Lee et al. [1]. They have generated a numerical model from the trial information generated which in the future could be utilized for accessing the mechanical properties of laminates delivered utilizing distinctive types of fibers without additionally tries. They have also concluded that GFRP laminates when subjected to flexural loading the stiffness after initial failure does not drop promptly but rather decays step by step. Sang-Young Kim et al. [2] have used vinyl ester epoxy resin and glass fiber reinforced fillers to analyse mechanical properties such as tensile strength, compressive strength and in-plane shear properties using the vacuum infusion process and hand layup process samples and found the vacuum infusion processed GFRP samples showed better mechanical properties than the hand layup technique due to handlayup's technique increased porosity.Mouritz et al. [3] analysed the development of fibre-reinforced polymer composites in naval ships and submarines. The high strength to weight ratio finds applications in superstructures, decks, bulkheads, advanced mast systems, propellers etc. which were meant to reduce weight and in turn increases the speed and decreases fuel costs in the long run, as well as contributes to corrosion resistance and less maintenance costs. Selvaraju et al. [4] analysed the application of Aramid fiber reinforced polymer (CFRP) and glass fiber reinforced Polymer (GFRP) in marine applications such as naval patrol boats, submarine diesel storage tanks, lube tanks, utility tanks, low pressure pipes, cable ladders and trays with the intention of reducing weight, improving corrosion resistance and increasing the strength to weight ratio. Vallbo [5] has analysed the effects of the mechanical properties with manufactured transport containing vinyl ester matrix and polyvinyl chloride (PVC) as a core material with carbon fibre reinforcement. This research found improvement in the mechanical properties, electrical properties, as well as lighter in weight, which in turn will reduce fuel consumption, thus being more environmentally friendly and finally having an increased payload.

Chalmers [6] studied the different types of resins such as polyester resins, vinyl ester and epoxy resin with different combinations of reinforcements such as E-glass, carbon and aramid fiber to reduce the stiffness, reduced maintenance load and manufacturing costs. Nagalingam et al. [7] conducted an analysis of different combinations of polyester resin, fiber and nano-clay (MMT) with varying percentages of combinations and observed the ultimate tensile strength decreased with an increase in the percentage of nano-clay and at the same time impact strength increased with the increase in the percentage of nano-clay.

Isao Kimpara [8] has studied the different composite materials used in marine construction such as GFRP, CFRP, AFRP and hybrid composite (FRP with two or more different type of fibers) materials with their current and potential applications. The research concluded that GFRP has been mainly utilized in the marine environment with the key considerations; reduction in weight, increased speed of the boat with an improvement in the mechanical strength and dynamic strength such as creep, impact and fatigue. Md Ekramul Islam et al. [9] concluded in their study FRP material was mostly utilised for marine applications and conducted experimental analysis using carbon fiber reinforced epoxy composites, which was modified with 2 % by weight of Montmorillonite (MMT) nanoclay and 0.3 % by weight of multi-walled carbon nanotubes (MWCNTs). The results were compared with unmodified carbon/epoxy composites and found that reinforcement with both nanoparticles significantly improved the mechanical and thermo-mechanical properties of carbon fiber reinforced (CFRP) composites. Anbusagar N.R.R et al. [10] determined FRP and Sandwich FRP has been mostly used in the marine sector and experimental studies were done with different combinations of FRP and sandwich FRP modified with Montmorillonite (MMT) nanoclay by the hand lay-up technique. It was observed that tensile strength and impact strength were significantly increased over the range of nanoclay loading. SEM images showed the adhesion of glass fibers to the polyester matrix was also greatly increased with an increase in nanoclay content.

2.2 Wear of Polymer Composite Materials

Law Mei Lin et al. [11] conducted an experimental study using epoxy resin and recycled carbon fiber, which was obtained from two conditions, as received and cryogenic treated condition. The tribological properties including the wear rate and the coefficient of friction were taken at different time intervals for both combinations and found the coefficient of friction decreased and wear resistance improved in the treated carbon fiber with epoxy resin when compared to the as received condition. Shakuntala Ojha et al. [12] studied the wear behaviour of both raw and carbon black wood apple shell particles filled with epoxy resin. It was found the carbon black particulate composites showed a lower wear rate as compared to the raw particulate composite. The effect of wood apple shell particle concentrations with different impingement angles of 30, 45, 60 and 90° and at a constant impact velocity of 48 m/s was observed and found the maximum erosion rate occurred at 60° impingement angle. The minimum erosion rate occurred for carbon black composite at 20 wt% as compared to raw 10 wt% of filler composite.

Gai Zhao et al. [13] analysed friction and wear of polyimides reinforced with carbon, glass and aramid fibers. The research showed reinforcement improves the tribological properties of composites and the best performance was achieved by inorganic fiber reinforced composites compared to glass fiber reinforcement due to the effective load sharing between surfaces in contact. Anu Gupta et al. [14] conducted research using epoxy resin reinforced with relatively inexpensive bamboo fibers to determine the mechanical and erosion behaviour of epoxy composites. Fiber composition ranged from 0-40 % with increments of 10% by weight. The results demonstrated the erosion wear rate improved quite significantly by the addition of bamboo fibers and mechanical properties, such as tensile strength shown maximum value at 40 wt%, flexural strength and impact strength increased with 0-20% inclusion of fibers. The test results have also shown decrease in properties with 30 % inclusion of fibers, furthermore mechanical properties increased above 30 % inclusion of bamboo fiber reinforcement. Refer graphs in Figure 4,5 and 6 for details.

Mohit Sharma et al. [15] conducted a study using polyethersulfone (PES) reinforced with carbon fabric. A comparison was made between carbon fibers treated with cold remote nitrogen and untreated carbon fibers reinforced with PES. It was observed that the plasma treatment of carbon fabric increased the fiber-matrix adhesion and enhanced the coefficient of friction and wear resistance properties when compared to untreated fibers. There was a correlation between the tribo-performance with molecular weight and tensile strength. Suresha et al. [16] carried out a study using epoxy resin with reinforced E-Glass and graphite filled epoxy composites. The wear resistance was observed for three different combinations and it was found that epoxy with E-Glass filler reduced the wear properties and the epoxy composites with E-Glass and graphite showed the least wear when compared with neat epoxy. The mechanical properties, such as tensile strength, tensile modulus and hardness increased with the addition of E-Glass and graphite filled composites.

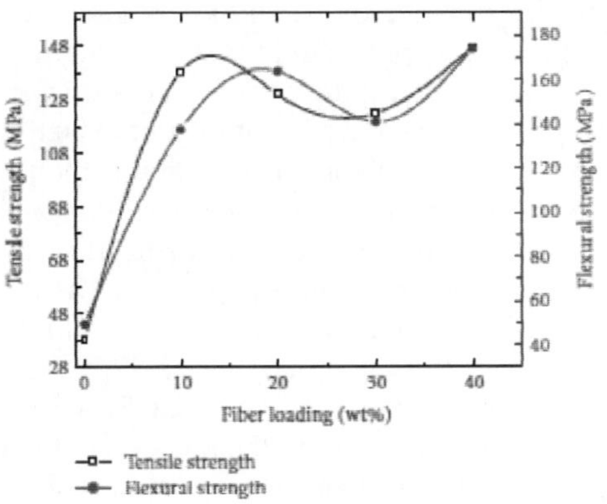

Figure 4. Effect on tensile and flexural strength with inclusion of bamboo fiber [14]

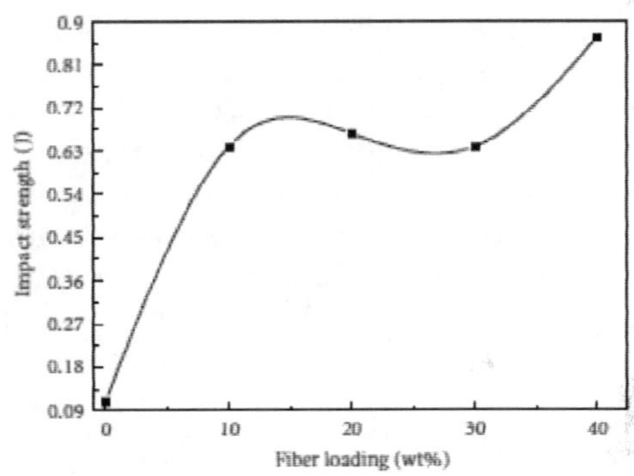

Figure 5. Effect on Impact strength of composites with inclusion of bamboo fiber [14]

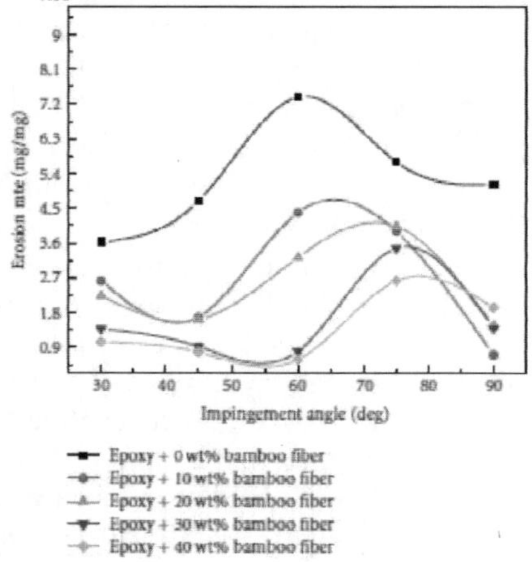

Figure 6. Effect on the erosion wear rate of composites with Impingement angle [14]

Mihail Botan et al. [17] analysed the tribological performance of polybutylene terephthalate (PBT) with an aramid fiber composite (10 % by weight) to measure the wear and frictional properties under different loads and velocities. The analysis showed the presence of aramid fibers increases the friction coefficient gradually with the applied load and the sliding speed. Jian [18] studied the friction and wear performance of polyvinylidene fluoride (PVDF) reinforced with carbon fibers of varying percentages ranging from 0, 10, 20 and 30 % under varying loads. The coefficient of friction and wear increased with the increase of load and the optimal wear resistance was obtained with a carbon fibre content of 20 %. All filled PVDF composites exhibited improved tribological properties. Singha et al. [19] utilized natural fibers of Hibiscus Sabdariffa with phenol-formaldehyde resin matrix to analyze the tensile strength, flexural strength, compressive strength and wear resistance properties. The results revealed that the mechanical and wear resistance properties increased up to 30 % with fiber reinforcement and then tended to decrease beyond that loading. Reza Arjmandi et al. [20] used the natural fiber of rice husks with different polymer matrices such as PE, PP, PVC and PLA to form polymer composites. The results revealed that inclusion of rise husk resulted in the reduction of tensile strength and the addition of other fillers such as MMT increased the mechanical properties. Also noted was that rice husks combined with PE showed the best results when compared to the other composites.

Pravuram Panda et al. [31] explored the mechanical and erosion wear of E-glass fiber reinforced 50 % by weight in epoxy resin. It was further reinforced with aluminum nitride of different percentages by weight (5, 10 and 15 %) using the hand lay-up technique. The results indicated that as the hardness increases with an increase in the percentage of aluminum nitride which is maximum at 15% inclusion, the tensile strength decreased which was contradictory. The maximum wear rate was observed at 60° impingement angle. Refer graphs 7 and 8 for details.J. Sudeepan et al. [32] utilized Acrylonitrile-butadiene-styrene(ABS) filled with micrometer sized titanium di-oxide of varying concentrations ranging from 5, 10 and 15 % by weight as filler content to determine the tribological (friction and wear) behavior of the polymer composite. It was also noted that the optimal conditions for the coefficient of friction was found to be with 15 % filler content and a specific wear rate with 10 % filler content. Hiral H Parikh et al. [33] has carried out a literature review for different fiber reinforced polymer composites and studied the effects of both material parameters such as type of filler, fiber length, filler content, fiber orientation and the operating parameters such as load, sliding distance, sliding velocity and temperature. Both parameters have a significant effect on the tribological properties of composite materials.

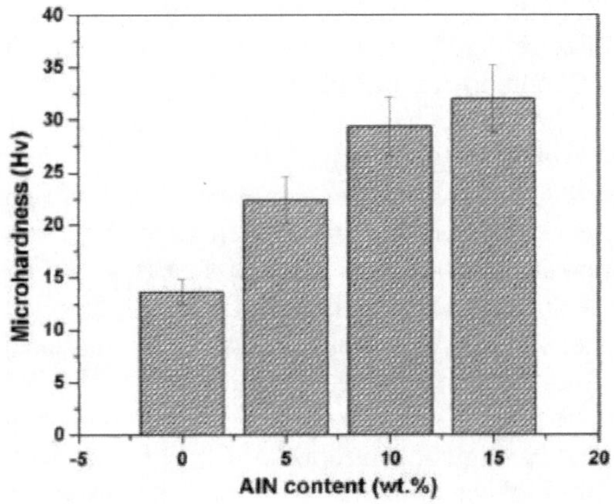

Figure 7. Variation of microhardness with AlN content [31]

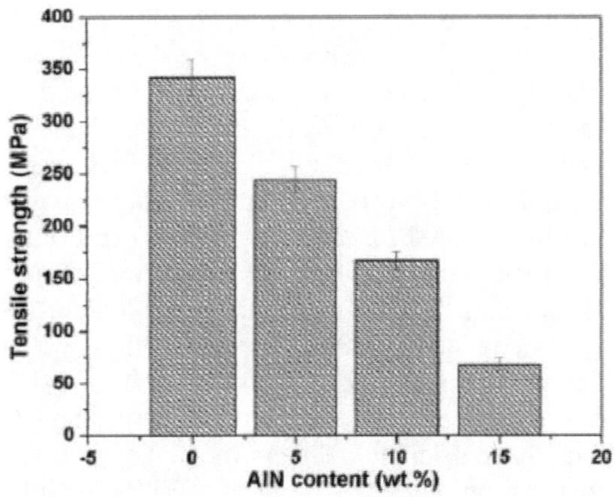

Figure 8. Variation of tensile strength with AlN content [31]

2.3 Tribological (Friction and Wear) behaviour of Nanocomposites

Tribology is related to the study of interacting surfaces in relative motion which in turn applies the principles of friction, lubrication and wear. Nano particles added to polymers tend to be effective in reducing the wear rate and the coefficient of friction of composites. Chun et al. [21] have conducted an analysis to study the mechanical behaviour of nanoclay in epoxy composites with varying percentages of nanoclay ranging from 0 to 4 % by weight. It was observed that wear resistance and hardness of nanoclay composites increased with the increasing percentage of nanoclay but this also depends on the degree of agglomeration of nanoclay clusters in the Nanoclay/Epoxy composites. Bhuyan et al. [22] studied the tribological behaviour of bio-based nanocomposites using soybean oil and functionalized organoclay (VMMT) with varying percentages of VMMT from 0, 1 and 5 %. The study revealed that the addition of clay of 1 % by weight resulted in a superior wear performance compared to results with lower and higher clay modifications. Wang et al. [23] have conducted a study to analyse the hardness and tribological behaviour of Polyacrylonitrile-methylmethacrylate-carbon nanotubes (AMMA-CNT) which are co-polymer nanocomposites, with varying compositions of CNT ranging from 0 to 3 % by weight in increments of 0.5 % by weight. The results indicated that nanocomposites with a 1.5 % CNT addition exhibited the lowest wear rate of 5.82×10^{-4} mm^3 N^{-1} $m^{-1}$and a lower coefficient of friction of 0.36. The significant improvement in tribological properties of AMMA-CNT composites have also attributed to excellent improvement in mechanical properties of composites.

Jawahar et al. [24] conducted an analysis using the pristine polyester, polyester with conventional clay-filled composite and polyester reinforced with nanoclay to study the flexural strength, flexural modulus, wear rate and coefficient of friction. It was found the nanocomposites possessed an improvement in flexural properties up to 20 % and wear properties up to 85 % when compared to the conventional clay-filled composites and pristine polyester. Lingaraju et al. [25] analysed the mechanical and tribological properties of polymer hybrid composites including nano particles. Results of the experiments have shown that the inclusion of nanoparticles such as silica and clay to the glass fiber reinforcement epoxy composite with varying percentages of 0 to 3 % by weight improved the mechanical properties such as tensile strength, impact strength, hardness and wear rate when compared to the pure composite material.

Kadhim et al. [26] conducted an analysis of the tribological properties using an Epoxy resin/Unsaturated polymer resin (80 % vol. /20 % vol.) reinforced with Nano-SiC and ultrafine graphite particles with varying percentages of SiC. Results indicated that the wear rate was higher for the blended resin when compared to the Nano-SiC composite. In addition, the wear and friction coefficient decreases with an increase in time and the percentage of Nano-SiC. Galetz et al. [27] conducted an analysis utilizing ultra-high molecular weight polyethylene (UHMWPE) reinforced with Carbon nanofibre (CNF) with varying percentages of CNF from 0, 5 and 10 % in weight. The test results revealed that the addition of nanofibers increased the mechanical properties such as modulus, yield stress and hardness. The wear rate of the composite material was reduced significantly from a 5 % inclusion of nanofibers although; an inclusion of 10 % nanofibers increased the wear rate.

Dong et al. [28] used polymethyl methacrylate/styrene and multiwall carbon nanotubes with varying composites of CNT ranging from 0-3 % with increments of 0.5 % and it was observed that the micro hardness increased with an increasing percentage of CNT and the tribological parameters such as friction coefficient and wear rate decreased with an increasing concentration of CNT. It was also noted that carbon nanotubes with 1.5 % CNT exhibited the highest wear resistance and the lowest friction coefficient. Zhang et al. [29] observed the wear resistance properties of epoxy fillers using short carbon fibers, graphite, Polytetrafluoroethylene (PTFE) and nano-TiO_2. The results indicated that each individual filler improves the wear resistance independently and it was also observed that the optimum wear properties was determined at 4-6 % nano-TiO_2 and the best wear resistance properties were achieved from nano-TiO_2 combined with conventional fillers.

Xin-Rui et al. [30] carried out an investigation on the friction and wear using different combinations of Polyimide (PI) with short carbon fibers (SCFs), micro SiO_2 and graphite particles. The results showed that incorporation of SCF and graphite particles to PI improved the tribological properties and the addition of SiO_2 to PI showed an adverse effect. The most effective results were observed with a combination of PI, SCF, graphite and micro SiO_2 particles.

3. Conclusion of literature review analysis

The careful review of published literature indicates that there has been no comprehensive analysis carried out so far and it is clear that further research on the mechanical and tribological behavior of polymer nano-composite materials is needed to investigate. Most of the research was carried out without the use of lubrications; therefore the use of lubrication in fabricating composite materials and the effect of lubrication on the wear rate [33] could be studied. Also it is rare to find the tribological properties of nano polymer composite materials having nano metal fibers with process optimization [18, 19, 22, 28 & 32]. The literature study is carried out to fill the gap in the literature, for the study of fiber reinforces polymer metal nano composites. Based on the literature study, the research gap is identified and they are presented as follows:

- Very limited mechanical property study has been carried out in fiber reinforced polymer metal nano composites so far and especially for carbon fiber reinforced polymer-SiC composites.
- Literature review indicated that the polymer composites with glass fiber reinforcement have a higher erosion rate than those without glass [33], which can reveal the need to use carbon fibers as reinforcement in this research.

- Epoxy resin LY556 and hardener HY951 are used so far in the previous researches for making composites. New grade is used in this study as a difference in epoxy resin as B-11 (3101) and hardener as K-6 (5205).
- Very limited study is carried out on the modeling of tribological properties for fiber reinforced polymer-metal nano composites.
- The review of literature indicated that the application oriented studies for the fiber reinforced polymer-metal nano composites are very few.
- Further the review of literature indicated that the optimization studies carried out on the fiber reinforced polymer-metal nano composites are very limited.

References

[1] Ya-Jung Lee and Cheng-Hsien Chung. Study on the Mechanical Properties of Marine FRP Laminates, *Journal of Composite Materials, Vol. 37, No. 11/2003.*

[2] Sang-Young Kim, Chun Sik Shim, Caleb Sturtevannt, Dave (Dae-Wook) Kim and Ha Cheol Song. Mechanical Properties and production quality of hand-layup and vacuum infusion processed hybrid composite materials for GFRP marine structures, *Int.J.Nav.Archit. Ocean Eng. (2014) 6:723~736.*

[3] Mouritz.A.P, Gellert.E, Burchill.P and Challis.K. Review of advanced composite structures for naval ships and submarines, *Composite structures 53(2001) 21-41.*

[4] Selvaraju.S and Illaiyavel.S. Applications of Composites in Marine Industry, *Journal of Engineering Research and Studies/Vol. II/Issue II/April-June 2011/89-91.*

[5] Vallbo.S. Material Selection Considerations for Polymer Composite Structures in Naval Ship Applications, *Journal of Sandwich Structures and Materials, Vol. 7 – September 2005.*

[6] Chalmers.D.W. The Potential for the Use of Composite Materials in Marine Structures, *Marine Structures 7(1994) 441-456.*

[7] Nagalingam.R, Sundaram.S and Stanly Jones Retnam.B. Effect of nanoparticles on tensile, impact and fatigue properties of fiber reinforced plastics, *Bull. Mater. Sci., Vol. 33, No.5, October 2010, pp. 525-528.*

[8] Isao Kimpara. Use of Advanced Composite Materials in Marine Vehicles, *Marine Structures 4 (1991) 117-127.*

[9] Md Ekramul Islam, Tanjheel H. Mahdi, Mahesh V.Hosur, Shaik Jeelani. Characterization of Carbon Fiber Reinforced Epoxy Composites Modified with Nanoclay and Carbon Nanotubes, *Procedia Engineering 105(2015)821-828.*

[10] Anbusagar N.R.R. and Giridharan P.K. Effect of Nanoparticles on Tensile and impact properties of FRP and Sandwich FRP Laminates by Hand – Layup Technique, *National Journal on Advances in Building Sciences and Mechanics , Vol.2 No.2 October 2011.*

[11] Law Mei Lin, Qumrul Ahsan, Hairul Effendy Ab Maulod, Noraiham Mohamad and Sivaraos. Wear Behaviour of Cryogenic Treated Recycled Carbon Fibers Filled Epoxy Composite, *Applied Mechanics and Materials Vol 761(2015) pp 489-493.*

[12] Shakuntala Ojha, Samir Kumar Acharya, Raghavendra Gujjala. Characterization and Wear Behavior of Carbon Black filled Polymer Composites, *Procedia Materials Science 6 (2014) 468-475.*

[13] Gai Zhao, Irina Hussainova, Maksim Antonov, Qihua Wang, Tingmei Wang. Friction and Wear of fiber reinforced polyimide composites, *Wear 301(2013) 122-129.*

[14] Anu Gupta, Ajit Kumar, Amar Patnaik, Sandhyarani Biswas. Effect of Different Parameters on Mechanical and Erosion wear behavior of Bamboo Fiber Reinforced Epoxy Composites, *Hindawi Publishing Corporation, International Journal of Polymer Science, Volume 2011, Article ID 592906.*

[15] Mohit Sharma, Jayashree Bijwe, Kuldeep Singh. Studies for Wear Property correlation for Carbon Fabric-Reinforced PES Composites, *Tribol Lett (2011) 43:267-273.*

[16] Suresha.B, Chandramohan.G, Siddharamaiah, Jayaraju.T. Three-Body Abrasive wear Behavior of E-Glass fabric reinforced/ Graphite-filled Epoxy composites, *Polymer Composites – 2008.*

[17] Mihail Botan, Igor Roman, Constantin Georgescu, Alina Cantaragiu and Lorena Deleanu. Tribological performance of the blend PBT and short aramid fibers, *INCAS Bulletin, Volume 7, Issue 1/2015,pp, 29-36.*
[18] Jian.L. The sliding friction and wear properties of poly (vinylidene fluoride) composites filled with carbon fibres, *Proc. IMechE Vol.223 Part J:J. Engineering Tribology.*
[19] Singha.A.S and Thakur.V.K. Synthesis, Characterisation and Analysis of Hibiscus Sabdariffa fibre Reinforced Polymer Matrix Based Composites, *Polymers & Polymer Composites, Vol. 17, No. 3, 2009.*
[20] Reza Arjmandi, Azman Hassan, Khaliq Majeed and Zainoha Zakaria. Rice Husk Filled Polymer Composites, *Hindawi Publishing Corporation, International Journal of Polymer Science, Volume 2015, Article ID 501471.*
[21] Chun Ki Lam, Kin Tak Lau. Tribological behavior of nanoclay /epoxy composites, *Material Letters 61(2007)3863-3866.*
[22] Bhuyan.S, Sundararajan.S, Lu.Y, Larock.R.C. A study of the physical and tribological properties of biobased polymer –clay nanocomposites at different clay concentrations, *Wear 268(2010) 797-802.*
[23] Ce Wang, Bin Dong, Guo-Yu Gao, Mao-Wen Xu, Hu-Lin Li. A study on microhardness and tribological behavior of carbon nanotubes reinforced AMMA-CNTs copolymer nanocomposites, *Materials Science and Engineering A 478(2008)314-318.*
[24] Jawahar.P, Gnanamoorthy.R, Balasubramanian.M. Flexural and tribological properties of polyester-clay nanocomposites, *Journal of Material Science 40(2005)4391-4393.*
[25] Lingaraju.D, Ramji.K, Pramila Devi.M and Rajya Lakshmi.U. Mechanical and tribological studies of polymer hybrid nanocomposites with nano reinforcements, *Bull. Mater. Sci., Vol. 34, No.4,July 2011, pp.705-712.*
[26] Bahjat B.Kadhim, Zainab A.Jabur, and Fadhil K.Farhan, Tribological properties of SiC filled polymer blend nanocomposites. *International journal of advances in Engineering & Technology, Apr.,2015.*
[27] Galetz.M.C, Blab.T, Ruckdaschel.H, Sandler.J.K.W, Altstadt.V, Glatzel.U. Carbon Nanofibre-Reinforced Ultrahigh Molecular Weight Polyethylene for Tribolological Applications, *Journal of Applied Polymer Science, **Vol. 104**, 4173-4181 (2007).*
[28] Dong.B, Wang.C, He.B.L, Li.H.L, Preparation and Tribological Properties of Poly(methyl methacrylate)/Stryrene/MWNTs Copolymer Nanocomposites, *Journal of Applied Polymer Science, Vol. 108, 1675-1679 (2008).*
[29] Zhang.Z, Breidt.C, Chang.L, Haupert.F, Friedrich.K. Enhancement of the wear resistance of epoxy: short carbon fibre, graphite, PTFE and nano-TiO_2, *Composites: Part A 35(2004) 1385-1392.*
[30] Xin-Rui Zhang, Xian-Qiang Pei, Qi-Hua Wang. Friction and wear studies of polyimide composites filled with short carbon fibers and graphite and micro SiO_2, *Material and Design 30(2009) 4414-4420.*
[31] Pravuram Panda, Sisir Mantry, S.Mohapatra , SK Singh and A. Satyapathy. A study on erosive wear analysis of glass fiber–epoxy–AlN hybrid composites. *Journal of Composite Materials 2014, Vol 48(1) 107–118.*
[32] J.Sudeepan, K.Kumar, T.K.Barman and P.Sahoo. Tribological behavior of ABS/TiO2 polymer composite using Taguchi statistical analysis. *Procedia Materials Science 5 (2014) 41-49.*
[33] Hiral H Parikh and Piyush P Gohil. Tribology of fiber reinforced polymer matrix composites- A review. *Journal*
[34] Petra Potschke, Sven Pegel and Andreas Janke. Melt compounding of thermoplastic polymers with carbon nanotubes, *Nanopoly workshop, Dredsen, 2005.*
[35] William Gacitua E. Aldo Ballerini A. Jinwen Zhang. Polymer Nanocomposites: Synthetic and Natural Fillers a review, Maderas. *Ciencia y tecnología 7(3): 159-178, 2005.*
[36] Palanikumar. K, and Anbu Sagar NRR. Sandwich nano composites for naval applications, *Naval research board Report, 2015.*

Design and Simulation of a PID Controller for Motion Control Systems

Zakariyya Hassan Abdullahi[1, a], Bashir Ahmed Danzomo[1, b], Zainab Suleiman Abdullahi[2, c]

[1] Department of Computer Engineering Hussaini Adamu Federal Polytechnic, Kazaure, Jigawa State Nigeria
[2] Department Mechanical Engineering, Hussaini Adamu Federal Polytechnic, Kazaure, Jigawa State Nigeria

[a]hassanzakariyya8@gmail.com, [b]zeeshazahm05@gmail.com , [c]bdzomo@gmail.com

Abstract. Motion control system plays important role in many industrial applications among which are in robot system, missile launching, positioning systems etc. However, the performance requirement for these applications in terms of high accuracy, high speed, insignificant or no overshoot and robustness have generated continuous challenges in the field of motion control system design and implementation. To compensate this challenge, a PID controller was design using mathematical model of a DC motor based on classical root-locus approach. The reason for adopting root locus design is to remodel the closed-loop response by putting the closed-loop poles of the system at desired points. Adding poles and zeros to the initial open-loop transfer function through the controller provide a way to transform the root locus in order to place the closed-loop poles at the required points. This process can also be used for discrete-time models. The Advantages of root locus over other methods is that, it gives the better way of pinpointing the parameters and can easily predict the fulfilment of the whole system. The controller performance was simulated using MATLAB code and a reasonable degree of accuracy was obtained. Implementation of the proposed model was conducted using-Simulink and the result obtained shows that the PID controller met the transient performance specifications with both settling time and overshoot less than 0.1s and 5% respectively. In terms of steady state error, the PID controller gave good response for both step input and ramp.

1. Introduction

Tool Motion control is a sub- discipline of automation with which the position or velocity of machine are regulated by using devices like hydraulic pump and electric motor, a servo. Motion control system plays significant role in many industrial applications among which are in robot system, missile launching, positioning systems etc. However, the performance requirement for these applications in terms of high accuracy, high speed, insignificant or no overshoot and robustness have generated continuous challenges in the field of motion control system design and implementation. The mechatronic approach to system design offers a solution to this problem by applying con-current engineering instead of the traditional sequential approach. In light of these, various techniques starting from classical to modern design approach have been developed and employed by researchers and industrialists in designing appropriate controller to meet the above demands [1] and [2].

IOP Conf. Series: Materials Science and Engineering **344** (2018) 012016 doi:10.1088/1757-899X/344/1/012016

Nowadays, modern control systems, like semiconductor manufacturing instruments, transportation, X-Y driving devices as well as robots have been important systems which normal require high speed and high efficiency linear motions [4]. These systems splitted into two components; which include the mechanical parts with servo drive systems and the servo controllers that regulate the multi axis agitation of the mechanical parts. In general, the system has consisted of an X-Y table, where each every motion axis can be driven by individual actuator arrangements, like DC or AC motors [5]. DC motors has been prominent in the industry because of their uncomplicated control manner. the armature voltage to differ the motors speed and because the motor carbon brush alongside the convertor were mechanical units, they tend to produce sparks and create ravage when the motor is operating, this is another dominant drawback of the DC Motors additionally, DC motor presents risks to the environment and have a short life cycle maintenance of DC motors is also costly, AC motors are normally classified into three groups, namely; the synchronous motors, the induction motor as well as the reluctance motor. Both the rotor and stator of an AC motor are the sole contact bearing units. The rotation of the motor is initiated by the stators magnetic field .this requires some control techniques' unlike the magnetic field quide control to regulate different moments. With the advent of semi conductor control devices, the estimation needed for AC motor regulation can be achieved with greater ease. Due to this advantage, AC motors becomes favourites this days [6].The movement mechanisms using motor mechanisms normally regulated to unknown dynamics linked interferences disturbances and friction among the relevant units, which usually damage the systems performance [3]. In an attempt to enhance the tracking performance in machining cutting, previous literatures have been presented in [9, 10].The versatility of numerical machines with output time delay and the dimensional error are presented in [9]. Frequency feedback and Phase margin analysis are operated for controller design to actuate regions of system steadiness on the other hand, cross coupled dynamics, and uncertainties among relevant axis components were not examined in the stability investigation. Additionally, the shift region nonlinearity as a result of mechanical dynamics was compensated by using adoptive fuzzy controller .with the determination of the stiff region nonlinearity [9] the major aims of this investigation is to examine the application of PID controller in the control of servo motor motion control system.

2. The PID Controller

A PID controller calculates an error value as the difference between a measured variable and a desired set point. The controller (shown in Figure 1) attempts to minimize the error by adjusting the process control inputs. In the absence of knowledge of the underlying process, a PID controller has historically been considered to be the best controller Bennett [8].

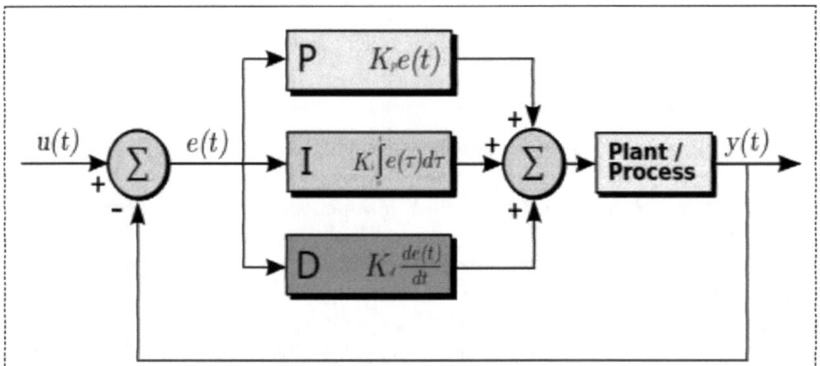

Figure 1. PID Controller Feedback System

The PID algorithm is the most popular feedback controller used within the process industries. It has been successfully used for over 50 years [5] and [6]. It is a robust easily understood algorithm that can provide excellent control performance despite the varied dynamic characteristics of a given system or process plant. The calculation algorithms given in (1) involved three separate constant parameters; P, I, and D. These parameters are interpretable in terms of time; where P relied on the present error, I on the buildup of past errors, and D is an indicator of impending errors, based on ongoing rate of change formula – where, Kp is the proper.

$$u(t)=MV(t) = K_P e(t) + K_i \int_0^t e(\beta)d\beta + K_d \frac{d}{dt}e(t) \quad [1]$$

Where, Kp represents the proportional gain, Ki represents the value integral gain, Kd is the derivative gain, e represents the error, t the period of the sudden time (the current and β are the parameters of integration (takes value from 0 - t). Equivalently, the transfer function in the Laplace domain L (s) of the PID controller is;

$$L(t)= K_P + K_i / s + K_d s \quad [2]$$

Where s is the complex frequency

2.1. *The proportional term, K_p*

The proportional term produces an output value that is proportional to the current error value. The proportional response can be adjusted by multiplying the error by a constant Kp, called the proportional gain constant given by;

$$P_{out} = K_P e(t) \quad [3]$$

A high proportional gain results in a large change in the output for a given change in the error (Figure 2). If the proportional gain is too high, the system can become unstable. In contrast, a small gain results in a small output response to a large input error, and a less responsive or less sensitive controller. If the proportional gain is too low, the control action may be too small when responding to system disturbances. Tuning theory and industrial practice indicate that the proportional term should contribute the bulk of the output change.

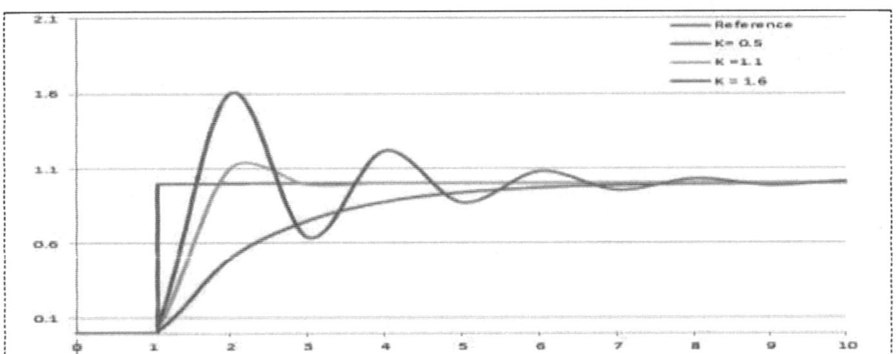

Figure 2. Plot of Time against PV for values of K_P, for K_p (K_i and K_d being held constant)

As a PID controller depends on the estimated process parameters, and not particularly on the knowledge of the process at hand. It is also widely applicable; through manipulating these three parameters of the model , the PID system works with specific process variables .The feedback of the controller can be explain in terms of its responsiveness to an error, the extend which the system pass over a set-point and the intensity of any system oscillation. The practice of the PID does not assured ideal regulation of the system or even its constancy. Some functions may need adopting only one algorithm to bring the correct system control. This is accomplishing by putting the other variables to zero. A PID controller is referred to as a PI, PD, P or I Controller in the absences corresponding regulation function. PI Controller is fairly familiar since derivative function is delicate to measurement noise, whereas lack of an integral term may restrict the system from meeting its target range.

2.2. The proportional term, K_p

The contribution from the integral term is similar to the magnitude of the variation and the period of the same. The integral in a PID Controller is a representation of the affected error variation overtime. Which presents the accumulate offset that may have been fixed earlier , the accumulated error variation hence multiplied by the integral gain (Ki) and included the controller output ,the integral function can be presented as;

$$I_{out} = K_i \int_0^t e(\beta)\, d\beta \qquad [4]$$

The integral Function pushes the motion of the process towards set point and avoids the resultant steady –state error variation that exists with pure preoperational controller (Figure 3).

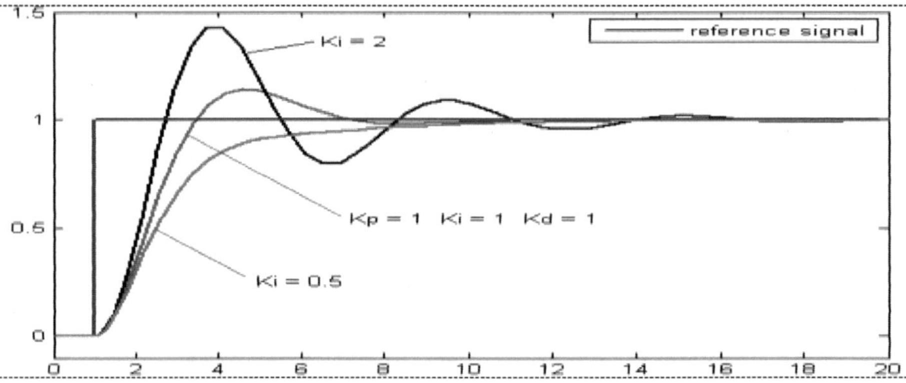

3.

Figure 3. Plot of Time against PV, for Three Values of K_i (K_p and K_d held constant

2.3 The Derivative term, K_d

The derivative of the process error is estimated by stabling the slope inclination of the error over time as well as multiplying the rate of change by the derivation gain kp (Figure 4). The size of the contribution of the derivation function to the entire control behaviors is referred to as the derivative gain kp. The derivative function therefore presented by

$$D_{out} = K_d \frac{d}{dt} e(t) \qquad [5]$$

Derivative response for see system performance and enhances the setting time and system balance , optimal derivative is not accidental so that operation of PID Controllers involve an added low pass filtering for the derivation function term to restrain the high frequency gain and noise. Derivation behaviour is sometimes used in reality even by one estimate in just 25% of deployed regulators by virtue of its variable encounter on system balance and other application.

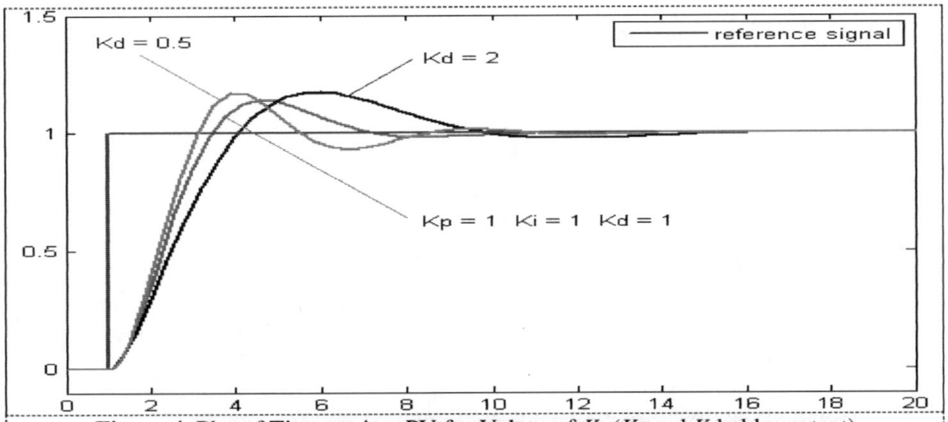

Figure 4. Plot of Time against PV for Values of K_d (K_p and K_i held constant)

4. Materials and Method

4.1. Mathematical Modelling of the Motion Control System

In this research, a Quanser rotary motion system driven by DC motor was considered. Basically, in motion control system, the DC motor acting as power actuator converts electrical energy into rotational mechanical energy as shown schematically in Figure 5. The major motor parameters are; armature resistance Rm, armature inductance Lm, moment of inertia of the motor system Jeq, damping friction of the mechanical system Beq and other motor constants in K subscripts. The input to the system is the armature voltage in volts driven by a voltage source. And the measured variable is the angular position of the shaft, θm in rad which is a function of the angular velocity of the motor shaft, ω in rad/s, $\dot{\theta} = \omega$

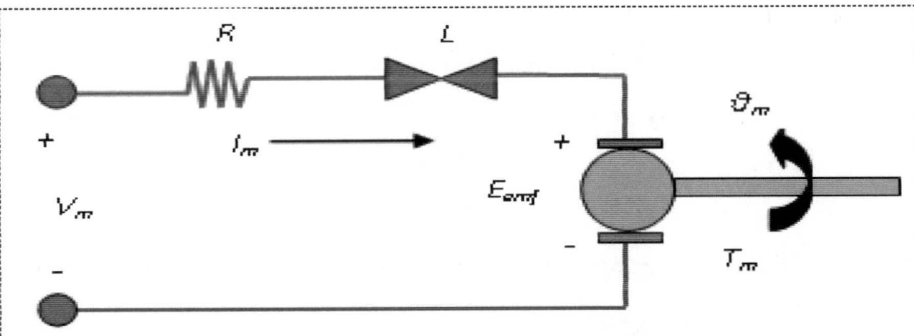

Figure 5. Schematic Representation

Using Kirchoff's law, the voltage equation has been obtained as follows:

$$V_m - R_m I_m - L_m \frac{dI_m}{dt} - E_{emf} = 0$$

[6]

IOP Conf. Series: Materials Science and Engineering **344** (2018) 012016 doi:10.1088/1757-899X/344/1/012016

Given that $L_m << R_m$, the motor inductance is disregarded, therefore;

$$I_m = \frac{V_m - E_{emf}}{R_m} \qquad [7]$$

The Drive emf is proportional to the drive shaft velocity $\dot{\theta}_m$ yields;

$$I_m = \frac{V_m - K_m \dot{\theta}_m}{R_m} \qquad [8]$$

By putting Newton's Law to the mechanical part of the system (second law of motion) the torque relation is given by:

$$J_m \ddot{\theta}_m = T_m - \frac{T_l}{\eta_g K_g} \qquad [9]$$

Where $\frac{T_l}{\eta_g K_g}$ is the load torque seen through the gears, and η_g is the efficiency of the gear box. Applying the second law of motion at the load of the motor, we have:

$$J_l \ddot{\theta}_l = T_l - B_{eq} \dot{\theta}_l \qquad [10]$$

Substituting;

$$J_l \ddot{\theta}_l = \eta_g K_g T_m - \eta_g K_g J_m \ddot{\theta}_m - B_{eq} \dot{\theta}_l \qquad [11]$$

Given that $\dot{\theta}_m = K_g \dot{\theta}_l$ and $T_m = \eta_m K_t I_m$ where η_m the motor efficiency, (11) becomes;

$$J_l \ddot{\theta} + \eta_g K_g^2 J_m \ddot{\theta}_l + B_{eq} \dot{\theta}_l = \eta_g \eta_m K_g K_t I_m \qquad [12]$$

The transfer function (TF) of the system (plant) is presented below as;

$$\frac{\theta_l(s)}{V_m(s)} = \frac{\eta_g \eta_m K_g K_t}{J_{eq} R_m s^2 + (B_{eq} R_m + \eta_g \eta_m K_m K_t K_g^2) s} \qquad [13]$$

Where J_{eq} represents the equivalent moment of inertia of the motor system as seen at the output described by the equation.

$$J_{eq} = J_l + \eta_g J_m K_g^2 \qquad [14]$$

These system parameters and their values is described in Table 1.

IOP Conf. Series: Materials Science and Engineering **344** (2018) 012016 doi:10.1088/1757-899X/344/1/012016

Table 1: System Parameters [5]

Parameter	**Symbol**	0.00767
Back EMF Constant	K_m	0.00767
Motor Torque Constant	K_t	3.87×10^{-7}
Motor Moment of Inertia	J_m	2.0×10^{-3}
Equivalent Moment of Inertia at the Load	J_{eq}	4.0×10^{-3}
Equivalent Viscous Damping Coefficient	B_{eq}	0.9
Gearbox Efficiency	η_g	0.69
Motor Efficiency	η_m	14x5=70
System Gear Ratio (For High Gear)	K_g	2.6
Armature Resistance	R_m	

Substituting these values in Table 1 into Equation 13, we have:

$$\frac{\theta_l(s)}{V_m(s)} = \frac{0.1282}{s(0.002s+0.0728)} = \frac{64.1}{s(s+36.42)}$$

Therefore the system open loop transfer function is given by:

$$G_p(s) = \frac{\theta_l(s)}{V_m(s)} = \frac{64.1}{s(s+36.42)} \qquad [15]$$

The system's open-loop response to a unit step input was observed using MATLAB code described graphically in Figure 6.

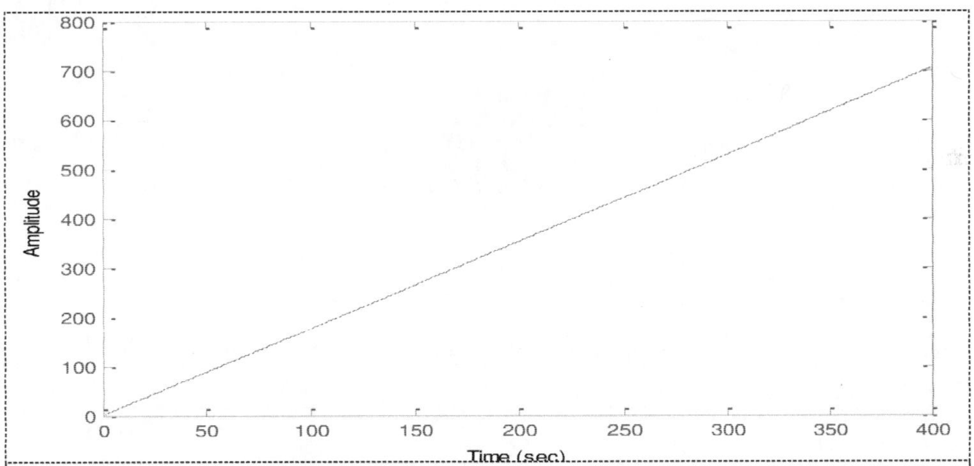

Figure 6. Open-Loop Response to a Unit Step

From the response in Figure 6, the open-loop response is unstable and cannot meet any realistic specification, therefore a PID controller with the required settling time, overshoot and velocity error constant has been designed so as to achieve the objective of the research.

IOP Conf. Series: Materials Science and Engineering **344** (2018) 012016 doi:10.1088/1757-899X/344/1/012016

4.2. PID Controller Design

The PID controller consists of a proportional plus derivative (PD) compensator cascaded with a proportional plus integral (PI) compensator. The PID controller configuration is of the form:

$$\frac{K_D s^2 + K_P S + K_I}{s} \Rightarrow \frac{K_D(s^2 + as + b)}{s} \qquad [16]$$

Where,

$$a = {K_P}/{K_D} \text{ and } b = {K_I}/{K_D}$$

The controller can then be expressed with two zeros and one pole at origin.

$$\frac{K_D(s + z_1)(s + z_2)}{s} \qquad [17]$$

The two zeros used to be located anywhere at left side of the root locus plot to achieve the desired response. Considering the root locus plot of figure 3, a pole at origin introduced by the controller doubled the pole at origin, thus introduce additional -140 degree. In line with the objective of this work, a PID controller with the following specifications is to be designed with Settling time, T_s as 0.09s, Overshoot of 2.5% and Velocity error constant, K_v given as 200. Based on these the system response parameters, ω_n and ξ are obtained as 58.38 and ±0.7613 respectively.

4.3. Simulation of the PID Controller and Result

The proposed PID controller was simulated with the modelled plant in MATLAB –Simulink with 10 degree step and ramp input and with aid of root-locus command the gain which was fine tuned to optimize the system performance as follows. The MATLAB- Simulink set-up for the PID controller is shown in Figure 7 – Figure 8.

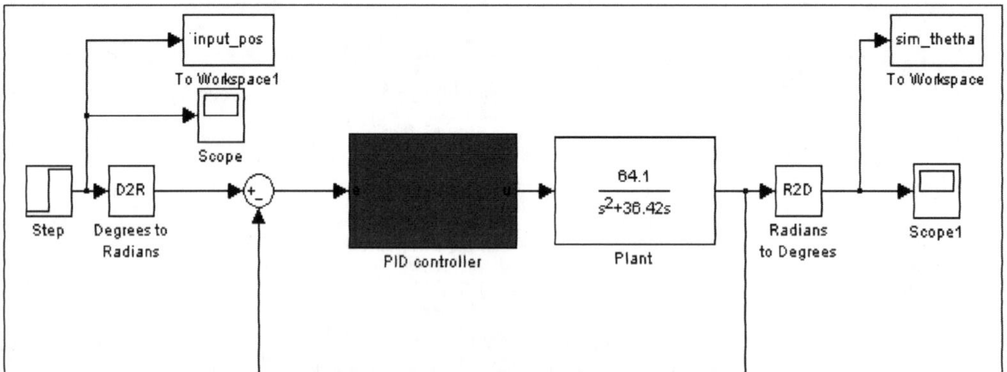

Figure 7. PID simulation Set-up with Step input

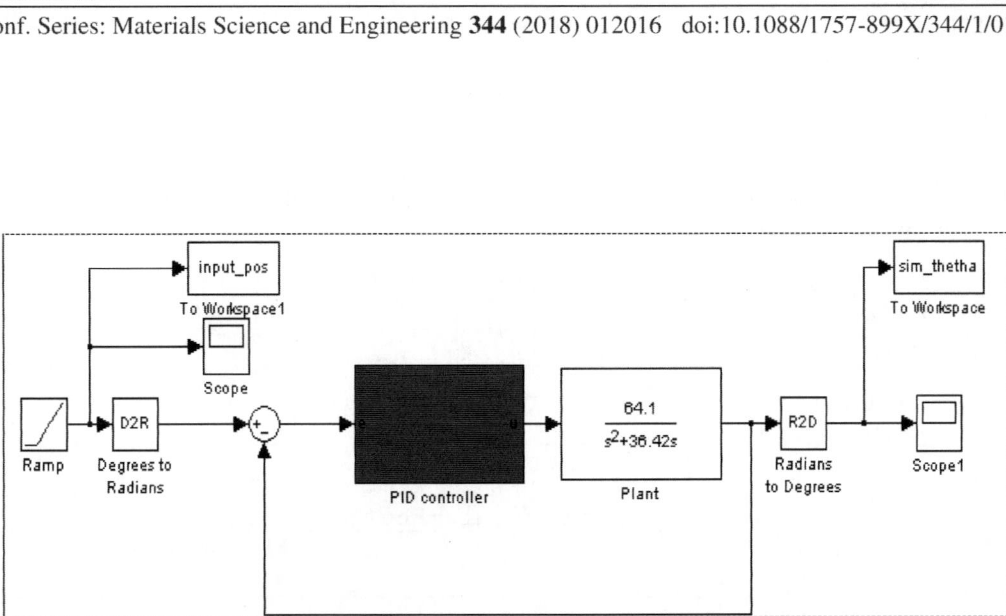

Figure 8. PID simulation Set-up with Ramp input

The response of each set-up for both step input and ramp are given in Figures 9 - 10.

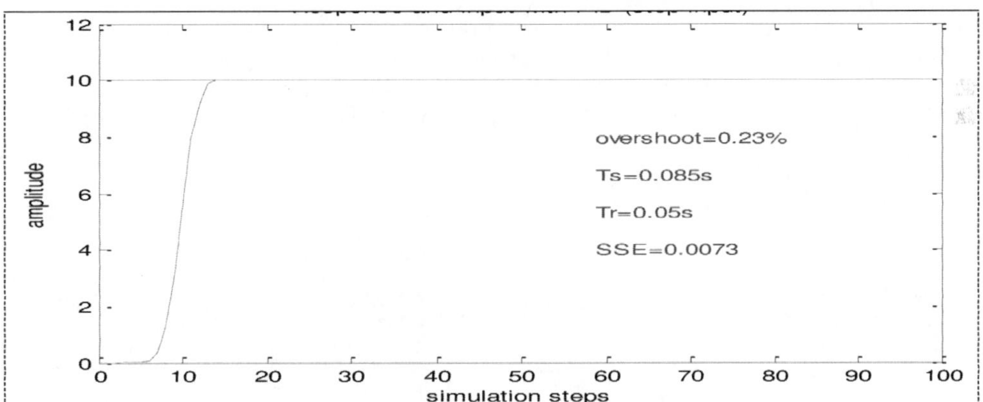

Figure 9. Simulation response of the plant with PID controller (Step input)

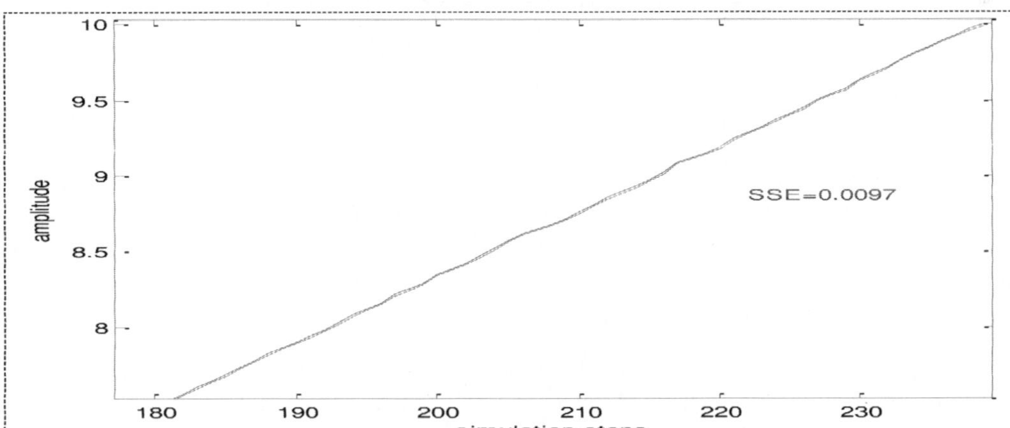

Figure 10. Simulation response of the plant with PID controller (Ramp input)

Table 2 and Table 3 show the overview of the simulation results for the step input and ramp input respectively.

Table 2: Overview of the Simulation Result for Step Input of 10 Degree

Performance	PID Controller
Overshoot %	0.23
Settling Time (s)	0.085
Rise Time(s)	0.049
Steady state error	0.0073

Table 3: Overview of the Simulation Result for Ramp Input

Performance	PID Controller
Steady state error (SSE)	0.0097

5. Hardware Implementation

5.1. Experimental Set-Up and Results

The PID controller was implemented experimentally on one of the available motion control system shown in Figure 12, 13 and 14 .Thus the experimental set-up consists of rotary motion system driven by DC motor, Encoder position sensor and Simulink-based controller implementation with Winton compiler. The Simulink based set-up of the experiment for the PID controller is giving in Figure 11.

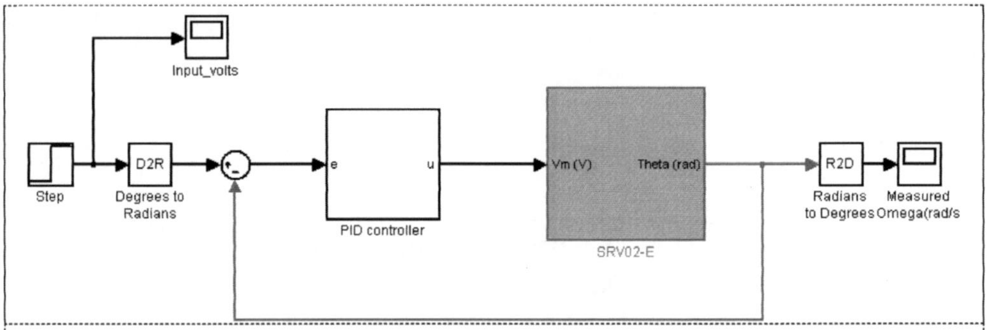

Figure 11. Experimental Set-up for the PID Controller

The experimental response of the plant with PID controller for step input and ramp input is shown in Figure 12 - Figure 15 respectively.

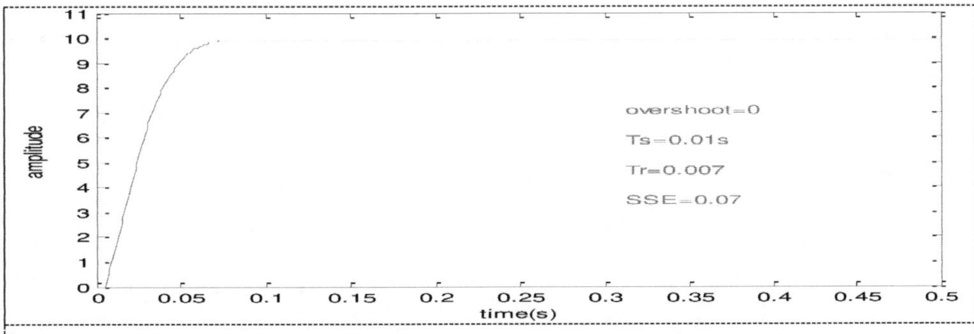

Figure 12. Response obtained by plant to a 10 degree step input with PID

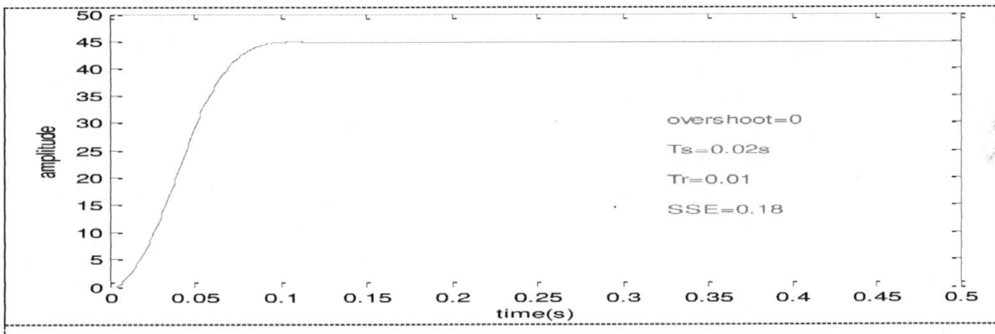

Figure 13. Response obtained by plant to a 45 degree step input with PID

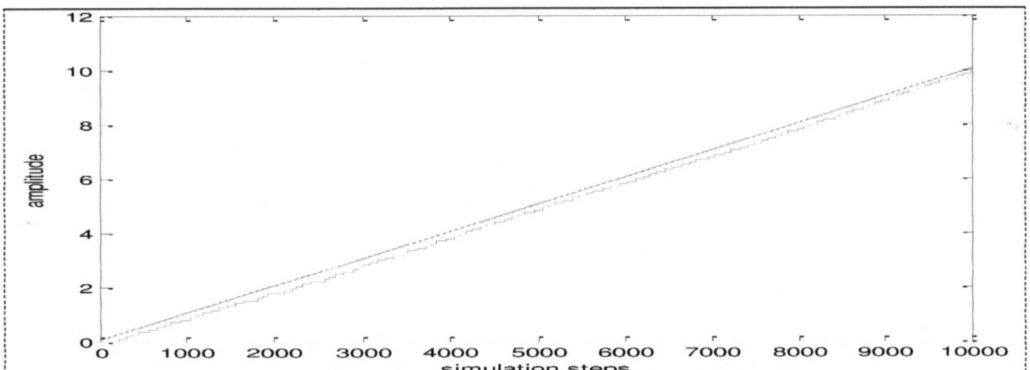

Figure 14. Response of the plant to a Ramp input with PID

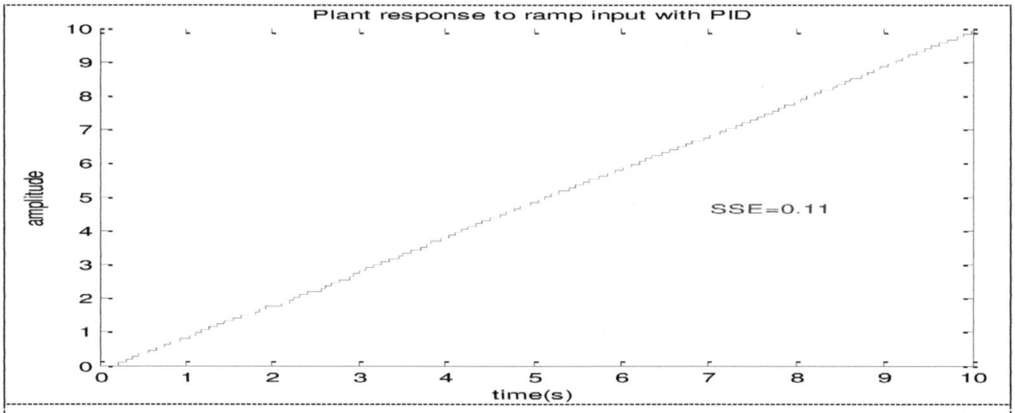

Figure 15. SSE for Response of the plant to a Ramp input with PID

5.2. *Fine -Tuning the PID Controller*

The controller gains were fine-tuned with the aid of root-locus plot to enhance the experimental response of the system and the resulting response obtained for the PID controller for 10 and 45 degree step input are shown in Figures 16 and 17 while an overview of the experimental results obtained for step and ramp input are shown in Tables 4 and 5.

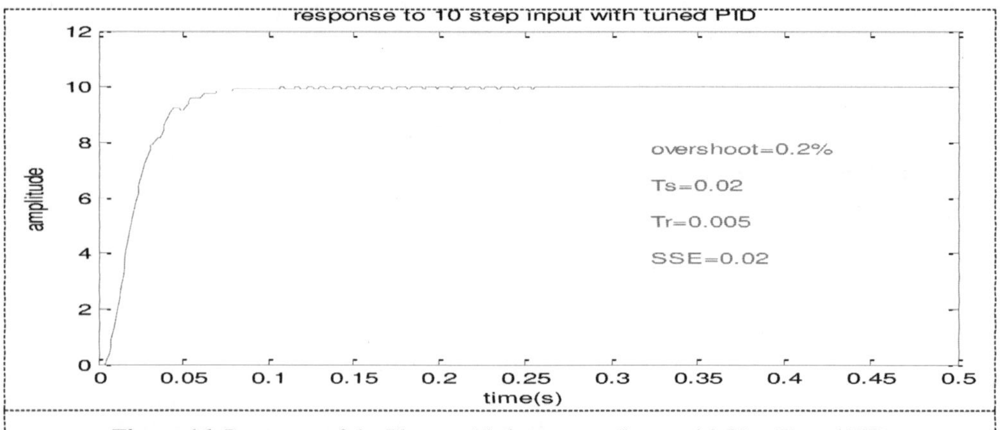

Figure 16. Response of the Plant to 10 degree step Input with Fine Tuned PID

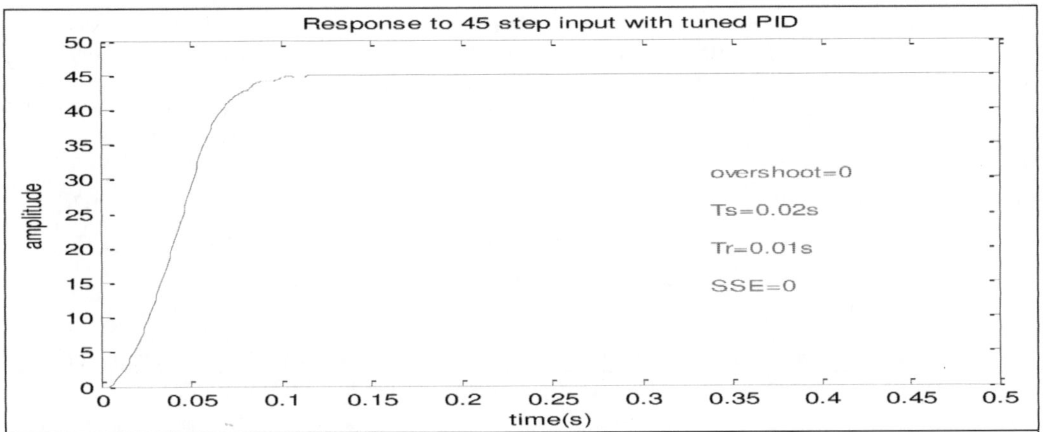

Figure 17. Response of the Plant to 10 degree step Input with Fine Tuned PID

Table 4: Overview of the Experimental Results with Step Input

Performance	PID (10 Degree Step)	PID (45 Degree Step)	PID Tuned (10 Degree Step)	PID Tuned (45 Degree Step)
Overshoot (%)	0	0	0.2	0
Settling Time (s)	0.01	0.02	0.02	0.02
Rise Time (s)	0.007	0.01	0.005	0.01
Steady State Error	0.07	0.18	0.02	0

Table 5: Overview of the Experimental Results with Ramp Input

Performance	**PID**
Steady State Error	0.11

5.3. Validation of the Result

The experimental result was validated by comparing result obtained for Lead-lag controller implementation with motion rotary system in Ismail (2008). From the summary of the comparative analysis conducted on the two controllers as shown in Table 6 and Table 7, the simulation results obtained for both controllers met the desired system specifications with PID given a better response in terms of overshoot and steady state error. However Lead-Lag compensator is seen to perform better in respect of transient response of settling and rise time. There are marginal differences between the simulation performance of the designed controller and experimental performance. An over damped responses were recorded for both controllers in the experiment. These could be attributed to the fact that the mathematical model used to design the controllers did not capture the inherent dynamic characteristics of the system which includes the nonlinearity effect of such factor as frictions.

Table 6: Comparison between the PID Controller and Lead-Lag Controller With Step Input

	PID CONTROLLER RESULT WITH STEP INPUT				LEAD-LAG CONTROLLER (Isma'il, 2008)	
PERFORMANCE	PID (With 10 Degree Step Input)	PID (With 45 Degree Step Input)	PID Tuned (With 10 Degree Step Input)	PID Tuned (45 Degree Step Input)	Lead-Lag Controller (With 10 Degree Step Input)	Lead-Lag Controller (With 45 Degree Step Input)
Overshoot (%)	0	0	0.2	0	0	0
Settling Time (s)	0.01	0.02	0.02	0.02	0.02	0.02
Rise Time (s)	0.007	0.01	0.005	0.01	0.003	0.01
SSE	0.07	0.18	0.02	0	0.24	0.18

Table 7: Comparison between the PID Controller and Lead-Lag Controller with Ramp Input

Performance	Lead-Lag Controller (Isma'il, 2008)	PID Controller
Steady state error	0.13	0.11

6. Conclusion

The present investigation has demonstrated the design of a PID controller using the classical root-locus approach. The simulation as well as experimental investigation use of this approach in handling simple positioning control system to a reasonable degree of accuracy for relatively simple situation where the mathematical model of the system is available. From the experimental investigation, the PID controller performs better for both step and ramp inputs with settling time and overshoot less than 0.1s and 5% respectively.

References

[1] Richard C D and Robert H B, 2001, Pren Hal Int'l
[2] Katsuhiko O, 2002 , Pear. Educ int'l
[3] Quans Rot Motion Syst Lab Manual. 2007
[4] Liu T H, Lee Y. C. Lee and Chang Y H 2004 IEEE Aero El Syst
[5] Lin F J and Shen P H, 2006, IEEE Ind Electron M
[6] Fuyu, D Xing'an C and Jiasheng C, 2005
[7] Groovier M P , Fund of Mod Manuf Mater
[8] Lim H, Seo, J W , Choi, C H ,2000
[9] Park E C , Lim H and Choi C H , 2003 IEEE Contr Syst Mag
[10] Gieras F and Piech Z J,2005 CRC Pres
[11] Chong K H, GCY and L Y ,2005, IEEE Contr Syst Mag

IOP Conf. Series: Materials Science and Engineering **344** (2018) 012017 doi:10.1088/1757-899X/344/1/012017

Toxicity prediction of ionic liquids based on Daphnia magna by using density functional theory

M N Nu'aim[1], M A Bustam

Chemical Engineering Department, Universiti Teknologi PETRONAS, 32610 Seri Iskandar, Perak, Malaysia

[1]Email: nizam.sza@gmail.com

Abstract. By using a model called density functional theory, the toxicity of ionic liquids can be predicted and forecast. It is a theory that allowing the researcher to have a substantial tool for computation of the quantum state of atoms, molecules and solids, and molecular dynamics which also known as computer simulation method. It can be done by using structural feature based quantum chemical reactivity descriptor. The identification of ionic liquids and its Log[EC50] data are from literature data that available in Ismail Hossain thesis entitled "Synthesis, Characterization and Quantitative Structure Toxicity Relationship of Imidazolium, Pyridinium and Ammonium Based Ionic Liquids". Each cation and anion of the ionic liquids were optimized and calculated. The geometry optimization and calculation from the software, produce the value of highest occupied molecular orbital (HOMO) and lowest unoccupied molecular orbital (LUMO). From the value of HOMO and LUMO, the value for other toxicity descriptors were obtained according to their formulas. The toxicity descriptor that involves are electrophilicity index, HOMO, LUMO, energy gap, chemical potential, hardness and electronegativity. The interrelation between the descriptors are being determined by using a multiple linear regression (MLR). From this MLR, all descriptors being analyzed and the descriptors that are significant were chosen. In order to develop the finest model equation for toxicity prediction of ionic liquids, the selected descriptors that are significant were used. The validation of model equation was performed with the Log[EC50] data from the literature and the final model equation was developed. A bigger range of ionic liquids which nearly 10^8 of ionic liquids can be predicted from this model equation.

1. Introduction

Salts which have a melting point temperature below the boiling point of water is also known as ionic liquids(ILs) which is now nearly 10^8 [1, 2]. Usually, ILs consist of an organic cation and combine with some inorganic anion such as Cl^-, BF_4^-, PF_6^- which will be matching to the organic cation. Since the ionic liquids have high capability to be used in many applications in the world such as fuel cells [3, 4], batteries [5-7], sensors [8-10], extractants [11, 12], capacitors [13, 14], ionogels and solvents [11, 15-19] in analysis, synthesis [15, 20, 21], catalysis and separation [22], they have been one of the interest in the industry. Most of the ionic liquids can be considered as a "green", but nowadays the development of these compounds also may cause some accidental discharge and immoral pollution [16, 23].

The definition of toxicity is the measure of its harmful effects on living organisms and it can be classified into three categories which are chemical toxicity, biological toxicity and physical toxicity. Recently, the toxicity of ILs has been given more consideration. Studies have shown that ILs do give harmful effect to human [16, 23], fish [20, 24], bacteria [25, 26] and many more. Therefore, ionic liquids are not so being considered as a naturally "green" [2, 16, 18, 24, 26]. Since the interest on ILs has raised, it's only feasible to come up with a solution to measure its toxicity without the sacrifice of animals as lab tests. This could be done by a computational approach towards the properties of the ionic liquids and developing a model to analyze the toxicity levels of the liquid. The most important theory that needs to be implemented in this research paper will be the Density Functional Theory [27-29]. The density functional theory is used to compute the descriptors such as chemical potential, ionization potential, electrophilicity index, as well as electron affinity of those ionic liquids [30-32]. All these properties are later compared with the different experimental data done on laboratory animals. From these validation, the toxicity with just the information needed from the density functional theory can be determined.

In this project, the main apprehension will be to develop the theoretical model based toxicity descriptors such as HOMO, LUMO, electrophilicity index, electronegativity, refractive index, enthalpy, entropy, Gibbs free energy, polarization and dipole moment so that observation can be done on the characteristics of these descriptors and compare the results obtained with toxicity through experimentation. The objective of this research is to develop a theoretical model equation to predict the toxicity ILs based on Daphnia Magna. The scope of this project is to identify the theoretical model using density functional theory in Material Studio software specifically using a Dmol3 modules calculation for ILs based on Daphnia Magna.

2. Materials and methods

2.1. *Dataset*

Ionic liquids based on Daphnia Magna with different head group of cations such as (Imidazolium, Pyridinium, Ammonium, Sulfonium), there were 31 ionic liquids have been chosen. The log[EC_{50}] is considered as the direction for the toxicity which also being used for the model equation to develop.

2.2. *Computational Methods*

Determination of interaction between the cation and anion in term of geometric optimizations and electronic structure calculations were made through a density functional theory computations using $DMol^3$ module of Materials Studio software [33]. First, full structural optimization for cations and anions for each ILs were carried out; the generalized gradient approximation (GGA) and Perdew and Wang functional 91 (PW91) exchange-correlation functional in Dmol3 module calculation [34]. The basis set in Dmol3 module is set to the Double numerical plus polarization function (DNP) [35]. A real-space orbital global cut off of 3.7 Å was used for the geometry optimization and the self-consistent field (SCF) convergence was set to 10^6. The convergence calculation conditions were compute as follows: energy = $1x10^{-5}$ Ha; Force = $2x10^{-3}$ Ha/ Å; displacement = 0.005 Å. The calculation process includes both geometry optimizations and energy calculation.

At the PW91/GGA level of theory in Dmol3 module, the HOMO and LUMO were calculated from density functional theory results. The other molecular descriptor which are Electrophilicity Index (ω), Energy Gap (ΔE), Chemical Potential (μ), Electronegativity (χ), and Hardness (η), were obtained basically from the formulas that involve HOMO and LUMO value as shown in Table 1.

Table 1. Descriptors Calculation

Descriptors	Formula
Energy Gap (ΔE)	$\Delta E = E_{LUMO} - E_{HOMO}$
Chemical Potential (μ)	$\mu = (E_{HOMO} + E_{LUMO}) / 2$
Hardness (η)	$\eta = (E_{LUMO} - E_{HOMO}) / 2$ $\eta = (\Delta E) / 2$
Electrophilicity Index (ω)	$\omega = (\mu^2) / (2\eta)$
Electronegativity (χ)	$\chi = (- E_{HOMO} - E_{LUMO}) / 2$

2.3. Statistical Analysis

To develop a model equation for ILs, a data analysis known as multiple linear regressions (MLR) were conducted on selected descriptors that are significant to the $Log_{10}[EC_{50}]$ from literature data. The reason in doing this MLR is because to obtain the best correlation between one dependent $Log_{10}[EC_{50}]$ quantity (toxicity) with selected significant descriptors as variables. The multiple linear regressions (MLR) were conducted using the Data Analysis in Microsoft Excel 2016. Toxicity prediction model was developed based on the statistical information from the data analysis.

2.4. Data Training and Test Sets

To develop a precise and accurate model equation, all the molecular descriptors were performed by MLR and from the results of MLR, the best correlated descriptors were selected. 80% of the ILs from based on Daphnia Magna were randomly picked to develop the MLR model and another 20% of ILs were used for the validation of the model. Flow chart below in Figure 1 shows the details of the model development.

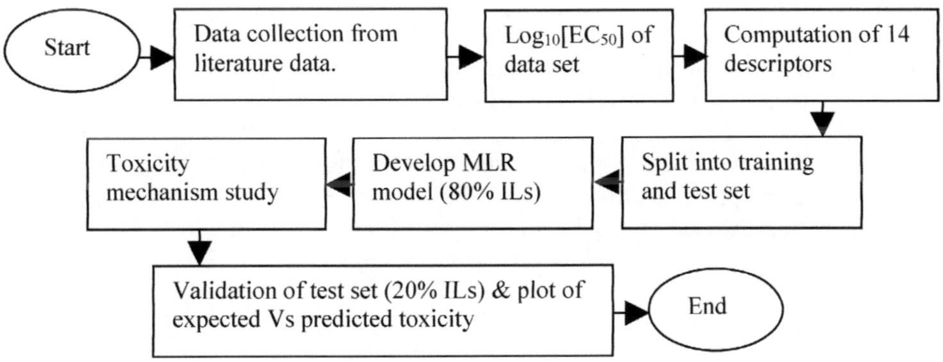

Figure 1 : Flow chart of methodology

IOP Conf. Series: Materials Science and Engineering **344** (2018) 012017 doi:10.1088/1757-899X/344/1/012017

3. Results and discussion

The studied descriptors were lowest unoccupied molecular orbital energy (E_{LUMO} in kJ/mol), highest occupied molecular orbital energy (E_{HOMO} in kJ/mol), hardness (η in kJ/mol), chemical potential (μ in kJ/mole), electrophilicity index (ω in kJ/mol), energy gap (ΔE in kJ/mol), and electronegativity (χ in kJ/mol). After the optimization of crystals structure of 31 ILs based on Daphnia Magna in Material Studio software, all the descriptors have been calculated after the HOMO and LUMO value was obtained from the computation of Dmol3 module in Material Studio software.

3.1. Statistical Analysis

3.1.1 Multiple Linear Regressions (MLR)

All the data sets consist of 14 descriptors were subjected to multiple linear regressions (MLR) to develop the model equation and the best relationship between one dependent variable log[EC_{50}] and significant variables (descriptor) has been made for ILs based on Daphnia Magna. The significant descriptors were HOMO and LUMO for both cation and anion, and the electrophilicity index (ω in kJ/mol) of cation. The correlation is a linear equation is shown in Table 2.

Table 2. Model Equation

Category of ILs	**Equation of the model**
Daphnia Magna	$\log[EC_{50}] = -1.1263 - 0.00991(E_{HOMO,\,cation}) + 0.0039995(E_{LUMO,\,cation}) + 0.00005(\omega_{,\,cation}) + 0.00535(E_{HOMO,\,anion}) - 0.004511(E_{LUMO,\,anion})$

The computation of 14 descriptors were done after division of the descriptors for cations and anions. Therefore, there were 7 descriptors for cations and 7 descriptors for anions. The subject that being focused on is the log[EC_{50}]. From MLR, Anova results produced and from it, the descriptors that significant to the log[EC_{50}] from literature data can be determined. As can be seen in Table 2, the model equation, there are some descriptors that are correlated with log[EC_{50}] and significant. The descriptors that significant to the log[EC_{50}] of Daphnia Magna are 5 descriptors which are HOMO and LUMO of cation and anion, and electrophilicity index of cation. Only these 5 descriptors are significant to the log[EC_{50}] of literature data, while the other 9 descriptors are not significant and correlated.

For the organism, the HOMO and LUMO energy level feature to the pattern of reactivity and to the liability of electrons sharing in the ionic liquids. The lower the toxicity if the HOMO and LUMO value are lower. While for the electrophilicity index, it indicates the ability of ILs to accept electrons which means the lower the value of electrophilicity index, the lower the toxicity. The ILs which can be categorized as hard ILs also have a lower value of electrophilicity index.

After the result of MLR, the final equation for toxicity prediction of ILs based on Daphnia Magna was produced. By using this equation, the validation of the value of log[EC_{50}] for the ILs were done. As shown in Figure 2, the predicted log[EC_{50}] are the value of log[EC_{50}] by using the model equation and the experimental log[EC_{50}] are the value of log[EC_{50}] from the literature data.

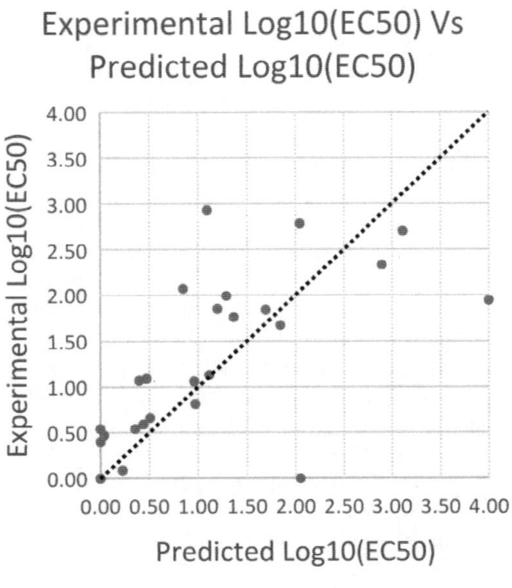

Figure 2 : Graphical representation of experimental vs predicted toxicity of ILs (Daphnia Magna)

Figure 2 shows graphical representation of experimental $Log_{10}[EC_{50}]$ data from literature against predicted $log_{10}[EC_{50}]$ from the model equation developed of Daphnia Magna. This graph plotted to show the consistency of the predicted data with the experimental data from literature. If the experimental data and predicted data lies on y=x line, it means that the predicted data are align and accurate with the experimental data from the literature. As can be seen in the graphs that, there are some inconsistency of predicted data against experimental data. This is because of there are some literature data that have different value of $log[EC_{50}]$ because it is from lot of reference. These outliers may be due to variability in the measurement or it may indicate as an experimental error.

4. Conclusion

The toxicity prediction technique without experimental work based on density functional theory on ionic liquids based on Daphnia Magna help to find risk of ionic liquids that have high toxicity to environment and human. The discovery of this density functional theory based determination on toxicity prediction of ionic liquids based on Daphnia Magna can be used to design ILs for wide applications such as (solvents, catalyst etc. that harmful as toxic for environment and human) without sacrificing animals, plant, fish and some sort of experimental organism. A lot of sets of data can also be featured and computed in the future works.

Acknowledgement

This work done in Universiti Teknologi PETRONAS, Malaysia. The software called Material Studio was used to perform the computation in high performance computer in research lab of Universiti Teknologi PETRONAS, Malaysia.

References

[1] F. Dommert, J. Schmidt, C. Krekeler, Y. Y. Zhao, R. Berger, L. Delle Site, *et al.*, "Towards multiscale modeling of ionic liquids: From electronic structure to bulk properties," *Journal of Molecular Liquids,* vol. 152, pp. 2-8, 2010.

[2] M. H. Fatemi and P. Izadiyan, "Cytotoxicity estimation of ionic liquids based on their effective structural features," *Chemosphere,* vol. 84, pp. 553-563, 2011.

[3] R. F. De Souza, J. C. Padilha, R. S. Gonçalves, and J. Dupont, "Room temperature dialkylimidazolium ionic liquid-based fuel cells," *Electrochemistry Communications,* vol. 5, pp. 728-731, 2003.

[4] B. Lin, L. Qiu, J. Lu, and F. Yan, "Cross-linked alkaline ionic liquid-based polymer electrolytes for alkaline fuel cell applications," *Chemistry of materials,* vol. 22, pp. 6718-6725, 2010.

[5] L. J. Hardwick, M. Holzapfel, A. Wokaun, and P. Novák, "Raman study of lithium coordination in EMI-TFSI additive systems as lithium-ion battery ionic liquid electrolytes," *Journal of Raman Spectroscopy,* vol. 38, pp. 110-112, 2007.

[6] C. Sirisopanaporn, A. Fernicola, and B. Scrosati, "New, ionic liquid-based membranes for lithium battery application," *Journal of Power Sources,* vol. 186, pp. 490-495, 2009.

[7] C. Y. Wang, V. Mottaghitalab, C. O. Too, G. M. Spinks, and G. G. Wallace, "Polyaniline and polyaniline–carbon nanotube composite fibres as battery materials in ionic liquid electrolyte," *Journal of Power Sources,* vol. 163, pp. 1105-1109, 2007.

[8] M. Elyasi, M. A. Khalilzadeh, and H. Karimi-Maleh, "High sensitive voltammetric sensor based on Pt/CNTs nanocomposite modified ionic liquid carbon paste electrode for determination of Sudan I in food samples," *Food chemistry,* vol. 141, pp. 4311-4317, 2013.

[9] H. Khani, M. K. Rofouei, P. Arab, V. K. Gupta, and Z. Vafaei, "Multi-walled carbon nanotubes-ionic liquid-carbon paste electrode as a super selectivity sensor: application to potentiometric monitoring of mercury ion (II)," *Journal of hazardous materials,* vol. 183, pp. 402-409, 2010.

[10] D. Wei and A. Ivaska, "Applications of ionic liquids in electrochemical sensors," *Analytica Chimica Acta,* vol. 607, pp. 126-135, 2008.

[11] S. Dai, Y. Ju, and C. Barnes, "Solvent extraction of strontium nitrate by a crown ether using room-temperature ionic liquids," *Journal of the Chemical Society, Dalton Transactions,* pp. 1201-1202, 1999.

[12] G.-T. Wei, Z. Yang, and C.-J. Chen, "Room temperature ionic liquid as a novel medium for liquid/liquid extraction of metal ions," *Analytica Chimica Acta,* vol. 488, pp. 183-192, 2003.

[13] A. Lewandowski and M. Galiński, "Carbon–ionic liquid double-layer capacitors," *Journal of Physics and Chemistry of Solids,* vol. 65, pp. 281-286, 2004.

[14] T. Sato, G. Masuda, and K. Takagi, "Electrochemical properties of novel ionic liquids for electric double layer capacitor applications," *Electrochimica Acta,* vol. 49, pp. 3603-3611, 2004.

[15] E. R. Cooper, C. D. Andrews, P. S. Wheatley, and P. B. Webb, "Ionic liquids and eutectic mixtures as solvent and template in synthesis of zeolite analogues," *Nature,* vol. 430, p. 1012, 2004.

[16] M. J. Earle and K. R. Seddon, "Ionic liquids. Green solvents for the future," *Pure and applied chemistry,* vol. 72, pp. 1391-1398, 2000.

[17] R. D. Rogers and K. R. Seddon, "Ionic liquids--solvents of the future?," *Science,* vol. 302, pp. 792-793, 2003.

[18] V. Tsarpali, A. Belavgeni, and S. Dailianis, "Investigation of toxic effects of imidazolium ionic liquids,[bmim][BF 4] and [omim][BF 4], on marine mussel Mytilus galloprovincialis with or without the presence of conventional solvents, such as acetone," *Aquatic Toxicology,* vol. 164, pp. 72-80, 2015.

[19] A. Zakari, S. Waziri, B. Aderemi, and S. Mustapha, "Computational study of environmental fate of ionic liquids using conductor-like screening model for real solvents (COSMO-RS) method," *Journal of Environmental Chemistry and Ecotoxicology,* vol. 5, pp. 90-95, 2013.

[20] M. I. Hossain, "Synthesis, characterization and quantitative structure toxicity relationship of imidazolium, pyridinium and ammonium based ionic liquids," Chemical Engineering Department, Universiti Teknologi PETRONAS, IRC, UTP, 2013.

[21] J. Sun, S.-i. Fujita, and M. Arai, "Development in the green synthesis of cyclic carbonate from carbon dioxide using ionic liquids," *Journal of Organometallic Chemistry,* vol. 690, pp. 3490-3497, 2005.

[22] H. Olivier-Bourbigou, L. Magna, and D. Morvan, "Ionic liquids and catalysis: Recent progress from knowledge to applications," *Applied Catalysis A: General,* vol. 373, pp. 1-56, 2010.

[23] T. P. T. Pham, C.-W. Cho, and Y.-S. Yun, "Environmental fate and toxicity of ionic liquids: a review," *Water research,* vol. 44, pp. 352-372, 2010.

[24] M. I. Hossain, B. B. Samir, M. El-Harbawi, A. N. Masri, M. A. Mutalib, G. Hefter, *et al.*, "Development of a novel mathematical model using a group contribution method for prediction of ionic liquid toxicities," *Chemosphere,* vol. 85, pp. 990-994, 2011.

[25] R. J. Bernot, M. A. Brueseke, M. A. Evans-White, and G. A. Lamberti, "Acute and chronic toxicity of imidazolium-based ionic liquids on Daphnia magna," *Environmental Toxicology and Chemistry,* vol. 24, pp. 87-92, 2005.

[26] Y. Zhao, J. Zhao, Y. Huang, Q. Zhou, X. Zhang, and S. Zhang, "Toxicity of ionic liquids: Database and prediction via quantitative structure–activity relationship method," *Journal of hazardous materials,* vol. 278, pp. 320-329, 2014.

[27] H. Eschrig, *The fundamentals of density functional theory* vol. 32: Springer, 1996.

[28] S. R. Pilli, T. Banerjee, and K. Mohanty, "HOMO–LUMO energy interactions between endocrine disrupting chemicals and ionic liquids using the density functional theory: Evaluation and comparison," *Journal of Molecular Liquids,* vol. 207, pp. 112-124, 7// 2015.

[29] M. Appell and W. B. Bosma, "Assessment of the electronic structure and properties of trichothecene toxins using density functional theory," *Journal of Hazardous Materials,* vol. 288, pp. 113-123, 5/15/ 2015.

[30] B. Elidrissi, A. Ousaa, M. Ghamalia, S. Chtitaa, M. Ajanaa, M. Bouachrineb, *et al.*, "Toxicity in vivo of nitro-aromatic compounds: DFT and QSAR results," *J Comput Aided Mol Des,* vol. 4, pp. 28-37, 2014.

[31] M. Ghamali, S. Chtita, A. Adad, R. Hmamouchi, M. Bouachrine, and T. Lakhlifi, "Combining DFT and QSAR results for predicting the cytotoxicity of a series of orthoalkyl substituted 4-X-phenols."

[32] M. Zhu, F. Ge, R. Zhu, X. Wang, and X. Zheng, "A DFT-based QSAR study of the toxicity of quaternary ammonium compounds on Chlorella vulgaris," *Chemosphere,* vol. 80, pp. 46-52, 2010.

[33] K. Z. Sumon, A. Henni, and A. L. East, "Predicting p K a of Amines for CO2 Capture: Computer versus Pencil-and-Paper," *Industrial & Engineering Chemistry Research,* vol. 51, pp. 11924-11930, 2012.

[34] J. P. Perdew, J. Chevary, S. Vosko, K. A. Jackson, M. R. Pederson, D. Singh, *et al.*, "Atoms, molecules, solids, and surfaces: Applications of the generalized gradient approximation for exchange and correlation," *Physical Review B,* vol. 46, p. 6671, 1992.

[35] B. Delley, "From molecules to solids with the DMol3 approach," *The Journal of chemical physics,* vol. 113, pp. 7756-7764, 2000.

IOP Conf. Series: Materials Science and Engineering **344** (2018) 012018 doi:10.1088/1757-899X/344/1/012018

Traffic condition with road upgrading during construction and operation stages based on level-of-service (LOS)

A C Tay[1] and **H H Lee**

Faculty of Engineering and Science, Curtin University Malaysia, CDT 250, 98009 Miri, Sarawak, Malaysia

[1]Email: tayaichen@curtin.edu.my

Abstract: In this study, traffic survey was done at signalized intersection of both ends of an upgrading road during morning and afternoon peak hours throughout construction period. The intersections of Tudan Road located at Miri City in Sarawak State, Malaysia were chosen as study sites. The aim of this research was to investigate the traffic condition for the signalized intersections of a R3 secondary road that being upgrading to standard U5 dual carriageway during peak hours for construction and operation stages. Data collected from these sites were analyzed according to Highway Capacity Manual (HCM) 2010. From the results, the level-of-services (LOS) of Kuala Baram By Pass Road intersection improved from Level F to Level E at operation stage for weekday during peak hours in the morning. LOS of Lutong-Kuala Baram Road intersection at construction stage were Level C during peak hours in the morning and Level D during peak hours in the afternoon for weekends. However, the LOS during both peak hours for weekends were projected to improve to Level B at the operation stage. It proved that the road construction has affected LOS, and this temporary issue can be solved after the upgraded road is in operation.

1. Introduction

Traffic assessment is a process that used to identify the traffic condition of the roads. Over the years, the population in Miri is rising. Increase in population could result in high density of traffic causing traffic congestion. Hence, the road construction activity is growing rapidly in order to accommodate the traffic volume. With a deep investigation of proposed road construction on the road network system, the traffic impacts such as congestion and road safety could be greatly enhanced.

Traffic control condition can be explained as the saturation flow of an approach. Saturation flow of an approach is expressed as the maximum number of vehicles which are able to pass through the intersection during the green indication, which is also known as the maximum flow [1].

Delay could be explained as the discrepancy between the additional time that a vehicle experiences while passing through an intersection with and without traffic signal control [2]. The major factors that may influence control delay are cycle length, effective green time, traffic volume (demand) and capacity. In fact, there are also some minor factors which may also affect control delay such as signal control, grade of approach and etc. [3]. The volume-to-capacity ratio is considered one of the major factors to identify the traffic condition at the intersection. The volume-to-capacity (v/c) ratio is known as the degree of saturation. It stands for an ability of an intersection to accommodate the traffic demand in terms of ratio [4]. Normally, the volume-to-capacity ratio which is smaller than

0.85 implying that the capacity is sufficient enough to overtake the traffic demand. For this, the drivers suffer delays and queues.

A traffic impact assessment was carried out at proposed Hypermarket in Skudai Town of Malaysia, it reported that LOS were D for weekday and C for weekend at the pre-construction stage, but the projected LOS was F for the year 2025 after the completion of Hypermarket [5]

A case study from Istanbul, Turkey shows that adaptive control does improve the performance at signalized intersection. It avoids the problem for growing at exit road and prevent sending excessive vehicles from the entry by implementing a proper signal timing [6]. Weaving effect may also influence the safety of other drivers. Therefore, it is important to encourage drivers not to change their travel lane too often [7]. In semi actuated operation, the detectors are placed on the minor approaches only. It indicates that the major approach of an intersection is having a green indication at all times until a vehicle is detected on any other minor approaches [8].

Vehicle queue acts as a very important factor to measure the traffic performance and it should be considered as a major evaluation criteria of signalized intersections. Lane groups that are having queues are probably facing serious delay [9]. A pre-timed control at an intersection is suitable for an intersection that is having a high concentrated volume such as city center [10]. Even though pre-timed control is able to provide an efficient operation during peak hour, it can still result an excessive delay to drivers.

A signal phase is one part of the cycle length distributed to different traffic movement that guide their direction way at the same time in one or more intervals. A simple intersection should have a two phase cycle. However, there are also some intersections which consist of more than three phase cycle. Intersections with more than three phase cycle may cause more delay as the green time for other phases may decrease and the intersections will be consisting of longer cycle length and additional time intervals. The excessive phases may decrease the progression level and increase the percentage of queue vehicles and fuel consumption. A two phase cycle is then result in higher lane capacity and provide a better intersection performance [11]. However, from the point of view in discussing the road safety, more signal phase cycle will be favorable to the intersection as it leads to less traffic accidents in the intersection.

In this study, the signalized intersections of Jalan Tudan were chosen as study sites as Jalan Tudan was under and upgrading construction. In order to accomplish the goal of this project, the objectives are stated below:

- To understand the traffic condition of the both signalized intersections which allocated along Jalan Tudan
- To determine the delay and level-of-service (LOS) of the intersections during peak hours for road construction stage and projected completion stage.
- To study the improvements that can used to enhance the signalized intersections.
- This study focuses the control delay result from the signalized intersections. It also provides the survey data to show the traffic condition in terms of level-of-service (LOS) which are having 6 different of levels, from level A to level F. Additionally, this study will also assess the traffic significance relating to the civil engineering work construction to determine whether the intersections be able to handle the traffic demands in acceptable level-of-service.

2. Methodology

The signalized intersections along Tudan Road, Miri City in Sarawak, Malaysia were chosen as study site as it was a road upgrading project. Therefore, the intersections were investigated to compare the traffic condition for road construction stage and completion stage. This location provided a good access for survey members to collect data. The study area was located between the longitude of E 114° 0’ to E 114° 3’ and latitude of N 4° 28’ to N 4° 29’.

To carry out the traffic survey, a tally sheet was prepared. The traffic volume data in one-hour duration with 15 minutes interval was recorded. In this study, three survey members were recruited for the traffic survey manual counting. It is to improve the accuracy of the data obtained if the manual traffic count was carried out by more people. Each survey member owned a stopwatch and stayed at the spot where vehicle flow and signal time could be clearly observed. The time for data collection was implemented during morning peak (7am to 8am) and evening peak (5pm to 6pm).

Performance of the signalized intersections were analyzed by the rules and formulas provided from Highway Capacity Manual 2010 [12]. To analyze the traffic condition of signalized intersections, the parameters were determined in the sequence, which are movement groups, lane groups, adjusted saturation flow rate, proportion arriving during green, signal phase duration, capacity and volume-to-capacity ratio, delay and level-of-service (LOS). The road geometric condition of Jalan Lutong-Kuala Baram intersection and Kuala Baram By Pass intersection are shown in figure 1 and figure 2, respectively. The approaches for both intersections were identified as shown in figure 3 and figure 4, respectively. The descriptions for all the directions of the approaches are listed in table 1.

The level-of-service (LOS) was determined by the control delay and volume-to-capacity (v/c) as listed table 2. LOS A is described as the best performance in road condition while LOS F is the worst. LOS A describes the condition with less than 10 seconds control delay and v/c ratio that is less than 1.0.

Table 1. Descriptions for approaches direction.

Jalan Lutong–Kuala Baram Intersection	
Directions	Descriptions
NB(RT)	Northbound (Right Turn)
NB(TH)	Northbound (Through)
SB(TH)	Southbound (Through)
WB(RT)	Westbound (Right Turn)
Kuala Baram By Pass Intersection	
Directions	Descriptions
SB(TH)	Southbound (Through)
SB(RT)	Southbound (Right Turn)
NB(TH)	Northbound (Through)
EB(RT)	Eastbound (Right Turn)

Table 2. Level-of-service threshold.

Control Delay (s/veh)	LOS by V/C Ratio	
	V/C ≤ 1.0	V/C ≥ 1.0
≤ 10	A	F
□ 10 - 20	B	F
□ 20 - 35	C	F
□ 35 - 55	D	F
□ 55 - 80	E	F

□ 80	F	F

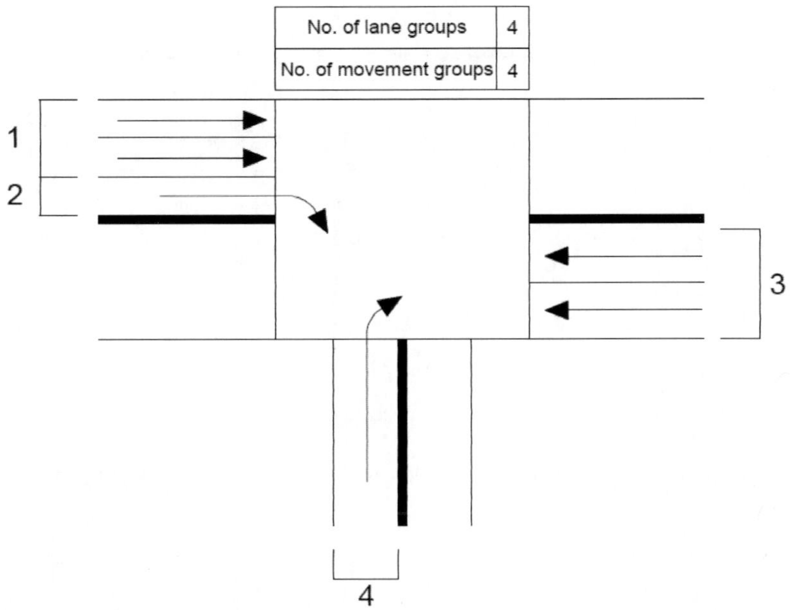

Figure 1. Road geometry of Jalan Lutong-Kuala Baram intersection.

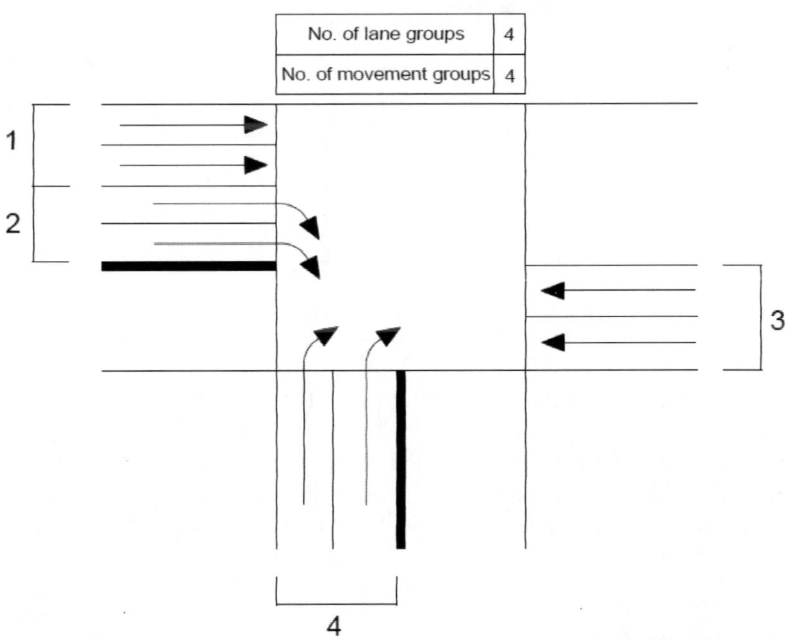

Figure 2. Road geometry of Kuala Baram by pass intersection.

IOP Conf. Series: Materials Science and Engineering **344** (2018) 012018 doi:10.1088/1757-899X/344/1/012018

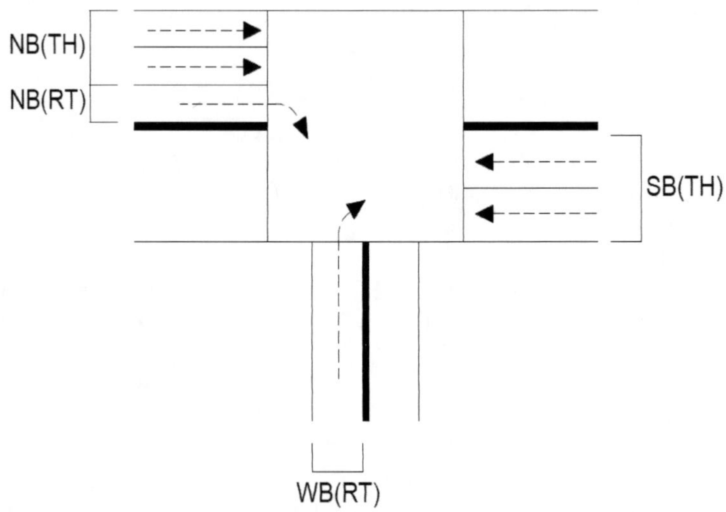

Figure 3. Approaches directions of Jalan Lutong-Kuala Baram intersection.

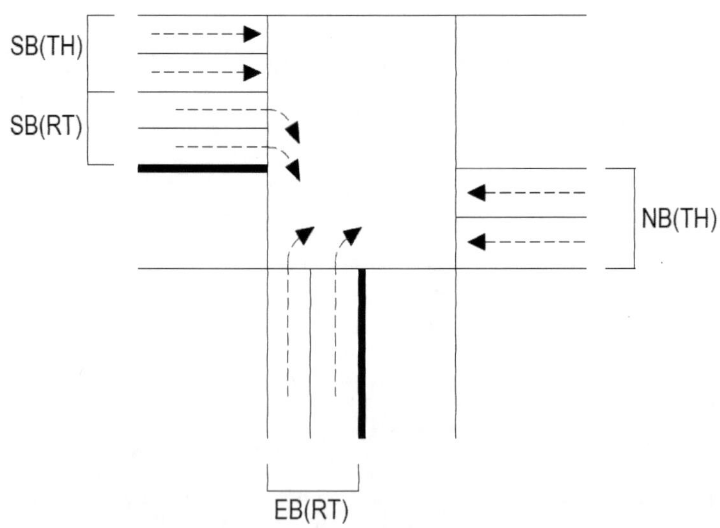

Figure 4. Approaches directions of Kuala Baram by pass intersection.

To project the traffic condition for completion stage, condition without heavy vehicles volume and road geometry enhancement were used for analysis. Amount of heavy vehicles obtained during road construction stage was excluded for projected completion stage. It is to investigate the effect of heavy vehicles to the intersection performance for completion stage.

3. Results and Discussions

Based on the data collected, NB(TH) approaches at Jalan Lutong-Kuala Baram and Kuala Baram

By Pass Intersections showed the highest average green ratio, which are 0.652 and 0.561, respectively. Among the green ratio at both intersections, the green ratio at Jalan Lutong-Kuala Baram Intersection is greater than the one at Kuala Baram By Pass Intersection.

In terms of passenger cars, buses and heavy vehicles, the traffic volume of passenger cars showed the highest distribution ratio at both intersections, followed by heavy vehicles and buses. The delay caused by heavy vehicles and buses to traffic condition at intersections may not be as significant as expected. It was due to the low distribution ratio of heavy vehicles.

For road construction stage, NB(RT) approaches at Jalan Lutong-Kuala Baram Intersection showed the highest volume-to-capacity ratio at weekday evening peak, 1.62. During weekday morning peak, SB(TH) at the same intersection showed the highest volume-to-capacity ratio, 1.50. For projected completion stage, the volume-to-capacity ratio for all approaches were reduced with road geometry improvement and without heavy vehicles.

The intersection delay is the average delay time of every up-stream approaches linked to the intersection. Intersection delay during road construction stage and projected completion stage is shown table 3. From the analysis, it showed that intersection delay was reduced in projected completion stage. The delay reduced at Jalan Lutong-Kuala Baram Intersection and Kuala Baram By Pass Intersection were around 49 percent and 19 percent, respectively.

The LOS for both intersections were proved to be enhanced for projected completion stage, except for weekday morning peak at Jalan Lutong-Kuala Baram Intersection (Table 4). It is due to high control delay at SB(TH) approaches at Jalan Lutong-Kuala Baram Intersection, which was affected by low capacity and high volume-to-capacity ratio.

Table 3. Intersection delay during road construction stage and projected completion stage.

Intersection Delay during Road Construction Stage and Projected Completion Stage (s)					
		Weekday		Weekend	
Intersection	Stage	Morning Peak (7am – 8am)	Evening Peak (5pm – 6pm)	Morning Peak (7am – 8am)	Evening Peak (5pm – 6pm)
Jalan Lutong–Kuala Baram	Road Construction	142.48	96.70	23.53	39.78
	Projected Completion	102.62	19.58	19.41	12.00
Kuala Baram By Pass Road	Road Construction	100.48	40.38	20.00	34.97
	Projected Completion	68.97	36.21	19.47	34.06

Table 4. LOS for both intersections during road construction and projected completion stage.

Level-of-service (LOS)					
		Weekday		Weekend	
Intersection	Stage	Morning Peak (7am – 8am)	Evening Peak (5pm – 6pm)	Morning Peak (7am – 8am)	Evening Peak (5pm – 6pm)
Jalan Lutong–Kuala Baram	Road Construction	F	F	C	D
	Projected Completion	F	B	B	B

Kuala Baram By Pass Road	Road Construction	F	D	B	C
	Projected Completion	E	D	B	C

4. Conclusion

Jalan Lutong-Kuala Baram Intersection indicated the low performance due to insufficient of lane capacity and green indication time during road construction stage. Therefore, high control delay occurred. During weekday, the LOS for road construction stage indicated to be level F. It proved that the traffic was the worst due to heavy traffic congestion. During weekend, the LOS determined were level C at morning peak and level D at evening peak, respectively. The LOS was improved to level B at projected completion stage. For projected completion stage, the delays at weekday morning and evening peak were analyzed to be reduced 102.62 and 19.58 seconds, while delays at weekend morning and evening peak were reduced to 19.41 and 12.00 seconds. From the analysis, it showed that Jalan Lutong-Kuala Baram Intersection was 49.22% improved from road construction stage to projected completion stage.

Kuala Baram By Pass Intersections showed the performance that is in average level. During weekday morning peak, the result showed that this intersection performed better at projected completion stage than road construction stage (from level F to level E). During weekday evening peak, it was found that the LOS was level D. During weekend morning and evening peak, both LOS were analyzed to be level B and level C, respectively for road construction and projected completion stages. Although the LOS was improved from level F to level E during weekday morning peak, it was not an ideal traffic condition for a signalized intersection because every driver was suffering a delay between 55 and 80 seconds at the intersection. The LOS was affected the road construction, However, this temporary issue can be solved after the upgraded road is in operation

References

[1] Gao L, Bhuiyan M A, Brian W R, Douglas K N and Richard S M 2016 *J. Traffic and Transportation Engineering* **3** 336
[2] Dion F, Rakha H and Kang Y A 2004 *Transportation Research Part B: Methodological* **38** 99
[3] Chen P, Liu H, Qi H S and Wang F J 2013 *J. Zhejiang University Science A* **14** 691
[4] Zhu Z H, Zheng J F, Gao Z Y and Du H M 2014 *Physica A: Statistical Mechanics and its Applications* **400** 200
[5] Minhans A, Zaki N H and Belwal R 2013 *Jurnal Teknologi* **65** 1
[6] Gündoğan F, Karagoz Z, Kocyigit N, Karadag A, Ceylan H and Murat Y S 2014 *J. Traffic and Logistic Engineering* **2** 198
[7] Kusuma A, Liu R, Choudhury C and Montgomery F 2014 *Transportation Research Procedia* **3** 51
[8] Lee C K, Yun I, Choi J H, Ko S J and Kim J Y 2013 *KSCE J. Civil Engineering* **17** 1749
[9] Wu N and Giuliani S 2016 *Transportation Research Procedia* **15** 63
[10] Park S Y, Kim D N, Lee S K and Jung J H 2016 *KSCE J. Civil Engineering* **20** 421
[11] Yang X F and Cheng Yao 2017 *Transportation Research Part C: Emerging Technologies* **74** 306
[12] Transportation Research Board 2010 *Highway Capacity Manual* vol 3 chapter 18 pp 1-107

IOP Conf. Series: Materials Science and Engineering **344** (2018) 012019 doi:10.1088/1757-899X/344/1/012019

Fundamental investigation on influence of external heat on chip formation during thermal assisted machining

A U Alkali[1], T L Ginta[2], A M Abdulrani[3], N M Elsiti[4]

[1]Department of Mechanical Engineering, School of Technology, Kano State Polytechnic, Kano – Nigeria
[2, 3]MOR, Biomedical Technology, Universiti Teknologi PETRONAS, Bandar Seri Iskandar, Tronoh, Ipoh, Malaysia
[4]Deptartment of Materials, Manufacturing and Industrial Engineering, Faculty of Mechanical Engineering Universiti Teknologi Malaysia, UTM, Skudai, Malaysia

[1]Email: adamalkalee@yahoo.com

Abstract. Various heat sources have been investigated by numerous researchers to reveal machinability benefits of thermally assisted machining (TAM) process. Fewer engineering materials have been tested. In the same vein, those researches continue to demonstrate effective performance of TAM in terms of bulk material removal rate, improved surface finish, prolong tool life and reduction of cutting forces among others. Experimental investigation on the strain-hardenability and flow stress of material removed with respect to increase in temperature in TAM has not been given attention in previous studies. This study investigated the pattern of chip morphology and segmentation giving close attention to influence of external heat source responsible for strain – hardenability of the material removed during TAM and dry machining at room temperature. Full immersion down cut milling was used throughout the machining conditions. Machining was conducted on AISI 316L using uncoated tungsten carbide end mill insert at varying cutting speeds (V) of 50, 79, and 100 m/min, and feed rates (f) of 0.15, 0.25, and 0.4 mm/tooth while the depth of cut was maintained at 0.2mm throughout the machining trials. The analyses of chip formation, segmentations and stain hardenability were carried out by using LMU light microscope, field emission microscopy and micro indentation. The study observed that build up edge is formed when a stagnation zone develops in front of tool tip which give rise to poor thermal gradient for conduction heat to be transferred within the bulk material during dry machining. This promotes varying strain – hardening of the material removed with evident high chips hardness and thickness, whereas TAM circumvents such impairment by softening the shear zone through local preheat.

1. Introduction

Recent studies undertaken on thermally assisted machining (TAM) have widely investigated various machinability factors but the phenomenological approach of investigation of chips formation received little attention to several researchers. Previous investigations reported by Alkali et al [1] have shown that the causes of variation of chips morphology can depend on the workpiece and cutting tool material and cutting conditions among other parameters. The formation of chips during metal cutting is further described by many scholars, including Ning et al [2] and Kovac and Sidjanin [3]. In another development, Yang and Lei [4] investigated chip morphology while cutting silicon nitride ceramic using laser assisted machining (LAM). Their investigation revealed semi continuous chips with various lengths that have been obtained at preheat temperatures of 1480^{0}C and 1595^{0}C. They further reported that larger chip dimensions indicate bigger plastic deformation. Their investigation did not report the morphological segmentation of the chips produced during the cutting process. Sun et al [5] reports that the chips thickness and serration in conventional milling was higher and presents cutting instability when compared to machining at elevated temperature (TAM) using laser beam as a heat source during milling titanium alloy (Ti-6Al-4V). Ding et al [6] study the chip formation and temperature effect on titanium alloy workpiece. Their report revealed that deformed chip thickness grew thicker during the initial 30µs cutting time irrespective of the parametric dimension of the undeformed chip thickness. Amin et al [7] reported that chips produced during TAM on carbon steel are more elongated and thinner than those obtained while conventional machining at room temperature. They described this occurrence to lower cutting forces and stresses on the cutting tool during the heat assisted machining. This significantly influences machinability benefit in terms of lower vibration and tool wear. The impulse of the present investigation is the formation of serrated chips reported in various TAM studies Viharos et al [8]. Currently, there is lacking literatures on chips formation of 316L stainless steel during TAM with oxyacetylene flame as a heat source mechanism. As a result, this research is aimed to reveal the mechanism involve in the chips formation during end milling 316L stainless steel using uncoated WC-Co insert. In particular, the study investigated the pattern of chip morphology and segmentation giving close attention to influence of external heat source responsible for strain – hardenability of the material removed during TAM and dry machining at room temperature.

2. Oxyacetylene as a Heat Source in TAM

The oxyacetylene gas flame has high penetration to width ratio. This enables its penetration into thicker work piece by using a single pass. These potentials make it suitable for use as heat source in TAM. Previous researches like the work of Davami and Zadshakoyan [9] and Mukherjee and Basu [10] have shown the feasibility of using TAM in terms of machinability improvement like longer tool life and improved surface finish using oxyacetylene as a heat source. Whereas the latter reported significant machinability improvement, the former reported that about 15% reduction of yield stress was experimentally observed while cutting AISI 1060 steel on centre lathe. These researchers concluded that tool life increment during thermally assisted machining is due to decrease of the yield stress of the work material. Interestingly, the mixed gas flow rate that eventually produces oxyacetylene heat flux can numerically be quantified. The flow rate of oxygen and acetylene gases should individually be determined. This gives the ability over thermal control of the heat source which has been a major challenge in TAM. Details of oxyacetylene parametric control could be obtained from Alkali et al [11]. It may be interesting to note that ability to control heat source in TAM process eliminates the tendencies for undesirable structural changes in the workpiece and improves machinability.

3. Experimental Set up

Machining was conducted on AISI 316L using uncoated tungsten carbide end mill insert at varying cutting speeds (V) of 50, 79, and 100 m/min, and feed rates (f) of 0.15, 0.25, and 0.4 mm/tooth while the depth of cut was maintained at 0.2mm throughout the machining trials. RSM technique was used to design the machining tests conducted both at room temperature and at elevated temperatures, in order to achieve valid and reliable conclusions in an effective and economical manner. The workpiece material composition is presented in Table 1.

Table 1. Nominal Composition of 316L Austenitic Stainless Steel

C	Mn	Si	P	S	Cr	Mo	Ni	N	Fe
0.03	2.00	0.75	0.045	0.03	18.67	3.16	10.00	0.10	Bal

The typical cutting tool signature of uncoated WC-Co insert is given in Figure 1. The inserts have neither a hole nor chip breaker.

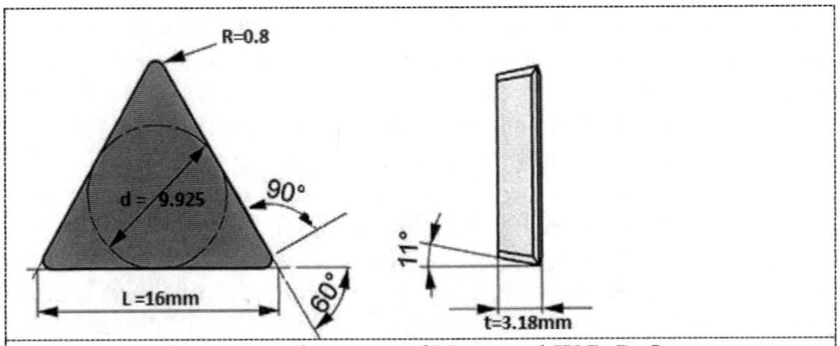

Figure 1. Signature of Uncoated WC-Co Insert

The samples have been sized 60mm x 50mm x 12mm with non-conventional electric discharge machining (EDM) wire cut using wire size of 0.25mm and tensile strength of 1000N/mm2. Experimental trials were obtained by using response surface (RSM) design to investigate the influence of external heat source responsible on segmentation and strain – hardenability of the formed chips during TAM and dry machining at room temperature. Full immersion down cut milling was used throughout the machining conditions in which the end teeth of the tool are predominantly used on Excel FU 281universal milling machine. The universal milling machine has main drive spindle speed in the range of 40 – 2000 rpm. It has table work surface of 300 x 1250mm. The table can swivel to left and right of 45inch. The table moves longitudinally traverse: 845mm, crosswise traverse: 230mm and vertically traverse: 390mm. ME-40TMT3x12 type cutter with 40mm diameter and three insert slots was used to securely hold one uncoated WC-Co insert at a time. The experimental set up is shown in Figure 2.

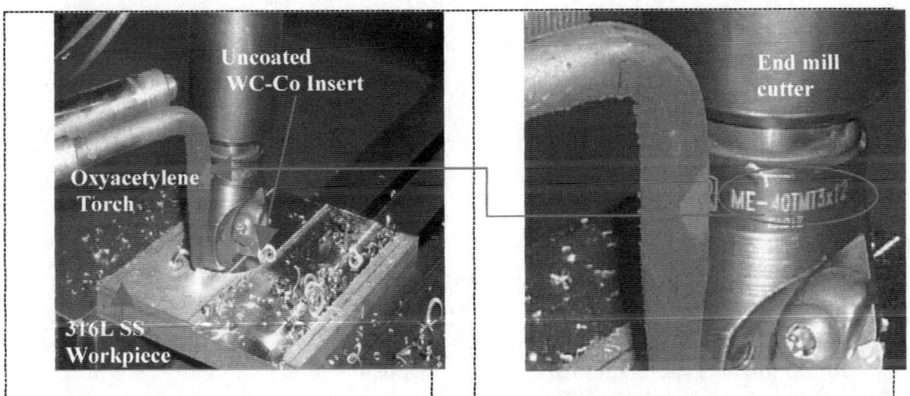

Figure 2. Typical TAM Set up with WC-Co Insert Secured on 40mm End Mill Cutter

3.1. Chips Preparation

Induced temperature during TAM was recorded using T640 Flir thermal image camera with program version 1.2. The objective of incorporating the thermal imager is to accurately examine the impact of external heat during preheat with oxyacetylene so as to identify specific preheat range associated with morphological segmentation and strain hardenability of the produced chips. The chips that have been produce from end milling 316L stainless steel have been collected and analyzed by using digital camera, optical microscope and scanning electron microscope (SEM). The chips have been prepared for microscopic investigation by hot mounting using Buehler SIMPLIMENT 1000, followed by grinding polishing and etching to be able to reveal the structural details of the chips. From the successive investigations, micro indentation, chip thickness, chip forms and chip serration have been critically studied to report their influence on the machinability of 316L stainless steel during TAM and conventional machining at room temperature. Chip images were captured with Leica light microscope using 100X magnification to study different segments of the chips produced and justify scientific hypothesis.

3.1.1. Machining Variables

The selection of the variables is restricted within the machine specifications except for the axial depth of cut. This study used three different ranges of both cutting speed and feed while the axial depth of cut was kept constant at 1mm as shown in table 2. Other heat source components like heat source height, lead distance mixed gas pressures are also kept constant except for the speed of the travelling heat source (in case of TEM).

Table 2. Machining Variables used for this Study

S/N	Parameters	Range		
1	Cutting Speed (m/min)	50	79	100
2	Feed (mm/tooth)	0.15	0.25	0.4
3	Heat source Height (mm)	7.5		
4	Lead Distance (mm)	40		
5	Acetylene gas Pressure (Psi)	10		
6	Oxygen gas pressure (Psi)	25		

4. Result and Discussion

Investigations were conducted on the chips formed during end milling cutting. The chips were collected, prepared through grinding, polishing and etching. The scanning electron microscopy (SEM) was used for micro-examination of the structured morphology. Lengthwise micro-sections of the chips at various cutting conditions were identified and viewed under the optical microscope to capture the structure of the chip and also examined the nature of primary and secondary serrated elements within a given length of the chips.

Accordingly, the chips obtained during the entire investigations are evaluated based on ISO 3685 classifications and the chips micrographs were examined to identify the following effects;

1- If the chip is burned or melted.
2- If the chips edges are rough or have straight edges in the cutting direction.
3- The morphology of the chips.
4- The chip serration

Micro sections of chips at stated selected cutting speeds are shown side in Figure 3 for TAM and conventional cutting at room temperature. It was generally observed that the 316L stainless steel (SS) chips are characterized with primary and secondary serrated teeth formed as a result of the adiabatic shear process. The study found out that built-up edge was formed when a stagnation zone develops in front of the tool tip during machining at room temperature condition while end milling at cutting speed V=50m/min and f=0.4 mm/tooth.

Room Temp. Machining	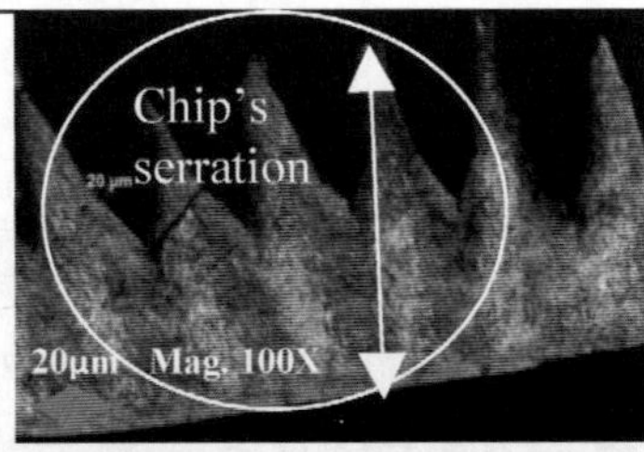
TAM	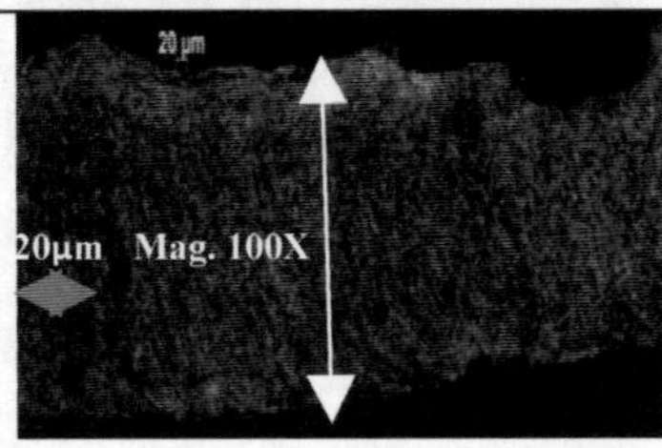

Figure 3. Micrograph of Chips Produced at 120seconds Cutting Time V=50m/min f=0.4mm/tooth while Machining at Room Temperature and Thermally Enhanced Machining at 253^0C

Generally, chips generated during machining at room temperature conditions are characterized with saw tooth shape, implying instability during the cutting. The serrations of the chips in this regard are mostly stable secondary type. It was believed that the serration was developed when the shear plane rotates in the direction of reduction of the shear angle during the phase of compression stages. Konig [12] concluded that the chip is formed by cracking and plasticisation as a result of high compressive stresses occurring from negative cut.

Room Temp. Machining	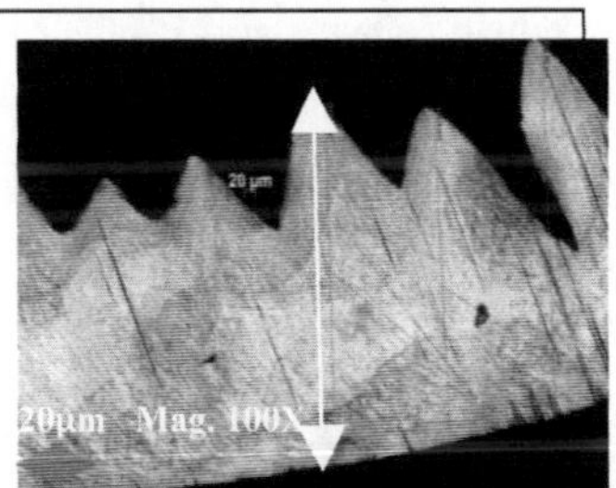
TAM	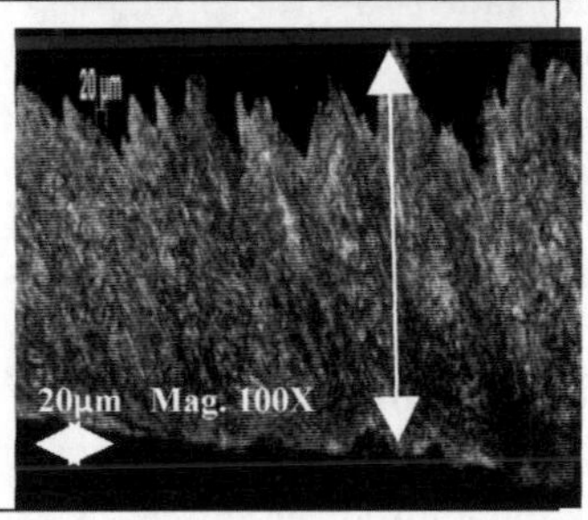

Figure 4. Micrograph of Chips Produced at V=79m/min f=0.4mm/tooth while Machining at Room Temperature and Thermally Enhanced Machining at 191^0C

IOP Conf. Series: Materials Science and Engineering **344** (2018) 012019 doi:10.1088/1757-899X/344/1/012019

Although there are obvious dissimilarities shown in Figure 4 when chips produced during TAM developed a primary serration form. This was believed to be influenced by low preheat temperature of 191^0C and high feed of f=0.4mm/tooth that was used. Figure 5 shows build up edge (BUE) phenomenon that was observed to initiate after three minutes cutting time until the tool failure criterion was attained. The BUE was believed to be due to local weld developed as a result of poor thermal gradient which also govern strain hardenability and restrain conduction heat transferred within the bulk material.

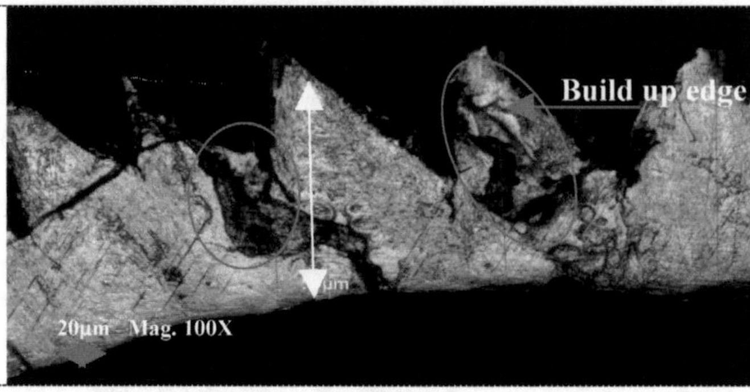

Figure 5. A Built up Edge Developed on Formed Chips during End milling Type 316L Stainless Steel At Room Temperature Condition while Cutting Speed of V=50m/min and Feed rate f=0.4mm/tooth using Uncoated WC – Co Insert

The chips obtained during machining at room temperature were observed to have high thickness and hardness. It was observed that the chips thickness varied when the feedrate was purposely increased while the cutting speed was maintained fixed at 79m/min. Amin et al [7] attributed thinner chips to lower cutting forces and stresses expended on the cutting tool. The chips obtained during both conventional machining at room temperature and while thermally assisted machining shows that chips thickness was influenced by cutting speed and preheat temperature, and that, with increase of preheat temperature, the chips thickness increases with decreases of saw tooth shape. Rao [13] reported that some of the ideal conditions that promote continuous chips in metal cutting included small chip thickness and high cutting speeds. Figure 6. Show an evident influence of feed rate on chip thickness during TAM and conventional machining at room temperature condition using uncoated WC – Co insert on type 316L stainless steel. It is generally concluded from the scope of the parameters investigated in this study that while the vibration amplitude decreases during thermally assisted machining, the surface roughness of type 316L decreases. The chips formed during thermally assisted machining are less hard with faded saw tooth shape. This implies that the resulting cutting force and the stress expended on the cutting tool are minimum as compared to when machining at room temperature conditions. This phenomena was believed to play a significant role lowering the tool wear and machine chatter amplitude as well, as reported in Amin et al [7] and Masood et al [14]. Hence, decreased machine vibration during end milling type 316L stainless steel using uncoated WC-Co insert substantiates that machinability improvements can be attained using oxyacetylene gas flame as a heat source for thermally enhanced machining process. In conclusion, the chips that are obtained during TAM are continuous type that exists in the classes of washer type – helical chips (long and snarled) and connected arc chips. However, both continuous and discontinuous chips were observed while machining at room temperature conditions. The chips type during this condition exists in the classes of arc chips (connected and loose) and snarled washer type – helical chips.

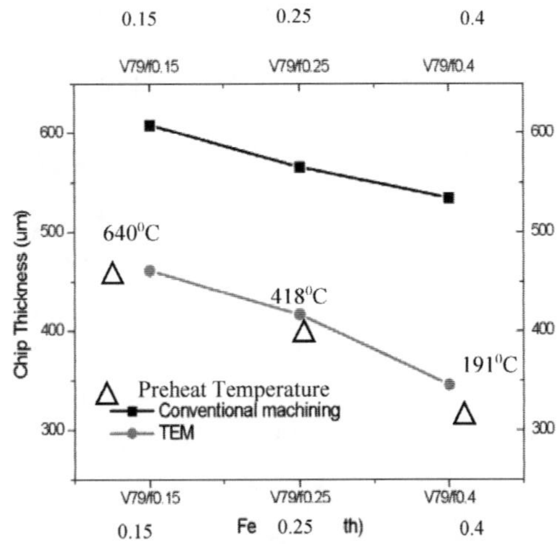

Figure 6. Influence of Feed Rate on Chip Thickness during Thermally Enhanced Machining and Conventional Machining at Room Temperature Condition using WC – Co Insert on Type 316L Stainless Steel (V = 79m/min)

5. Conclusions

From the scope of the study presented here, it is reliably concluded that the cutting parameters significantly influences the morphology of chips formed. This chips resulting from machining 316L stainless steel using WC-Co insert are characterized in form of saw tooth during TAM and conventional cutting at room temperature. However, chips formed during TAM are less hard with faded saw tooth shape. This implies that the resulting cutting force and the stress expended on the cutting tool are minimum as compared to when machining at room temperature conditions where the saw tooth shape is evidently characterized in form of both primary and secondary serration.

Acknowledgment

The authors wish to thank the Universiti Teknologi Petronas (UTP) in Malaysia, for providing the necessary facilities and grant to undertake this resourceful investigation. Additionally, the authors wish to acknowledge the support of Kano state polytechnic in Nigeria for encouragement and sponsorship of this publication.

References

[1] Alkali A U., Noordin M Y., Sa'ad Elmunafi M H , and Fawad H , 2013 Int. J. Mater. Mecha. Manu. 1 76-79.
[2] Ning Y, Rahman, M and Wong Y, 2001 J. Mater. Process Tech 113 360-367
[3] Kovac P and Sidjanin L, 1997 Int J. of Prod. Econs. 51 149-153
[4] Yang B and Lei S, 2008 Int. J. Mechatronic and Manuf. System, 1 116-130
[5] Sun S, Brandt M, Barnes J and Dargusch M, 2011 J. of Eng.Manu. 225 1512-1527
[6] Ding H, Shen, N and Shin, Y C, 2012 J. Mater. Process Tech (212): 601-613
[7] Amin A, Abraham I, Khairusshima N and Ahmed M I, 2007 J. Mater. Process Tech 185 97-105
[8] Viharos Z J, Markos, S and Szekeres, C. 2003 World Cong.–Metro. Millen.1469-1473
[9] Davami M and Zadshakoyan M, 2008 WASET 22.
[10] Mukherjee P and Basu S, 1979 Int. J. Prod. Res.11 21-36

IOP Conf. Series: Materials Science and Engineering **344** (2018) 012019 doi:10.1088/1757-899X/344/1/012019

[11] Alkali A U, Ginta T L and Abdul-rani A M, 2014 Adv. Mater. Res. 845 790-794
[12] Konig W 1990 Annals of CIRP 39 1 413-425.
[13] Rao P, 2009 Manufacturing Technology 2 — Metal cutting and Machine Tools (New Delhi, India: Tata McGraw-Hill Publishing)
[14] Masood S, Armitage K and Brandt M, Int. J. Mach. Tool Manu 51 450-456

IOP Conf. Series: Materials Science and Engineering **344** (2018) 012020 doi:10.1088/1757-899X/344/1/012020

Experimental investigation into effect of cutting parameters on surface integrity of hardened tool steel

K Bashir*[1], A U Alkali[1], M H S Elmunafi[2] and **N M Yusof[2]**

[1] Department of Mechanical Engineering, Kano State Polytechnic, Nigeria
[2]Faculty of Mechanical Engineering, Universiti Teknologi Malaysia

*Email: kbmaisallah405@gmail.com

Abstract. Recent trend in turning hardened materials have gained popularity because of its immense machinability benefits. However, several machining processes like thermal assisted machining and cryogenic machining have reveal superior machinability benefits over conventional dry turning of hardened materials. Various engineering materials have been studied. However, investigations on AISI O1 tool steel have not been widely reported. In this paper, surface finish and surface integrity dominant when hard turning AISI O1 tool steel is analysed. The study is focused on the performance of wiper coated ceramic tool with respect to surface roughness and surface integrity of hardened tool steel. Hard turned tool steel was machined at varying cutting speed of 100, 155 and 210 m/min and feed rate of 0.05, 0.125 and 0.20mm/rev. The depth of cut of 0.2mm was maintained constant throughout the machining trials. Machining was conducted using dry turning on 200E-axis CNC lathe. The experimental study revealed that the surface finish is relatively superior at higher cutting speed of 210m/min. The surface finish increases when cutting speed increases whereas surface finish is generally better at lower feed rate of 0.05mm/rev. The experimental study conducted have revealed that phenomena such as work piece vibration due to poor or improper mounting on the spindle also contributed to higher surface roughness value of 0.66Ra during turning at 0.2mm/rev. Traces of white layer was observed when viewed with optical microscope which shows evidence of cutting effects on the turned work material at feed rate of 0.2 rev/min

1. Introduction

Tool steels are one of the major consumed materials for engineering products because of their good thermo-mechanical properties. Tool steels presents various machining challenges like frequent tool failure, poor surface finish and surface integrity. It is also classified among the difficult-to-cut material because of their inherent hardened characteristics. Tool steels presents poor machinability factors like high surface roughness which impacts negative effects on the component that is produced like decreased fatigue life and corrosion resistance and high coefficient of friction. Fewer literatures have been published so far on the investigations on surface finish of tool steel during cutting with wiper coated ceramic insert. Alkali et al [1] have investigated the influence of cutting parameters on chip morphology using dry turning of AISI O1 tool steel. Their investigation does not report the influence of the mentioned parameters on surface finish of the AISI O1 tool steel. Yalçın [2] conducted an investigation on hardened tool steel (HRC 52) during hard turning with mixed ceramic cutter. Their investigation revealed that surface roughness values that were obtained while cutting with ceramic tools were found higher than those obtained with cubic boron nitride (CBN) tool. The study further reported that the cutting force components obtained while cutting with CBN tool were found higher than those obtained with mixed ceramic tool. However, Christina [3] reported that the major factor responsible for machinability influence have been the cutting parameters like cutting speed, feed rates, cutting tool geometry, edge preparation and type of work material among others. Thus, the impetus of the current investigation is to study the influence of wiper coated ceramic cutting tool on surface finish and surface integrity of AISI O1 tool steel. Sudhansu et al [4] investigated the resulting surface roughness of AISI 52100 steel during finish turning with CBN tool. Their investigation revealed that feed rate was found to be the only significant factor on arithmetic average (Ra) surface roughness with percent contribution of 68.12%, followed by cutting speed which contributed 25.11%. Samardžiová et al [5] studied the surface roughness parameters Ra and Rz of hardened steel (100Cr6) using mixed ceramic insert. Their investigation revealed that it is possible to achieved more than two times lower values of surface roughness parameters Ra and Rz with wiper geometry cutting tool than conventional geometry cutters when they are use at the same cutting parameters. Their investigation further concluded that insert geometry influences surface quality, especially surface roughness parameters. The study further reported that it is possible to use wiper geometry (which is different from the conventional geometry in the shape of the minor cutting edge), when better surface quality is needed. This study will report the performance of wiper coated ceramic tool on surface integrity and surface finish of AISI O1 tool steel.

1.1. Wiper geometry turning insert

Another overwhelming drive for adopting the wiper geometry insert during this investigation is its ability to increase feed rate twice while keeping abreast the surface finish. Hence, cutting with wiper geometry is often regarded as high feed rate hard turning [6]. Another advantage attributed with high–feed cutting inserts like wiper geometry is their ability to maintain a low surface roughness of the machined surface and obtained super finished surfaces. An evident outcome of such inserts is improved productivity by increasing efficiency and lowering production costs [6, 7]. This investigation used rhomboid wiper geometry ceramic insert ANSI coded CNGA 120412EFW. The ceramic tool was mounted on left corner tool holder and that was geometrically coded Kennametal MCLNL 16161412.

2. Experimental Set up

2.1. Work piece material

Through hardened cylindrical tool steel bar obtained from AISI O1 was used as work piece material. Tool steel is a general purpose oil hardened tool steel suitable for a wide variety of cold-work applications. The steel was selected for this study because of its inherent machinability characteristics, good dimensional stability in hardened state, good combination of high surface hardness and toughness after hardening and tempering. The hardness of the material is 55HRC.

2.2. Cutting Tool Material

As reported in section 1.1, the rhomboid shaped coated wiper ceramic tool ANSI coded CNGA 120412EFW has a tool nose radius ($r\varepsilon$) of 1.2 mm. The geometrically coded Kennametal MCLNL 16161412 ceramic tool was mounted on left corner tool holder and has the following geometry;
Side rake angle $\gamma = -5°$

Back rake angle $\alpha = -5°$

Side cutting edge (SCEA) = -5°

Nose radius r = 1.2 mm

3. Experimental set-up

To obtain a proper shaped facial area, the work piece was machined on the faces and center drilled at both ends to enable support between centers both in order to minimize vibration as well as impact of forces that may result from shearing. Cutting tool was mounted and securely held in position and aligned with the work piece at both *X* and *Z* coordinates being references for the machine set up. To facilitate tool entry, a chamfer was made at the end of the work piece. The surface roughness (Ra) of machined part was measured using a Portable surface profilometer (Brand/Model: Taylor Hobson Surtronic 3+) three times per cut. A cut-off length of 0.8 mm was used to compensate for surface waviness while an average of these reading was taken. Thereafter overall average was eventually recorded for each experimental.

4. Result and discussions

4.1. Effect of Cutting Speed on the Surface Roughness

It was observed that the surface roughness improves as the cutting speed increases from 100m/min to 210m/min. At high cutting speed, the frictional force decreases due to the increase of temperature. Hence, this improves the surface finish. However, it is observed that at higher cutting speed, the surface finish was better. This is shown in Figure 1 which shows the surface roughness when the cutting speed is increased to 200m/min.

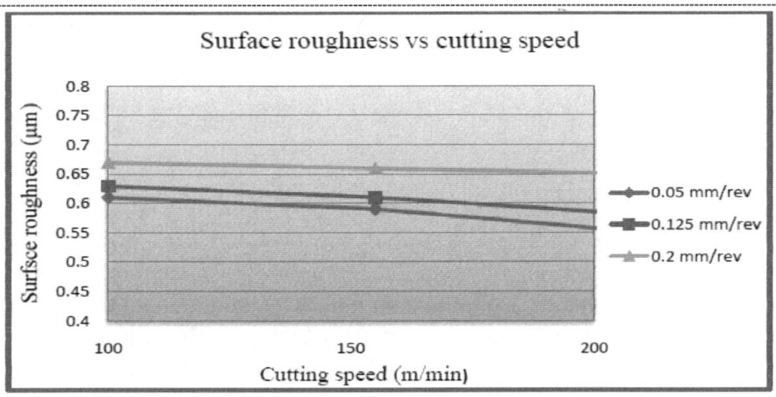

Figure 1. Surface Roughness of AISI O1 Tool Steel at Various Cutting Speed

4.2. Effect of Feed Rate on the Surface Roughness

The increase of feed rate results in an increase of the surface roughness values Ra, where the surface roughness at 210m/min with a feed rate of 0.05mm/rev appears to have a better Ra value compared to when feed rate is at higher value. This is to say, that surface finish is generally better at lower feed rate of 0.05mm/rev, whereas the poor surface roughness was recorded at 100m/min with a feed rate of 0.2mm/rev, as shown in Figure 2.

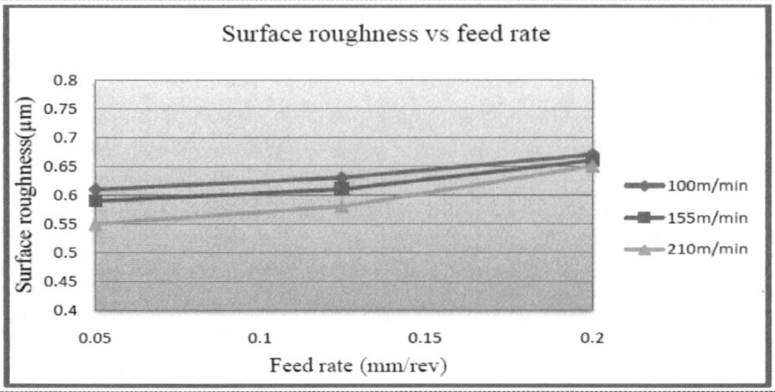

Figure 2 Surface Roughness of AISI O1 Tool Steel at Various Feed Rate.

4.3. Surface Roughness Propagation

Surface roughness propagation is effected by many conditions. The experimental study conducted revealed that phenomena such as work piece poor or improper mounting on the headstock spindle or mismatched in support of the work piece from the tail stock centre could result in higher surface roughness value. This in turn, affects the cutting tool, causing it to wear rapidly. Additionally, it is observed that the abrasive action of the deformed chip with the machined surface influences the surface roughness. It is also observed that the chips formed under the conditions investigated, were continuous chips which later curled and surrounded both the machined surface of the work material and of course the cutting tool rake face, hence contributing to crater wear on the cutting tool. At a feed rate of 0.05mm/rev, the surface roughness was found to be fair with the three sets of cutting speed investigated in this study. However, at 0.2mm/rev the roughness value was higher up to an average of 0.67µm with a corresponding cutting speed of 100m/min. This summarized the fact that, at lower feed rates the surface roughness always has a lower value which manifests in the form of good surface finish. These values gradually become better with increasing cutting speed. Figure 3 shows the surface roughness propagation.

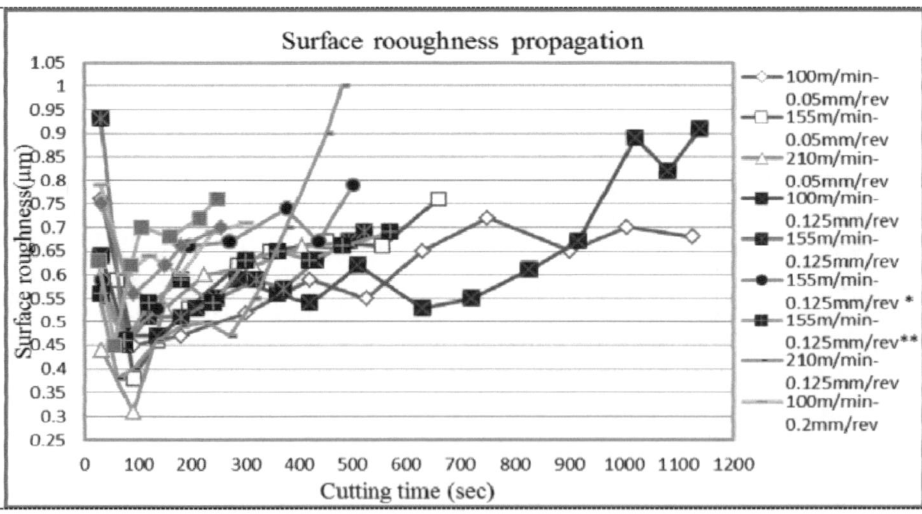

Figure 3. Surface Roughness Propagation when Turning AISI O1 Tool Steel

Although surface roughness value is imminent during the metal cutting, this study observed that, at every first cut in each experiment, the surface roughness value was found higher than the second cut. This is assumed to be due to the sharp cutting tool tip which becomes rounded after the first cut and before it is affected by a wear. The micro indentation obtained was in the range of 52 – 55 HRC. The micro indentation readings obtained were basically higher than those obtained on the machined work piece material. Guo and Sahni [8] also reported in his study that the change in hardness of a turned part was caused by the extremely small grain size and high dislocation density.

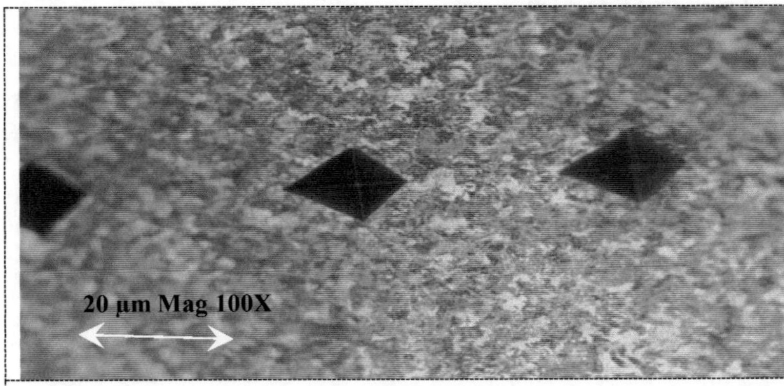

Figure 4. Micro Indentation on the Machined Work piece of AISI O1 Tool steel Spaced to avoid Interference

It is further observed that white layer exists with two layers. A dark layer beneath the white layer but above the bulk material as well as the white layer on top of the dark layer. The White layer appears white under an optical microscope after polishing and etching as evident and appeared smoother than bulk material as captured with SEM shown in Figure 5.

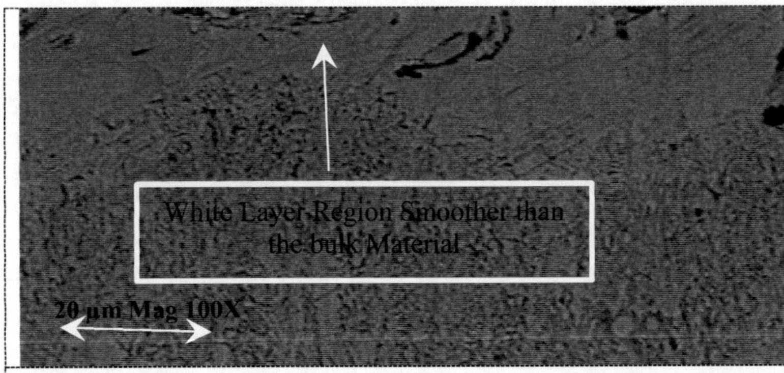

Figure 5. SEM Image Showing a White Layer Region Smoother than the Bulk Material

5. Conclusion

Based on the results and analysis carried out in this study during dry turning AISI tool steel with hardness 55 HRC using wiper edge preparation coated ceramic cutting tool, the study have revealed that the surface roughness improves as cutting speed increases from 100m/min to 210m/min. Surface finish is generally better at lower feed rate of 0.05mm/rev, whereas the poor value of surface roughness was recorded at 100m/min with a feed rate of 0.2mm/rev. Based on the presented result of the investigation obtained, it is recommended that the effect of cutting edge preparation and tool wear on the residual stresses should be investigated.

IOP Conf. Series: Materials Science and Engineering **344** (2018) 012020 doi:10.1088/1757-899X/344/1/012020

References

[1] Alkali A U, Noordin M Y, Elmunafi M S H, and Fawad H, 2013 Int. J. of Mater, Mech. Manuf, 1, 1
[2] Yalçın B, 2015 J. of the Trans. of the Can. Soc. Mech. Eng., 39, No. 2.
[3] Cristina B, 2015 Int. J. Sci & Eng Res, 6 5 2229-5518
[4] Sudhansu R D, Debabrata D and Amaresh K, 2014 Proceed. of Res., 5th Int. & 26th All India Manu. Tech., Design and Res. Conf.
[5] Samardžiová M, & Neslušan M, 2013 Facul. of Mater. Sci. and Tech. in Trnava Slovak Uni. of Tech.
[6] Ozel T, Karpat Y, Figueira L and Davim P, 2007 J. Mater. Process. Tech. 189 192–198.
[7] Stachurski W, Kruszynski B and Midera S L 2012 Mechanics and Mech. Eng'rng 16. 1 25–32
[8] GuoY B and Sahni J 2004 Int. J. Mach Tool Manu. 135 – 145

IOP Conf. Series: Materials Science and Engineering **344** (2018) 012021 doi:10.1088/1757-899X/344/1/012021

Development of Fire Detection Algorithm at Its Early Stage Using Fire Colour and Shape Information

Zainab Suleiman Abdullahi[1,a], Shehu Hamisu Dalhatu[2,b] and Zakariyya Hassan Abdullahi[1,c]

[1]Department of Computer Engineering, Hussaini Adamu Federal Polytechnic, Jigawa State Nigeria.
[2]Department of Electrical Engineering, Bayero University Kano, Nigeria

[a]zeeshahzahm05@gmail.com, [b]shdzango@yahoo.com, [c]hassanzakariyya8@gmail.com

Abstract. Fire can be defined as a state in which substances combined chemically with oxygen from the air and give out heat, smoke and flame. Most of the conventional fire detection techniques such as smoke, fire and heat detectors respectively have a problem of travelling delay and also give a high false alarm. The algorithm begins by loading the selected video clip from the database developed to identify the present or absence of fire in a frame. In this approach, background subtraction was employed. If the result of subtraction is less than the set threshold, the difference is ignored and the next frame is taken. However, if the difference is equal to or greater than the set threshold then it subjected to colour and shape test. This is done by using combined RGB colour model and shape signature. The proposed technique was very effective in detecting fire compared to those technique using only motion or colour clues.

1. Introduction

Fire can be defined as a state in which substances or materials combine chemically with oxygen from the air to give out heat, smoke and flame [1]. In general fire occurs only when 3 essential elements come together: heat (sufficient), material (Fuel), and oxygen as illustrated in Figure1. Fire plays very important role in human life. Human beings rely on fire for their cooking, warmth, blacksmith and industry applications. In spite of its numerous advantages fire can be dangerous if left uncontrolled. Such fire has the potential to cause physical damage like human burning, property/environmental destruction and water contamination. Fire can be characterized by the heat it generates, smoke, motion and irregular shape. This paper we restricted to class A and B fire formed by wood, papers, charcoal, rubber, petrol and kerosene etc. which are the most common fire occurring in our environment.

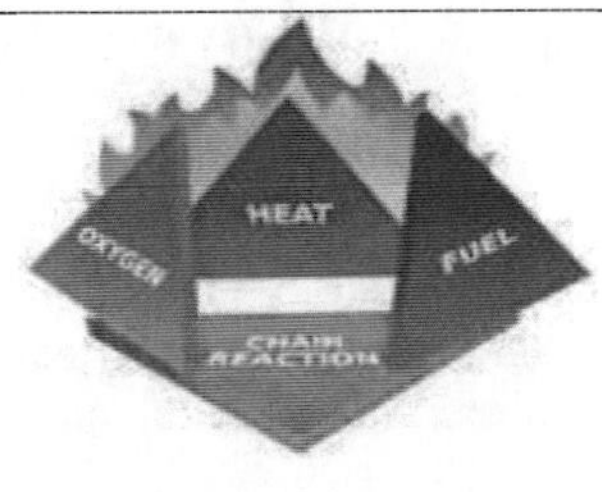

Figure 1. Triangle of a Fire

Fire can be characterized by the heat it generates, smoke, motion and irregular shape. This paper we restricted to class A and B fire formed by wood, papers, charcoal, rubber, petrol and kerosene etc. which are the most common fire occurring in our environment. In recent times many fire outbreaks were reported in Nigeria. For example [2] reported that about 80 fire disruption was cataloged in only 2015 in Kano state alone. Properties estimated to be valued more than to N185m were ruin by the fire disruption within the period and about 103 rescue teams were contacted with 10 false alarms available from members of the public. A total of 134 lives were saved from various incidents while 21 lives were lost. Reference [3] reported another Fire outbreak in Kano Sabon Gari Market where shops and properties worth millions of naira were destroyed and it spent more than 10 hours before it was finally put out. Conventional fire detection systems such as smoke and fire detectors are based on electronic sensors which are usually depend on heat and pressure. In the worst case scenario, if the sensors are damaged or not configured properly, these sensors can results false alarm. Also, the response time of proposed algorithm is faster compared to sensor-based fire detection system. However, the proposed technique does not require any type of condition to trigger and it has the ability to monitor a large area in both indoor and outdoor.

2. Related Work

This paper discusses the review work of fire detection using color models, motion or shapes to detect a fire. For example: A real-time video fire and smoke detection algorithm proposed by [4] which was based on foreground image accumulation and optical flow technique. Also, [5] proposed a system to detect a fire using color and shape through covariance matrix and send alarm using a mobile phone application. The mobile phone served as the receiving end and the user will be able to view and alert fire station. Reference [6] uses different color models to detect both fire and smoke .The color models are extracted using a statistical analysis of samples extracted from different type of video sequences and images. Reference [7] proposed a detection technique that uses YCbCr color space. In the event of fire outbreaks a frame is keep under consideration. If the area of the flame increases in a subsequent frame then an alarm is sounded. Reference [8] comes up with a technique using edge detection, color and motion to detect both fire and smoke. Reference [9] developed a method that extracts color and motion from video sequences to detect fire. The result of this paper is to identify color pixels in the scene and then identify fire region [10] focuses on optimizing the flame detection by identifying gray cycle pixels in the nearby flame, which is generated due to smoke in the area spread by flame.

3. Proposed Algorithm

The proposed technique uses background subtraction to detect changes in frames there by subjecting the result to colour and shape signature to detect a fire in both indoor and outdoor. Fig.3.1 shows the flow chart of the system.

3.1. Camera Details

In this work, a digital hand camcorder camera with resolution of 1280x720 pixels and 30 FPS frame rate with 5.1 megapixel CMOS image sensor having 2.0-inch LCD display was used for recording of the fire videos.

3.2. Detection of Fire Frames

Identifying moving objects from a video frame find to be very difficult in image processing technique. A common method used is to perform background subtraction that is to compare the current frame with reference frame. Usually, the first frame of the video which contain no object of interest is considered as the reference frame say frame (K).

IOP Conf. Series: Materials Science and Engineering **344** (2018) 012021 doi:10.1088/1757-899X/344/1/012021

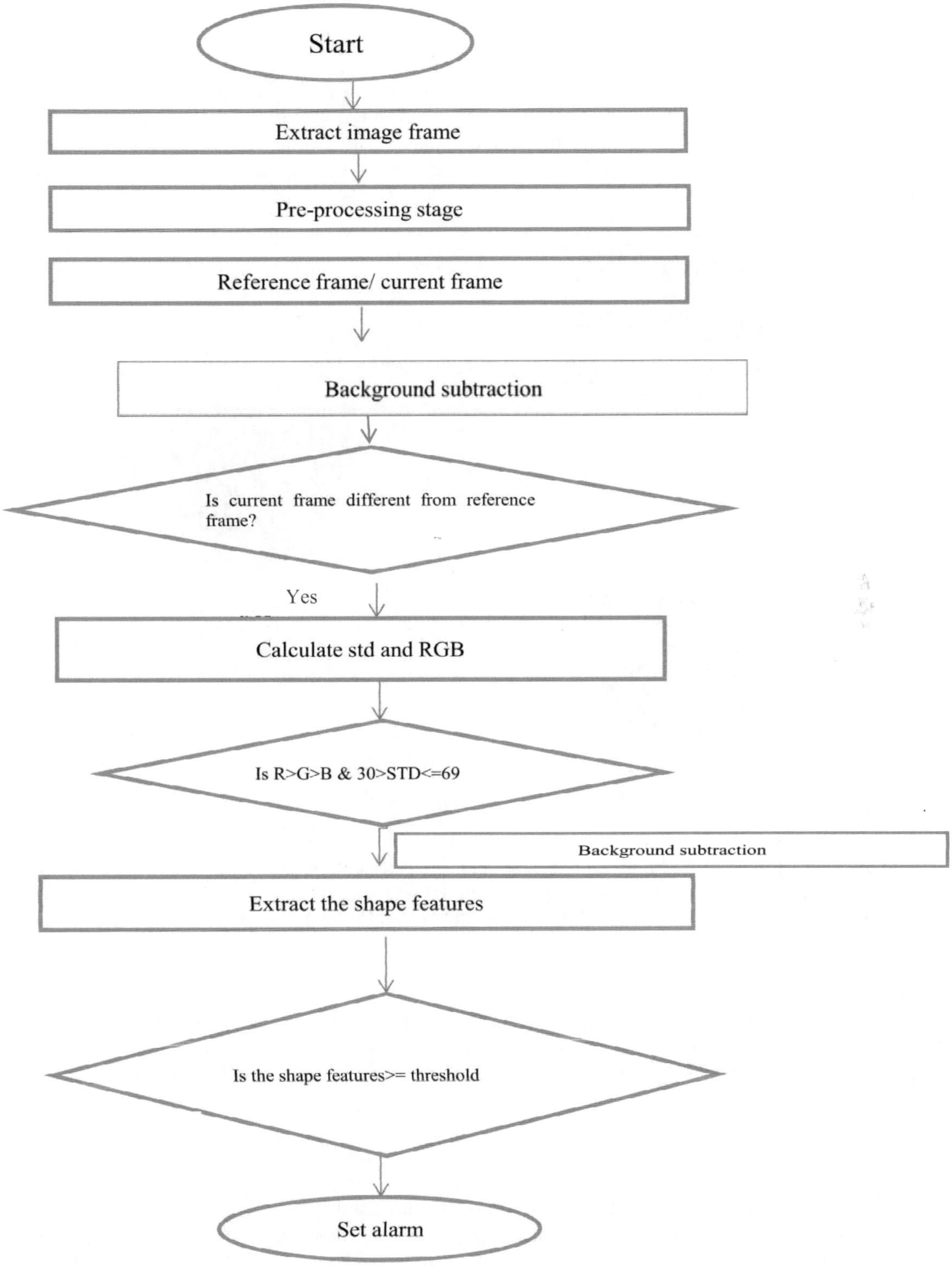

Figure 2: General Frame Work of the System

As the video advances, an object of interest starts appearing in the subsequent frames (K+1, K+2, K+3,…, K+N). In order to detect the object of interest, the current frame say frame (K+3), is subtracted from the reference frame that is ((K+3) – (K)), all the common pixels between the two frames will cancel out, the only pixel that will remain will be that belonging to the moving object in the current frame as illustrated in Figure 3.

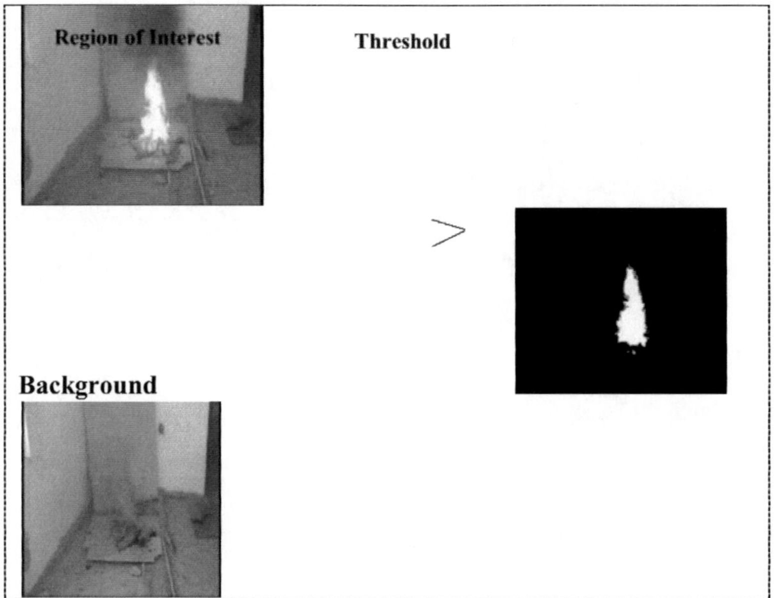

Figure 3. Frame Subtraction

The idea of background different method is to set the differences in pixel values between the current frame and background frame. If the result is less than the set threshold, the difference is ignored and the next frame is taken. However, if the difference is equal or greater than the set threshold, the different region is taken for segmentation.

4. Segmentation

Segmentation is a technique for discriminating the foreground image against the background (other regions in the image). In this paper, the threshold based segmentation method is used due to its simplicity and accuracy. The thresholding operation is a grey value remapping operation g defined by:

$$G(v) = \begin{cases} 0 & if\ v < t \\ 1 & if\ v \geq t \end{cases} \tag{1}$$

Where v represents a grey value and t is the threshold value. Thresholding maps a grey-valued image to a binary image. In the threshold approach once an appropriate threshold value t is selected (which was experimentally determined in this work as 80 (i.e. t = 80) the back ground pixels are forced to black while the foreground ones are converted into high intensity as shown in Figure 4.

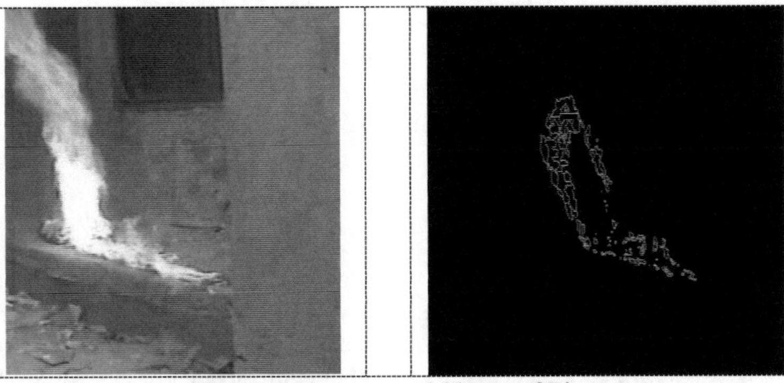

Figure 4: Segmented Shape of Fire

4.1. Fire Detection Using Colour

Color is very important attribute for object recognition. Fire can be described using its color properties which are resolved into three basic elements i.e. Red, Green and Blue (RGB). [10] decomposed RGB color model to detect color information. He further established the ratio of these color sand concludes that a fire has R-component greater than G- component and G-component greater than blue. R > G & G > B. Thus, fire color has R being the dominant color channel in an RGB image of fire. The correlation between the 3 channels is shown in Figure 5.

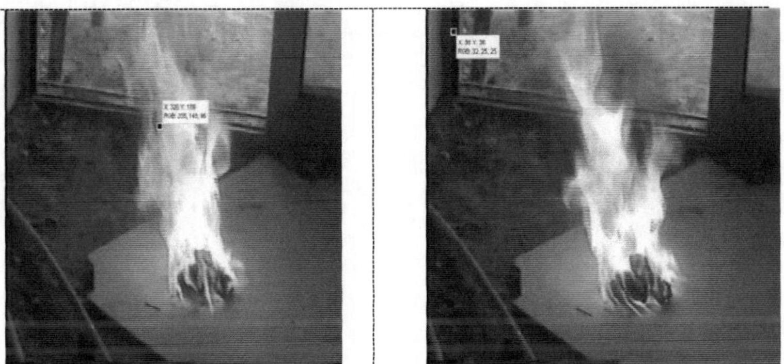

Figure 5: RGB Values of Same Frame at two Different Points

Figure 5a and b are the RGB values of same frame at two different points. Fig. 5 (a) shows the coordinate (x=326,y=189) with the corresponding values (255,148,96). This justifies the second RGB condition that R>G>B which qualify the region as a fire while fig.5(b) coordinates (x=56,y= 38) with the corresponding values (32,25,25) doesn't satisfy the RGB condition and therefore, it disqualifies the region as non-fire.

5. Fire Detection Test

Shape is a feature of object geometry that is variant with respect to time translation, rotation and scaling. Several algorithms were presented in the past by many researchers such as Box-counting method, Perimeter method, signature, skeleton, edge detection and contour extraction, etc. [11]. In this paper we employed a shape signature method to extract the shape of a fire. A shape signature is 1-D functional representation of a boundary and may be generated by plotting the distance from centroid to the boundary

IOP Conf. Series: Materials Science and Engineering **344** (2018) 012021 doi:10.1088/1757-899X/344/1/012021

as a function of angle. The basic idea is to reduce the boundary representation to 1-D function which presumably easier to describe the original 2-D boundary [12]. Based on the assumption of uniformity in scaling with respect to both axes, sampling is taken at equal interval of theta, changes in size of a shape result in the amplitude values of the corresponding signature. The signature of the given boundary is given as

$$[st, angle, x_0, y_0] = signature(b, x_0, y_0) \tag{2}$$

Where b is n_p by 2 array containing xy coordinates of the boundary ordered in clockwise or counterclockwise direction. The amplitude of the signature as a function of increasing angle and is output in st. The figures below show the signature of different shapes of objects. The basic idea of signature is to remove dependency on size while preserving the fundamental shape of the waveforms. This clearly differentiates the fire and non- fire frames by the uniqueness of the signature of an individual frames and yields better result.

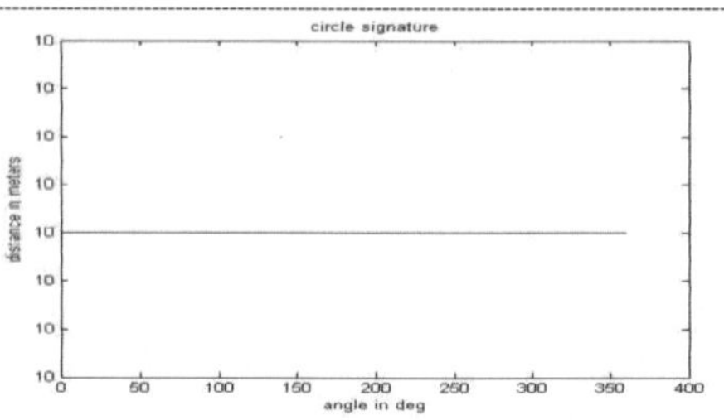

Figure 6. A Signature of a Circle

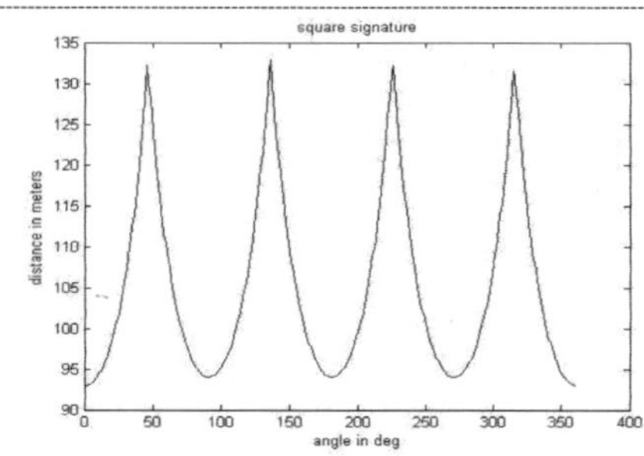

Figure 7. A Signature of a Square

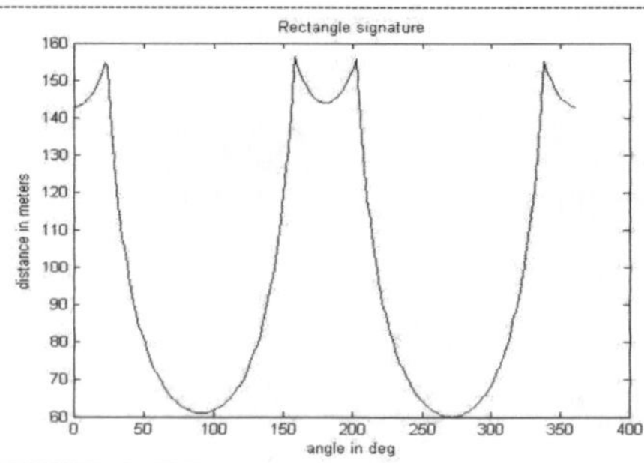

Figure 8. A Signature of a Circle

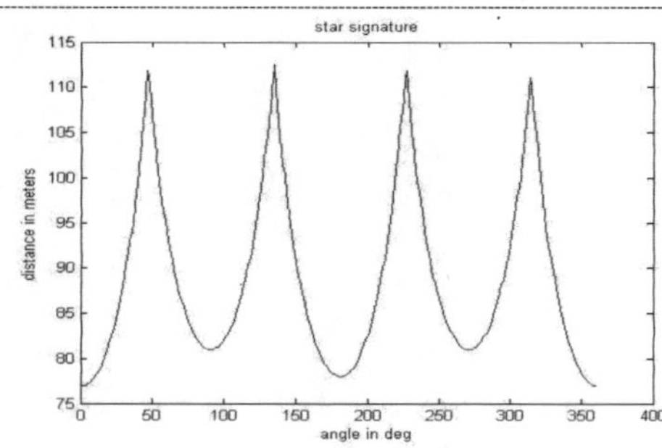

Figure 9. A Signature of a Square

6. Experimental Result

6.1. Fire Detection using Colour

Colour is very important tool in fire detection. Many algorithms were developed to detect a fire as discussed above. In this work, 10 video clips were sample from the database to test the algorithm using colour. Table 1 show the summary of the results obtained from colour.

Table 1 Summary Result of Colour Fire Detection

Video Clip	Total Number Of Frames	Rejected Frames	True Positive (TP)	False Positive (FP)	True Negative (TN)	False Negative (FN)
Video D1	275	68	124	48	19	16
Video D2	550	97	331	86	24	11
Video D3	425	52	318	25	11	19
Video D4	625	49	322	187	51	33
Video D5	500	68	376	18	21	17
Video D6	425	24	320	36	23	22
Video D8	500	69	359	51	13	8
Video D10	650	35	155	301	118	41
Video D11	550	216	118	161	25	30
Video D14	250	21	165	31	19	14
TOTAL	4750	699	2588	944	324	211

[13] Describe the Sensitivity, Specificity and Accuracy in terms of true positive (TP), true negative (TN), false negative (FN), and false positive (FP). Given that [13],

Sensitivity = $TP/(TP+FN)$ = (Number of true positive assessment) / (Number of all positive assessment) (3)

Specificity = $TN/(TN+FP)$ = (Number of true negative assessment) / (Number of all negative assessment) (4)

Accuracy = $(TN+TP)/(TN+TP+FN+FP)$ = (Number of correct assessments) / (Number of all assessments) (5)

From Table 1,
SENSITIVITY (True Positive Rate (TPR)) $= TP\ /\ (TP\ +\ FN)$

$$TPR = 2588\ /\ (2799) = 0.9246 = 92.461\%$$

SPECIFICITY (True Negative Rate (TNR)) $= TN\ /\ (FP + TN)$

$$SPC = 324\ /\ (1268) = 0.2555 = 25.552\%$$

FALSE DETECTION RATE $(FDR) \;=\; FP\,/\,(FP\;+TP)$

$$FDR = \;944\,/\,(3532)\; = \;0.2673\; = 26.73\;\%$$

The accuracy of the algorithm specifies the ability of the algorithm in detecting the region of interest (ROI).

Accuracy = $(TP + TN)/(TN + TP + FN + FP)$ 100% (6)

$$Accuracy \;=\; 2588\,/\,(4067)\;*\,100\%\;=\;63.634\%$$

6.2. Fire Detection Shape (Signature)

The concept of signature is to extract dependency on size while retaining the fundamental shape of the waveforms. Signature method helps in enhancing the technique by reducing the false detection in using colour. Table 2 shows the summary detection using a shape (Signature). From Table 2. [13] stated that :

SENSITIVITY $(TPR) \;=\; TP\,/\,(TP\;+\;FN) \;=\; 2793\,/\,2936 \;=\; 0.9513 \;=\; 95.13\%$

SPECIFICITY $(TNR) \;=\; TN\,/\,(FP + TN) \;=\; 245\,/\,690 \;=\; 0.3551 \;=\; 35.51\%$

False Detection Rate $(FDR) \;= FP\,/\,(FP\;+TP) \;=\; 445\,/\,3238 \;=\; 0.1374 \;=\; 13.74\%$

Accuracy= $((TN + TP)/(TN + TP + FN + FP)) * 100\% \;=\; 3038\,/\,3626 \;=\; 0.83784 \;=\; 83.784\%$

The accuracy of the algorithm using shape was found to be 83.784%.

Table 2 Summary Result of Fire Detection using Shape

Video clip	Total number of frames	Rejected frames	True positive (TP)	False positive (FP)	True negative (TN)	False negative (FN)
Video D1	275	68	163	6	16	7
Video D2	550	93	390	27	21	19
Video D3	425	41	356	5	12	11
Video D4	625	17	322	187	9	13
Video D5	500	68	378	16	23	15
Video D6	425	14	340	31	18	22
Video seven D8	500	65	412	11	5	7
Video eight D10	650	351	147	24	118	10
Video nine D11	550	226	98	131	19	30
Video ten D14	250	21	187	7	4	9
TOTAL	4750	964	2795	445	245	143

The proposed technique using both color and shape result in better detection despite the present of lightening and fire color objects by achieving 83.784% accuracy.

7. Conclusion

In this work, fire detection system was proposed, which is based on video image processing techniques. A number of sequential frames from original video clips were used which consists of fire and non-fire frames. The proposed algorithms differentiate the fire pixel frame from non-fire pixel frame using RGB colour model, and shape. The system has a better performance accuracy of 83.784% compare to [16] with 80.64%.

Thus application of the proposed fire detection system will ensure a better system performance in term of reducing the false alarm and better early fire detection. The performance of this technique can further be improved by detecting the growth of fire, along with the previous fire detection technique.

References

[1] $https://en.wikipedia.org > wiki > Fire$ 2016 28
[2] NAN 2015, Leadership New Paper Kano State 16
[3] Ahmad M, 2016, Daily Trust Kano State, 3
[4] Chunyu Y, Zhibin M and Zhang X, 2013, 9th Asia-Oceania Symp Fire Sci and Techno, 62, 891-898
[5] Chua A C, Leandicho, C L H, Magtibay A L C and Ortiz J T, 2013, Thesis, Mapua Inst of Techno, Malaysia
[6] Mengxin L, Weijing X, Ke X, Jingjing F, Dingding 2013, Rev of Fire Det Techno on Vid Image 49 2
[7] Saylee G, Sumeet, B, Sachin P, Rishi D and Rahul P, 2013, Int'l J Adv Res Comp And Comm Eng 2
[8] Bhargava S and Bhalla A V, 2015, Int'l J Emer Techno in Eng Res 1
[9] Nguyen T T, Nguyen-Phuc, T Do-Hong, 2013 Fire Det Based on Vid Proces Meth
[10] Gaurav Y 2012, J Comp Sci Eng, 2012 3 2
[11] Noda K. U, 1994 Veh Navig & Inform Syst Confer Proceed 57-62
[12] Gonzalez R C and woods R E, 2002 Digit Image Process
[13] Gonzaler R C, Richard E W and Steven I E, Prent Hal 2nd Edit 449- 452
[14] Nagavee N, Devaraj D and Kumar P, 2013, Int'l J of Sci and Techno
[15] NAN, 2015, Prem Time NewsPaper 5
[16] Kumarguru P and Siau-ChuinLiew 'Fire detection algorithm using image processing techniques' Proceeding of the 3rd International Conference on Artificial Intelligence and Computer Science (AICS2015), 12 - 13 October 2015, Penang, MALAYSIA.

Risk assessment framework on time impact: Infrastructure projects in soft soil during construction stage

W W Low[1], K S Wong[1] and J L Lee[1]

[1] Department of Civil and Construction Engineering, Curtin University Malaysia, CDT 250, Miri, Sarawak, Malaysia

Email: low.wai.wah@curtin.edu.my

Abstract. With the growth of economy and population, there is an increase in infrastructure construction projects. As such, it is unavoidable to have construction projects on soft soil. Without proper risk management plan, construction projects are vulnerable to different types of risks which will have negative impact on project's time, cost and quality. Literature review showed that little or none of the research is focused on the risk assessment on the infrastructure project in soft soil. Hence, the aim of this research is to propose a risk assessment framework in infrastructure projects in soft soil during the construction stage. This research was focused on the impact of risks on project time and internal risk factors. The research method was Analytical Hierarchy Process and the sample population was experienced industry experts who have experience in infrastructure projects. Analysis was completed and result showed that for internal factors, the five most significant risks on time element are lack of special equipment, potential contractual disputes and claims, shortage of skilled workers, delay/lack of materials supply, and insolvency of contractor/sub-contractor. Results indicated that resources risk factor play a critical role on project time frame in infrastructure projects in soft soil during the construction stage.

1. Introduction

The construction industry in Malaysia is developing steadily. A lot of rural areas are being developed to improve the living lifestyle and the overall development of the land. Infrastructure development in a country is influenced by a few factors, such as physical make-up of country, the growing economy, and the socio-economy disparities between each state [1]. Since the independence, Malaysia has been growing at a steadily good pace, this encourages more development in the country. New and old infrastructures are being developed, modernized and refurbished to serve the community even better. In addition, together with the congestion of population around the country, it increases the changes of construction work on soft soil area.

Infrastructure plays an important role in a country's development. According to MSG Experts [2], countries that are more advanced are seen to have better physical infrastructure such as better roads, stations, and other form of infrastructures as well. Infrastructure opens up the connectivity within the country as well as the connectivity to the outside world, thus encouraging more trades to occur.

There are several failures of infrastructures on soft soil throughout the years and there are a number of the cases are not reported to the authorities or released to the public as a fear of losing business and reputation [3]. It is believed that there are more actual cases than the number it is reported. Most of the failure cases are a result from the unaware of the excess pore water pressure from the soil when it is loaded [3]. According to an article written by [4], soil that are having low shear strength and permeability but high in compressive ability is considered as soft soil. Normally, there are a few situations that will be faced during the construction on soft soil. One of the situations is the site having insufficient bearing capacity. The other risks on the construction on soft soil are such as having continuous settlement after construction, unstable during excavation and embankment construction, and others as well [4].

Risk management is very vital in a construction project as it helps to reduce loss and maximize the profit [5]. Risk management system is used in many fields because it ensures a smooth operation of work and improves the efficiency as most of the possible risks that may happen are being identified and counter measures are implemented to prevent distuption to work as much as possible. As the construction process takes a long time to complete, thus it is very vulnerable of having time and cost overrun [5].

2. Research aim and scopes

2.1. Research aim

The research aim is to propose a risk assessment framework in infrastructure projects in soft soil during the construction stage based on time element.

2.2. Research scope

The scope of this research was focused on general infrastructures such as bridges, tunnels, roads, and sewerage system on soft soil. The risk factors are focused on time element only as this factor is one of the most influential element in a construction project. The research boundary is within the West and East Malaysia. The soft soil in this research refers to such as peat soil, loam soil, clay soil, and other soft soils available in Malaysia. Finally, this research focused on the construction stage.

3. Literature review

Based on the literature review, numerous studies focused on assessing risks in infrastructure projects. For examples, [6] focused on assessing risk and uncertainty inherent in Chinese highway projects and [7] focused on the application of risk assessment framework in a cross-sea route project in China. However, little or none of the studies focus on risk assessment in infrastructure projects in soft soil. Internal risks were covered in this research which is defined as risks that are caused by project team members. Table 1 indicates the internal risks extracted from some main literature. A total of 21 internal risks were identified and were categorized under 4 main groups, namely, resources risk, technical risk, project and client-related risk and legal risk as showed in Figure 1.

IOP Conf. Series: Materials Science and Engineering **344** (2018) 012022 doi:10.1088/1757-899X/344/1/012022

Table 1. Internal Risk Factors

Internal risks	Literature								
	A	B	C	D	E	F	G	H	I
Lack of special equipment	o								
Delay/lack of material supply	o								o
Shortage of skilled worker	o								
Poor/lack of soil treatment				o	o				
Poor/lack of site safety measures							o	o	
Poor site investigation		o							
Poor project planning		o							
Poor site management		o							
Delay due to contractor, sub-con, supplier, client, consultant			o						
Inexperienced contractor		o	o					o	
Bad workmanship	o							o	o
Inexperienced project manager			o						o
Technical difficulties in utilities diversions		o							
Complexity of project			o			o		o	
Project scope change/scope ambiguity		o							
Client's financial difficulties			o						
Tight project schedule			o					o	
Unrealistic/inadequate budgets			o						
Potential of contractual disputes and claims	o								
Problems in dispute settlement			o						
Insolvency of contractor/sub-contractor			o						

Notes:
A = [6]; **B** = [8]; **C** = [9]; **D** = [10]; **E** = [3]; **F** = [11]; **G** = [12]; **H** = [13]; **I** = [14]

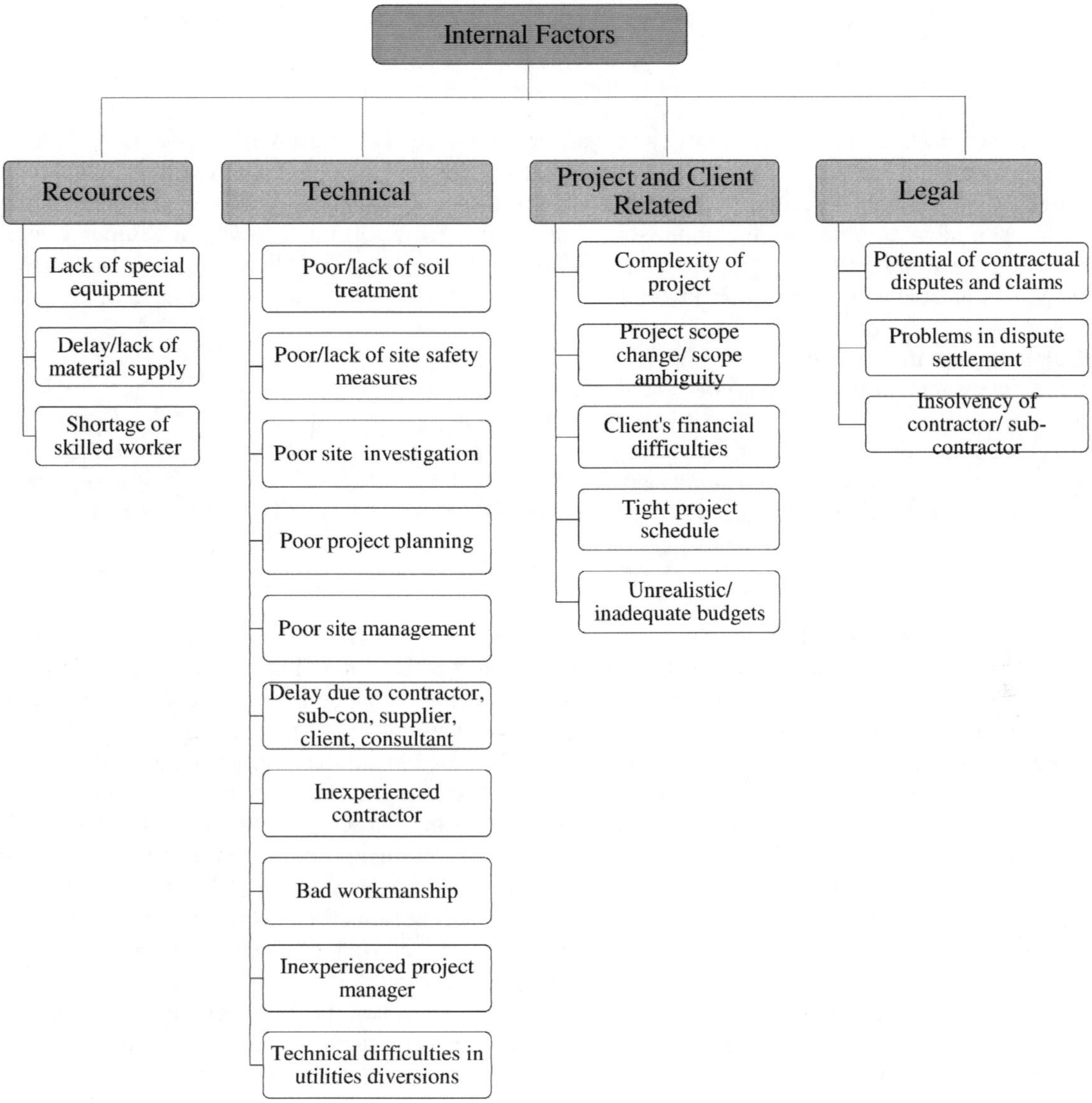

Figure 1. Internal risks hierarchy.

4. Research methodology

The research strategy used in this research is the quantitative research. The research method of this research was questionnaire survey. This is because the time is a limiting factor in this research. We need to complete this research within a year, thus making the survey method the most favorable. In this study, postal questionnaire technique was used as it is more convenient and economic.

4.1. Population and sample size

In this research, the population was civil engineers in the contractor organisations. There are some criteria for the targeted experts. Those criteria are: (1) the participants had to have civil engineering background; (2) have at least 15 years of working experience in the construction industry; and (3) have been involved in at least 1 local infrastructure projects in soft soil. Construction organisations were found through the Construction Industry Development Board (CIDB) website. The targeted experts were further identified from the respective companies' websites or through phone call to the human resource department. A total of 55 plus questionnaires were sent out and 13 valid responses were received.

4.2. Data analysis
The research analysis of this study was Analytic Hierarchy Process (AHP). AHP is a structured complex decision-making technique and developed by Saaty [15]. It is a method that compares the paired criteria and then derives the ratio scales from there. AHP basically follows five simple steps: construct a hierarchy, pair-wise comparison, consistency assurance, relative-weight computation, and lastly aggregation of relative weights. This was performed by using the Microsoft Excel. Consistency check was calculated in this research with a tolerance of 10% or less which is considered acceptable

5. Results and discussions
5.1. Demographic information of the respondents
All the respondents have more than 15 years working experience and have involved in at least 1 local infrastructure projects in soft soil. The responses were considered acceptable and this was supported by some previous studies. From an example, Khalil and his fellow researchers [16] shortlisted 22 potential respondents and obtained 12 valid responses. Also, according to [17], there is no fixed to the number of respondent for AHP analysis.

5.2. Anaylsis results
Table 2 represents the abbreviations of the 21 internal risks. Based on the analysis results, the consistency ratio was less than 0.10 (10%), within the acceptable limit. Table 3 indicates the local and global weightages among the identified internal risks. Based on Table 3, resource factor (R) is found out to be the most impacting risk factor among others when considering time element which gives a score of 0.39. The resources of a project include the manpower, material, and machineries.

Among the resources risks, the most impacting risk factor is lack of special equipment (R1) with a weightage of 0.4. While, among the technical risks, poor project planning (T4) has the highest weightage on time element. The planning of a project is very important as it ensures the smooth flow of a construction. If the project is not planned properly, one process of delay or problem could postpone the whole construction. This would result in a very serious time delay and late completion of construction project can cause the client to lose money. For the group of project & client related risks, the highest weightage is project scope change/project ambiguity (PC2). A project scope change means that the work under contract changed, this would result in a new project planning and this would require time to complete. For the legal factors, potential of contractual disputes and claims (L1) has the highest weightage. According to [6], the large number of variables and sections in the contracts that make it tough and risky to manage.

Based on Table 3, lack of special equipment (R1) has the highest impact on time element. This followed by potential of contractual claim (L1), shortage of skilled worker (R3), delay/lack of material (R2) and the insolvency of contractor/sub-contractor (L3). From here, we can see that generally resources (R) and legal (L) have the most impact on time element. According to [6], their research also stated that the lack of special equipment would have a large impact on the construction project as it takes time to tale in the special equipment as it may not be easily obtainable. On the other hand, the 5 least significant factors are poor/lack of site safety measures (T2), technical difficulties in utilities diversion (T10), bad workmanship (T8), inexperienced project manager (T9) and poor/lack of soil treatment (T1). All these factors are under the group of technical risks.

IOP Conf. Series: Materials Science and Engineering **344** (2018) 012022 doi:10.1088/1757-899X/344/1/012022

Table 2. Internal Factors Abbreviation

	Resources (R)
R1	Lack of special equipment
R2	Delay/lack of material supply
R3	Shortage of skilled worker
	Technical (T)
T1	Poor/lack of soil treatment
T2	Poor/lack of site safety measures
T3	Poor site investigation
T4	Poor project planning
T5	Poor site management
T6	Delay due to contractor, sub-con, supplier, client, contractor
T7	Inexperienced contractor
T8	Bad workmanship
T9	Inexperienced project manager
T10	Technical difficulties in utilities diversions
	Project & Client Related Risks (PC)
PC1	Complexity of project
PC2	Project scope change/scope ambiguity
PC3	Client's financial difficulties
PC4	Tight project schedule
PC5	Unrealistic/inadequate budgets
	Legal (L)
L1	Potential of contractual disputes and claims
L2	Problems in dispute settlement
L3	Insolvency of contractor/sub-contractor

Table 3. Local and Global Weightage for all Comparisons between Internal Factors

Risk Factors			Local Weightage	Global Weightage
Level 1		Level 2		
R	0.39	R1	**0.40**	**0.157**
		R2	0.28	**0.110**
		R3	0.31	**0.122**
T	0.15	T1	0.09	***0.013***
		T2	0.04	***0.006***
		T3	0.10	0.015
		T4	**0.17**	0.025
		T5	0.16	0.024
		T6	0.11	0.017
		T7	0.11	0.017
		T8	0.07	***0.011***
		T9	0.08	***0.012***
		T10	0.06	***0.009***
PC	0.16	PC1	0.17	0.028
		PC2	**0.31**	0.050
		PC3	0.19	0.030
		PC4	0.15	0.023
		PC5	0.19	0.030
L	0.30	L1	**0.43**	**0.128**
		L2	0.26	0.078
		L3	0.32	**0.095**

6. Conclusions

Throughout the research, it was found that resource factor has the highest weightage on time element. The 5 most significant risk factors are lack of special equipment, potential of contractual disputes and claims, shortage of skilled worker, delay/lack of material supply and insolvency of contractor/sub-contractor. Therefore, when a contractor company is contracting an infrastructure project in soft soil, they should be aware of the possible risks that are mentioned in this research and pay more attention on the top 5 listing risk factors and come out with an appropriate response plan so that when risks occur, they will have a countermeasure to deal with the risk. It is believed that this research can bring out awareness to the civil engineering industry as the construction on soft soil may poses different risks compared to other soil type. Engineers that take part in a soft soil construction need to be noted of the possible risks. Other than that, it can serve as a preliminary guideline for construction professionals during the planning, budgeting, and managing stages. This research was focused on the construction stage and did not cover other construction project stages such as designing stage, planning stage, and operational stage. Different risk factors may occur during different stages. Therefore, to fully understand how the risk factor changes throughout the whole project, future research on different project stages is recommended.

References

[1] Naidu G 2008 *Infrastructure Development in Malaysia* ed N Kumar International Infrastructure Development in East Asia-Towards Balanced Regional Development and Integration, no. March: 204–27. http://www.eria.org/publications/research_project_reports/images/pdf/PDF No.2/No.2-part2-7.Malaysia.pdf

[2] MSG Experts 2016 *Importance of Infrastructure in a Nation's Development* (Management Study Guide) http://www.managementstudyguide.com/importance-of-infrastructure-in-a-nations-development.htm

[3] Rahardjo P P 2001 Geotechnical failures case histories of construction on soft soils, forensic investigations and counter measures in Indonesia *International Journal of Integrated Engineering* **6** 2 11–23

[4] Mohamad N O, Razali C E, Hadi A A A, Som P P, Eng B C, Rusli M B and Mohamad F R 2015 Challenges in construction over soft soil – Case studies in Malaysia *IOP Conference Series: Materials Science and Engineering* **136** 1

[5] Khodeir L M and Mohamed A H M 2015 Identifying the latest risk probabilities affecting construction projects in Egypt according to political and economic variables *HBRC Journal* **11** 1 129–35

[6] Zayed T, Amer M and Pan J 2008 Assessing risk and uncertainty inherent in Chinese highway projects Using AHP *International Journal of Project Management* **26** 4 408–19

[7] Wang T, Wang S, Zhang L, Huang Z and Li Y 2015 A major infrastructure risk-assessment framework: Application to a cross-sea route project in China *International Journal of Project Management* **34** 7 1403-15

[8] Boateng P, Chen Z and Ogunlana S O 2015 An analytical network process model for risks prioritisation in megaprojects *International Journal of Project Management* **33** 8 1795–1811

[9] Malekitabar H, Ardeshir A, Sebt M H and Stouffs R 2016 Construction safety risk drivers: A BIM approach *Safety Science* **82** February 445–55.

[10] Bian X, Hong Z S and Ding J W 2016 Evaluating the effect of soil structure on the ground response during shield tunnelling in Shanghai soft clay *Tunnelling and Underground Space Technology* **58** September 120–132

[11] Qazi A, Quigley J, Dickson A and Kirytopoulos K 2016 Project Complexity and Risk Management (ProCRiM): Towards modelling project complexity driven risk paths in construction projects *International Journal of Project Management* **34** 7 1183–98
[12] Wang J, Zou P X W and Li P P 2015 Critical factors and paths influencing construction workers' safety risk tolerances *Accident Analysis and Prevention* **93** August 267–79.
[13] Zhang L, Wu X, Skibniewski M J, Zhong J and Lu Y 2014 Bayesian-network-based safety risk analysis in construction projects *Reliability Engineering and System Safety* **131** November 29–39
[14] Gue S S and Tan Y C 2002 Prevention of failures related to geotechnical works on soft ground Malaysian Geotechnical Conference
[15] Saaty R W 1987 The analytic hierarchy process-what and how it is used *Mathematical Modelling* **9** 3-5 161–76
[16] Khalil N, Kamaruzzaman S N and Baharum M R 2016 Ranking the indicators of building performance and the users' risk via analytical hierarchy process (AHP): Case of Malaysia *Ecological Indicators* **71** December 567-576.
[17] Saaty T L and Özdemir M S 2014 How many judges should there be in a group? *Annals of Data Science* **1** 3-4 359-368

IOP Conf. Series: Materials Science and Engineering **344** (2018) 012023 doi:10.1088/1757-899X/344/1/012023

Investigation of Boron addition and compaction pressure on the compactibility, densification and microhardness of 316L Stainless Steel

S Ali [1,2], A M A Rani [1], K Altaf [1], Z Baig [1]

[1] Mechanical Engineering Department, Universiti Teknologi PETRONAS, 32610 Bandar Seri Iskandar, Perak, Malaysia
[2] Department of Design and Manufacturing Engineering, National University of Sciences & Technology (NUST), H12, Islamabad, Pakistan

Email: majdi@utp.edu.my

Abstract. Powder Metallurgy (P/M) is one of the continually evolving technologies used for producing metal materials of various sizes and shapes. However, some P/M materials have limited use in engineering for their performance deficiency including fully dense components. AISI 316L Stainless Steel (SS) is one of the promising materials used in P/M that combines outstanding corrosion resistance, strength and ductility for numerous applications. It is important to analyze the material composition along with the processing conditions that lead to a superior behaviour of the parts manufactured with P/M technique. This research investigates the effect of Boron addition on the compactibility, densification, sintering characteristics and microhardness of 316L SS parts produced with P/M. In this study, 0.25% Boron was added to the 316L Stainless Steel matrix to study the increase in densification of the 316L SS samples. The samples were made at different compaction pressures ranging from 100 MPa to 600 MPa and sintered in Nitrogen atmosphere at a temperature of 1200°C. The effect of compaction pressure and sintering temperature and atmosphere on the density and microhardness was evaluated. The microstructure of the samples was examined by optical microscope and microhardness was found using Vickers hardness machine. Results of the study showed that sintered samples with Boron addition exhibited high densification with increase in microhardness as compared to pure 316L SS sintered samples.

1. Introduction

Powder Metallurgy (P/M) is one of the most diverse and continually evolving technologies among the various metal working methodologies for producing parts of various sizes and shapes. The main advantages of P/M include high productivity and dimensional accuracy of the parts with minimal use of energy and owing to material savings as there is minimal loss of raw material[1]. The attraction of P/M for producing functional parts originate from several attributes ranging from high volume production of precise parts, difficult to process metal materials and specialty alloys containing mixed phases.

The manufacturing of a sintered part using P/M starts with the compaction of the metal powder in a rigid die of the desired size and shape. In this process, the die cavity is filled with metal powder and a high compaction pressure is applied on the powder in the die cavity. The powder particles get squeezed and a certain amount of cold welding takes place. The powder particles get interlocked and remain intact with the surrounding particles leading to the development of interparticle bonds. Once the part is ejected from the die, the compact green part owns significant strength to withhold its shape and size. The next stage in this process is the sintering of the green compact parts where in the parts are placed in a furnace and heated to a temperature slightly below the melting point of the metal. The sintering cycle is designed accordingly depending upon the metal composition and the additives used. The heating rate is kept low to minimize the chances of crack initiation in the parts[2]. This step provides the thermal energy required to strengthen the powder particles and initiate the interparticle bond that helps in improving the mechanical properties of the parts.

Stainless Steel parts produced by P/M play an important role in engineering sector due to their low production cost and reduced need for post processing[3]. AISI 316L Stainless Steel is one of the promising materials used in P/M. It has low carbon concentration with high nickel and chromium contents that make it an attractive industrial material combining good corrosion resistance and ductility in a number of applications[4]. It's superior mechanical properties, high ductility, average strength and corrosion resistance makes it an ideal material for use in several applications including automotive, industrial, electrical and electronic, chemical, pharmaceutical, surgical and medical tools, aerospace, structural components, bio-applications and others[5-8]. Although P/M has been there for a quite long time but there has been research going on to improve the densification behavior of the P/M materials. To cope with, improvement of processing steps of compaction and sintering and modification of the powder has been developed to obtain high density P/M parts[5, 8].

Several Studies on sintered Stainless Steels have revealed that there are pores present inside the sintered parts which have a deleterious effect on the mechanical properties of the parts[9, 10]. This porosity is a cause of reduction in the tensile strength, impact resistance, ductility and corrosion resistance of the parts[11, 12]. Porosity thus hinders the above mentioned properties and researchers all over the world have tried to address this issue by incorporating additives in the stainless steel matrix to get fully dense parts. Several studies on improved densification of Stainless Steels have suggested the use of elemental Boron powder as an additive to get nearly fully dense parts for ferrous alloys[7, 13-19]. The analysis of Iron-Boride (FeB) diagram reveals that Boron is an active sintering additive for Iron based materials. The eutectic reaction takes place between Iron and Boron at a temperature higher than the temperature required for eutectic transformation. At this stage, a liquid phase appears which helps in improved sintering of the parts[1, 10]. Boron has also a good tendency to form complex borides with Iron resulting in FeB, with Nickel forming Ni2B and with Chromium it forms CrB and segregate at the grain boundaries if dispersed in the Stainless Steel matrix[10].

The addition of Boron depends not only on the density of the Stainless Steel but also on the mechanical properties and its microstructure[1]. To improve densification, microhardness and other mechanical properties, the amount of Boron addition plays an important role. In this article, 0.25% Boron was added to the 316L Stainless Steel matrix to investigate the effect of its addition on the compactibilty, densification, sintering characteristics and microhardness at different compaction pressures ranging from 100 MPa to 600 MPa.

2. Materials and methods

In this study, six samples from pure 316L Stainless Steel powder and another six samples from 0.25% Boron added Stainless Steel powder were produced by Powder Metallurgy method. The 316L Stainless Steel Powder used in the study was supplied by Zhe Jiang Bai Nian Yin Industry & Trade Co. Ltd Su Xi, Yongkang, Zhejiang, China and was produced by high-pressure gas atomization method. The chemical composition of 316L SS is given in Table 1. Nano size elemental Boron powder used in this study was supplied by Xuzhou Jiechuang New Material Technology Co. Ltd. Hong Kong and had particle size of 90 nm. The morphology of both the powders, observed by Scanning Electron Microscope (SEM) is shown in Figure 1.

IOP Conf. Series: Materials Science and Engineering **344** (2018) 012023 doi:10.1088/1757-899X/344/1/012023

Table 1. Chemical composition of 316L Stainless Steel powder

Element	Cr	Ni	Mo	Si	Mn	C	P	Fe
wt. %	17.04	12.01	2.4	0.9	1.5	0.028	0.045	Bal

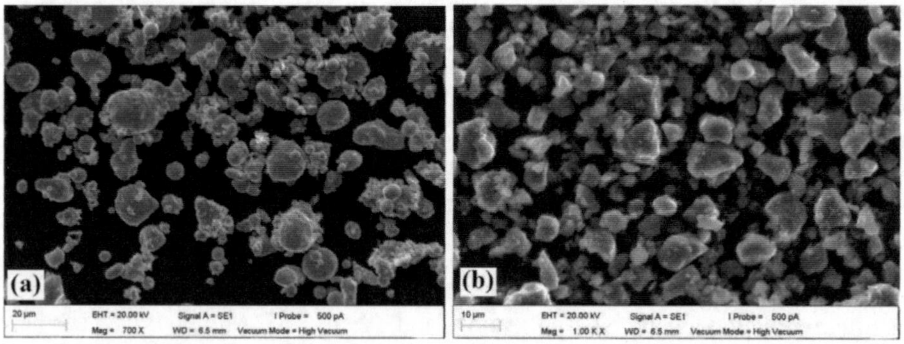

Figure 1. SEM micrograph of (a) 316L Stainless Steel powder and (b) Boron powder

The samples were made in disk shape at different pressures ranging from 100 MPa to 600 MPa as shown in Figure 2. In order to produce samples from Boron added Stainless Steel, the Boron and Stainless Steel powders were mixed together using the Turbula mixer device at 40 rpm for two hours. The Green Compact samples were then produced by pressing the powder uniaxially in a hardened steel die to obtain 30 mm diameter with 10 mm thick disk shape samples. The green compact samples were then sintered in Nitrogen atmosphere at 1200°C for 1 hour as shown in Figure 3.

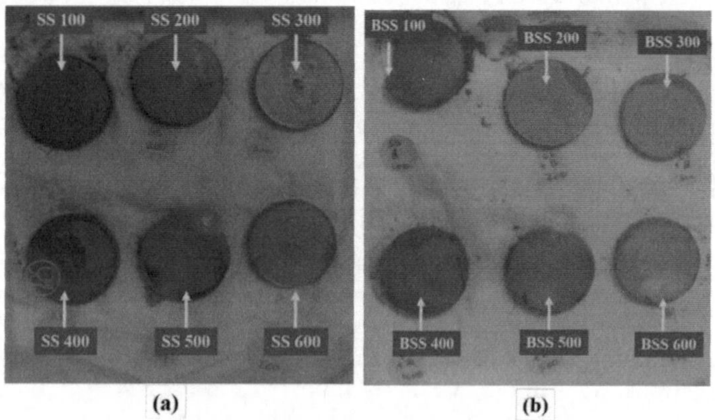

Figure 2. Green Compacts made from (a) pure 316L Stainless Steel (SS) powder and (b) Boron added Stainless Steel (BSS) powder

Figure 3. Sintered Samples in Nitrogen atmosphere at 1200°C

The density of the Green Compacts was measured by geometrical method and density of the sintered samples was measured by the water displacement method using Archimedes' Principle according to Standard ASTM B962-17. Optical Microscope (OM) studies were carried out to observe the microstructure and morphology of the sintered samples. For this, all the samples were prepared by mounting them in Bakelite and then polishing the surfaces. The samples were then placed in the Glycergia etching solution for 15 seconds to reveal the microstructure. Hardness of the sintered samples was measured using Vickers Hardness Tester according to Standard ASTM E92-82.

3. Results and discussion

3.1. Green density

The density of the Green Compact samples was measured from geometrical dimensions, applied load, weight of the sample and area of the used die. It was found that the green density increased with increasing the compaction pressure for pure Stainless Steel powder as well as the boron added Stainless Steel powder. The increase in density with the increase in compaction pressure has been depicted in Figure 4.

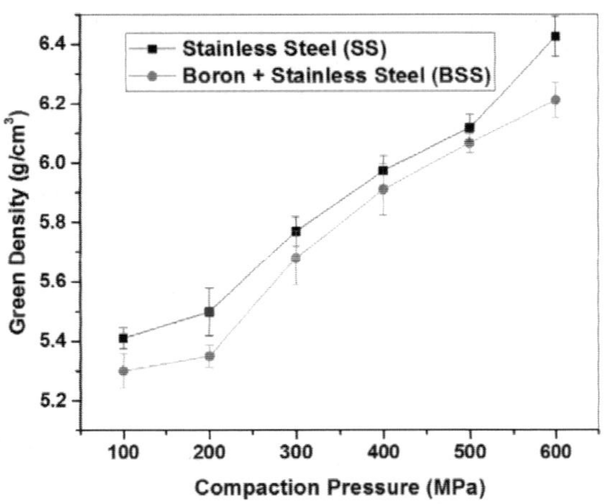

Figure 4. Graph representing Green Density as a function of Compaction Pressure for pure Stainless Steel and Boron added Stainless Steel

From Figure 4, it can be seen that the green density of the comacted samples is increasing with increasing the compaction pressure. However, the green density of Boron added Stainless Steel shows variation as compared to the pure Stainless Steel. It can be seen that Boron addition in stainless steel samples reduced the green density at all compaction pressure. This can be attributed to the presence of Boron particles between the SS particles that provide hindarance during compaction and avoid particles to come close even at high pressures.

3.2. Sintered density

Figure 6 demonstrates the sintered density of both pure Stainless Steel and Boron added Stainless Steel samples sintered under Nitrogen atmosphere. It can be clearly seen that all 316L Stainless Steel samples exhibited significant increase in sintered density at all compaction pressures. Moreover, the samples added with Boron powder show a notable increase in the densification of the sintered samples beyond 200 MPa. The schematic diagram of Boron addition to the Stainless Steel particles and the densification mechanism has been shown in the figure 5 below.

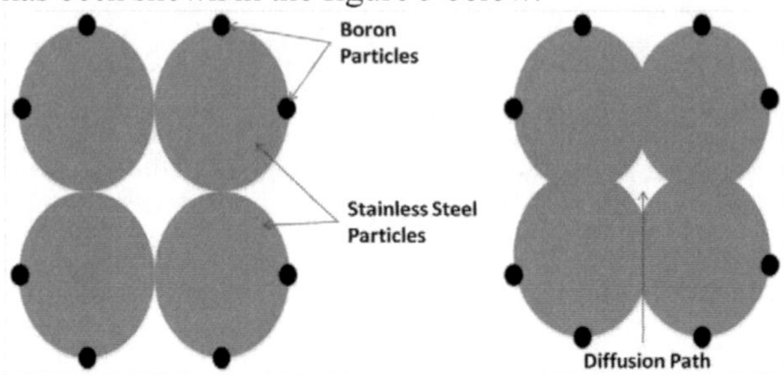

Figure 5. Schematic diagram of Boron diffusion and densification mechanism

A maximum densification of 92.5 % was achieved for Boron added Stainless Steel sample at 600 MPa. It thus validates that Boron addition promotes densification along with increase in compaction pressure for 316L Stainless Steel. The behaviour of the boron addition in Stainless Steel shows that it facilitates the sinterability of the samples by formation of liquid phase that is produced by the eutectic reaction that takes place between Iron and Boron. This reaction takes place at a temperature higher than the temperature required for eutectic transformation. At this stage, a liquid phase appears which helps in improved sintering of the parts by reducing the number of pores. Boron tends to segregate by forming a layer on the grain boundaries which helps in rapid densification of the powder by providing high diffusivity rate. This helps in reducing the number of pores inside the surface supporting improved densification. The amount of Boron addition and the sintering temperature also have a high impact on densification and if these parameters are correctly chosen, nearly full dense parts can be produced.

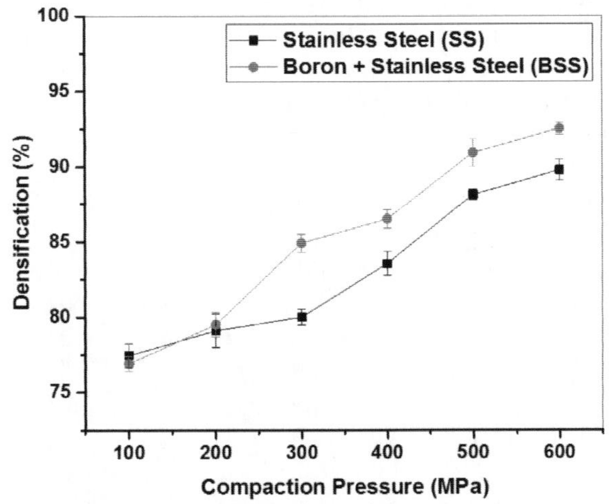

Figure 6. Graph representing Densification of Sintered Samples for pure Stainless Steel and Boron added Stainless Steel

3.3. Densification parameter

The compressibility of powder is an index of its densification behaviour under the application of an external pressure. The compressibility of powder compacts during compaction process is normally measured by green density which is the measure of its densification.

The fractional density of compacts can also be measured by Densification Parameter. It is an index of fractional densification achieved as a result of compaction as compared to theoretically maximum attainable densification. The Densification Parameter for all the samples is given in Table 2.

Table 2. Densification Parameter for pure Stainless Steel and Boron added Stainless Steel

Powder Composition	Densification Parameter at different Compaction Pressures					
	100 MPa	200 MPa	300 MPa	400 MPa	500 MPa	600 MPa
316L SS	0.30	0.32	0.33	0.345	0.49	0.52
Boron added 316L SS (BSS)	0.31	0.38	0.47	0.49	0.62	0.66

3.4. Microhardness

The microhardness measurements of the sintered samples were evaluated using Vickers Hardness testing to determine the effect of the sintering atmosphere. Figure 7 shows the hardness values calculated for all the specimens sintered at different compaction pressures ranging from 100 MPa to 600 MPa. It was observed that microhardness increased with increasing Compaction Pressure. Boron addition also favoured the increase in microhardness and the samples sintered with Boron added Stainless Steel showed a notable increase in the microhardness. A maximum hardness of HV 195.25 was observed for the Boron added Stainless Steel sample.

The Nitrogen atmosphere was found suitable for sintering all the samples because it favoured the dissolution of Nitrogen into the samples during the sintering process which had an impact on the improved densification and enhanced microhardness. It has been reported that Nitrogen is a better sintering atmosphere for austenitic Stainless Steels which helps in strengthening the grain size of the sintered samples[20]. Moreover, the addition of Boron helped in increasing the microhardness of the sintered samples. Boron has a tendency to form borides with Iron, Chromium and Nickel present in the Stainless Steel when dispersed in the matrix during the sintering process. These borides segregate themselves at the grain boundaries and are a source of increasing the strength of the resulting sample which can be distinguished in figure 7.

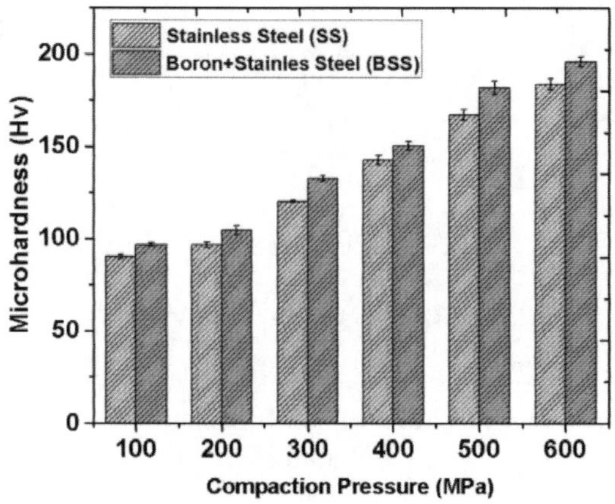

Figure 7. Microhardness of the sintered samples for pure Stainless Steel and Boron added Stainless Steel

3.5. Microstructure

The microstructures of the samples with and without the addition of Boron are shown in the figure 8 and 9. Figure 8 (a) to (f) shows the microstructure of the pure Stainless Steel samples compacted at different compaction pressures ranging from 100 MPa to 600 MPa. The figure 8 (a) shows that as the compaction pressure is too low (100 MPa), therefore, a number of empty spaces in the form of pores are present. As we increase the compaction pressure, the number of pores keep on decreasing and at compaction pressure of 600 MPa shown in figure 8 (f), there are very less pores left inside the sintered sample which represents good densification. A maximum densification of 89.5 % was achieved for the samples compacted at 600 MPa.

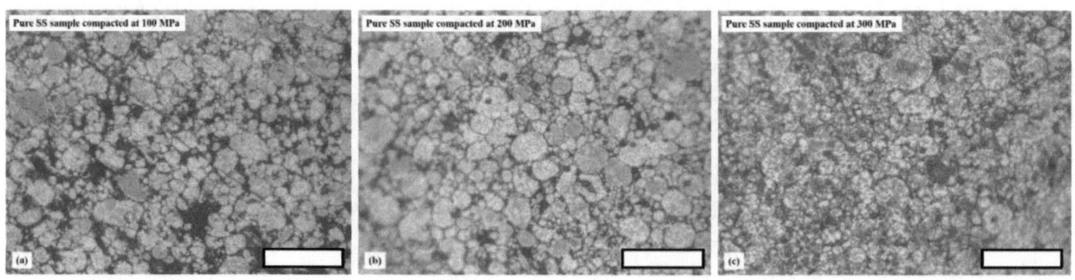

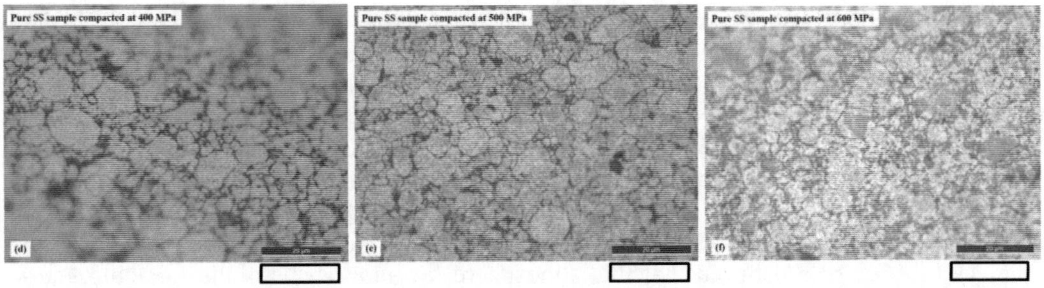

Figure 8. Microstructures of Pure Stainless Steel samples compacted from 100 MPa to 600 MPa

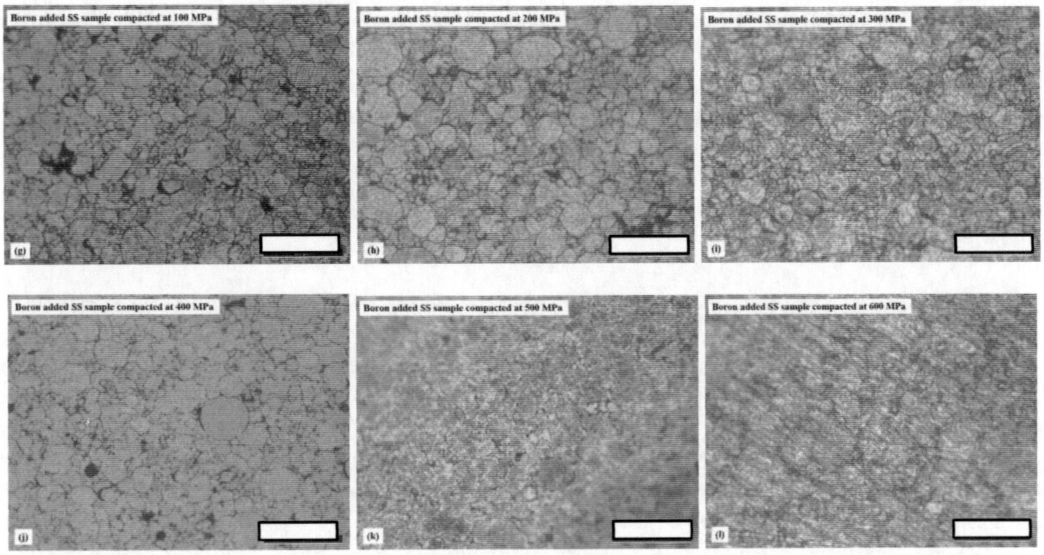

Figure 9. Microstructures of Boron added SS samples compacted from 100 MPa to 600 MPa

Figure 9 (a) to (f) show the microstructure of the Boron added Stainless Steel samples. It can be noted that with the addition of Boron, the number of pores start reducing and at 600 MPa compaction pressure, the Boron added Stainless Steel sample revealed 92.5 % densification. In the samples sintered with Boron added Stainless Steel, sufficient amount of liquid phase formed during sintering process which provided nearly full densification. The grains were well surrounded with eutectic phase indicating that high diffusivity layer was formed at the grain boundaries resulting in nearly full densification. The Boron also helped in smoothing the pore size from irregular to spherical shape. It thus concludes that the microstructure of the sintered samples has improved with the addition of Boron.

4. Conclusion

In this study, Stainless Steel was compacted at different Compaction Pressures ranging from 100 MPa to 600 MPa to study the effect of Compaction Pressure on the densification of the sintered Stainless Steel samples. It is concluded from the results that the densification increases as the Compaction Pressure is increased. A maximum densification of 89.5 % was achieved for the pure Stainless Steel samples compacted at 600 MPa. To improve the densification of the sintered samples, 0.25 % Boron was added to the Stainless Steel matrix. The study concluded that the Boron addition has a notable effect on the density of the sintered samples and a maximum of 92.5 % densification was achieved at Compaction Pressure of 600 MPa. Thus, we conclude that Boron aids in increasing the density of the sintered samples. Similar results of Boron aiding in increasing the densification of sintered samples using Powder Injection Method and Powder Metallurgy methods have been reported[1, 7, 12, 16, 19] and supports the fundamentals of increase in densification of sintered samples. The effect of Compaction Pressure on the microhardness of the samples was also studied using Vickers Hardness testing which confirmed that increase in Compaction Pressure increases the microhardness of the sintered samples. A maximum of HV 183.76 was achieved for the pure Stainless Steel samples compacted at 600 MPa. The microhardness for Boron added Stainless Steel was further improved to HV 195.25 at Compaction Pressure of 600 MPa. The sintering atmosphere also has an impact on the overall properties of the sintered samples and Nitrogen was found well suitable for sintering the samples at 1200°C. It is believed that this atmosphere favours the dissolution of Nitrogen into the samples during sintering process. This in turn has a notable effect on the densification, microhardness and strength of the sintered samples in terms of strengthening grain size, grain boundaries, pore geometry and porosity ratio. It is thus concluded that Compaction Pressure as well as Boron addition enhances density and microhardness of the sintered samples. It is now recommended to increase the Compaction Pressure to 800 MPa along with Boron addition upto 1.5 wt%. This will further increase the densification and microhardness of the sintered samples and can get nearly 98~99 % dense parts with improved strength.

References

[1]. Sulima, I., L. Jaworska, and J. Karwan-Baczewska, Effect of Boron Sinter-Aid on the Microstructure and Properties of Austenitic Stainless Steel-TIB2 Composites/Wpływ Dodatku Boru Na Mikrostrukturę IW Łaściwosci Kompozytów Stal Austenityczna-TIB2. Archives of Metallurgy and Materials, 2015. **60**(4): p. 2619-2624.

[2]. Erden, M.A., The Effect of the Sintering Temperature and Addition of Niobium and Vanadium on the Microstructure and Mechanical Properties of Microalloyed PM Steels. Metals, 2017. **7**(9): p. 329.

[3]. Kurgan, N., et al., Production of 316L stainless steel implant materials by powder metallurgy and investigation of their wear properties. Chinese science bulletin, 2012. **57**(15): p. 1873-1878.

[4]. Sander, G., et al., On The Corrosion and Metastable Pitting Characteristics of 316L Stainless Steel Produced by Selective Laser Melting. Journal of The Electrochemical Society, 2017. **164**(6): p. C250-C257.

[5]. Coovattanachai, O., et al., Performance improvement of P/M 316L by addition of liquid phase forming powder. Materials Science and Engineering: A, 2007. **445**: p. 440-445.

[6]. Padmavathi, C., A. Upadhyaya, and D. Agrawal, Corrosion behavior of microwave-sintered austenitic stainless steel composites. Scripta Materialia, 2007. **57**(7): p. 651-654.

[7]. Bayraktaroglu, E., et al., Effect of boron addition on injection molded 316L stainless steel: Mechanical, corrosion properties and in vitro bioactivity. Bio-medical materials and engineering, 2012. **22**(6): p. 333-349.

[8]. Federation, M.P.I., A.P.M. Institute, and A. International, Advances in powder metallurgy and particulate materials. Vol. 1. 2006: Metal Powder Industries Federation.

[9]. Schade, C., Processing, Microstructures and Properties of a Dual Phase Precipitation-Hardening PM Stainless Steel. 2010.

[10]. Szewczyk-Nykiel, A., The effect of the addition of boron on the densification, microstructure and properties of sintered 17-4 PH stainless steel. Czasopismo Techniczne, 2015.

[11]. Gülsoy, H., Production of injection moulded 316L stainless steels reinforced with TiC (N) particles. Materials Science and Technology, 2008. **24**(12): p. 1484-1491.

[12]. Bakan, H., D. Heaney, and R. German, Effect of nickel boride and boron additions on sintering characteristics of injection moulded 316L powder using water soluble binder system. Powder metallurgy, 2001. **44**(3): p. 235-242.
[13]. Bagliuk, G., Properties and structure of sintered boron containing carbon steels, in Sintering-Methods and Products. 2012, InTech.
[14]. Akhtar, F., Sintering behavior of elemental powders with FeB addition in the composition of martensitic stainless steel. Journal of Materials Engineering and Performance, 2007. **16**(6): p. 726-729.
[15]. Miramontes, J.C., et al., Effect of boron additions on sintering and densification of a ferritic stainless steel. Journal of materials engineering and performance, 2010. **19**(6): p. 880-884.
[16]. Skałon, M. and J. Kazior, Enhanced sintering of austenitic stainless steel powder AISI 316L through boron containig master alloy addition. Archives of Metallurgy and Materials, 2012. **57**(3): p. 789-797.
[17]. Mahathanabodee, S., et al., Effects of hexagonal boron nitride and sintering temperature on mechanical and tribological properties of SS316L/h-BN composites. Materials & Design, 2013. **46**: p. 588-597.
[18]. Gierl, C., I. Mohsin, and H. Danninger, Boron Activated Sintering of PM Steels–Alternative Boron Sources. Powder Metallurgy Progress, 2008. **8**(2): p. 135.
[19]. Lozada, L. and F. Castro, Controlled densification of boron-containing stainless steels. 2011.
[20]. Kurgan, N., Effects of sintering atmosphere on microstructure and mechanical property of sintered powder metallurgy 316L stainless steel. Materials & Design, 2013. **52**: p. 995-998.

IOP Conf. Series: Materials Science and Engineering **344** (2018) 012024 doi:10.1088/1757-899X/344/1/012024

The Qualitative Measurement towards Emotional Feeling of Design for Product Development

Fevi Syaifoelida[1*], M.A.M. Megat Hamdan[2], Murrad M.[2], Hazim Aminuddin[1]

[1] Mechanical Engineering Department, Universiti Tenaga Nasional (UNITEN)
[2] Mechanical Engineering Department, Universiti Pertahanan Nasional Malaysia (UPNM)

*Corresponding author: Fevilia@uniten.edu.my

Abstract. To compete in today's ever-growing technology market, a product needs to be well presentable to the customers. It is a challenge to design a product that is able to attract customer's attention and to build their loyalty towards the product. A product needs to be designed with focus to give the maximum level of satisfaction to the end user which is the customer. That is the focus of this paper, to achieve customer's satisfaction, by studying feelings and emotional value related to product designs using Kansei Engineering (KE) and to test how important that product element (level of satisfaction) by using Kano Method (KM). KE is a method of translating human emotions and feelings into product development. The method studies the human interaction and responses when a customer sees a product, then translates it into a new improved design. However, KE cannot stand on its own. It did not specify in which extends the feeling or emotions is important in a product. After we had the design appearance parameters from KE and existing design evaluation, it need to be classified which is more important than the other is. That is why Kano Method (KM) will also be used. Since this scope of study towards an emotional feeling of design (existing part/appearance) in Kano categories not in deep function of technical requirement, so KM will help to classify parts of product into categories, which part will give fully satisfaction while using it. It studies the more important attributes considered by the customers for improvement. The objective is to find out the design priority guide that can be used to maximize customer's satisfaction. Therefore, in order to apply the qualitative measurement idea into the real situation, the headphone product is chosen (popular among students) as the product (the appearance of part, feeling when use it) domain for this study. As the results progressed, it was found out that the headband part of headphone is the most important part of the product. It needs to be durable and comfortably designed to give full satisfaction while using (Kano-functional). The research and development (R&D), and designing process of a product can be improved greatly to increase the customer's satisfaction by capturing their emotional feeling in physiological design.

1. Introduction

When designing a product, our main target is to achieve customer satisfaction when the final product is being purchased and used by the consumer. Don Norman (1988) states that a design is a success only when the final product is successful in making customer buys it, uses it, and enjoys it, and spreads the word of the product to others. He claimed that designers need to ensure that the design satisfies people's needs, in terms of functions, usability, and the ability to deliver emotional satisfaction, pride, and delight. Customer satisfaction is based on the post choice evaluative judgment after a purchase decision is made (Oliver, 1980; Churchill and Suprenant, 1982; Bearden and Teel, 1983; Oliver and DeSarbo, 1988). Oliver (1997) also states satisfaction as a function of disconfirmation, which relates to both expectations and performances. Swan and Combs (1976) also provide the same argument that satisfaction can be achieved with product performance that fulfills the expectations. Both of these models agree that when the performance meets or is higher than the expectation, satisfaction has achieved, while when the expectation exceeds the performance, satisfaction is not achievable.

Based on the above reasons, what actually the initial "spark" for the demand in product design is a customer's own the process decision. First, the new products come out almost every day. Companies need to think ways to make their product stand out against the other products. One of the key factors is the product design. That is what will give the first impressions to the customers. Each day the industries faces increasingly higher competition in their business related to the customers' demands and product design requirement.

Second, what makes a design different from the others? One thing that is lacking in design community today is the needs to please clients first. The engine is as important as the car's exterior. Here, due to the emotions can change a person's behavior which is a response to environmental conditions (Hartono and Chuan, 2011), but how to unlock those customer decisions to be something that could be in the right interpretation? (Syaifoelida et al, 2016). Consequently, these ideas should translate feelings and emotional value (qualitative measurement) towards their level of satisfaction in each design element.

2. Decision Making Approach Theory

2.1. Kansei Engineering

Kansei Engineering is defined as "translating technology of consumer's feeling and image for a product into design elements" (Nagamachi, 1995). Kansei Engineering is a method of translating feelings, emotions, and impressions into product development (Watson 2011). Kansei Engineering's objective is to be able to capture and understand human's feelings and consideration, and translate those aspects into product design and development. Its main objective is towards customer's satisfaction in using the final product. Nagamachi (1995) realized that the industry desired to quantify customer's impression into their products. So, the developed Kansei Engineering can measure and capture feelings toward product properties.

There are some aspects and features in a product that customers are not aware of, but when they use it, they can feel it. It is difficult to be explained, but it is there. Therefore, the goal of Kansei Engineering is to study those features and use it to create a design and products. That is why the understanding of semantics is important in Kansei Engineering. Throughout the process, we will be collecting Kansei words. Kansei words are words that describe the product (Schütte, 2002). For example, when describing a car, adjectives such as fast, beautiful, and economical are used. Besides adjectives, verbs and nouns can also be used to express the Kansei words. Numbers of Kansei words can range from 50 to 600 words, (Nagamachi 1997). Words with similar meanings are compiled to a bank of words.

2.2. Kano Method

The design parameters from Kansei Engineering need to be classified which is more important than the other is. That is where the Kano Model plays its role. Kano Method is a theory developed by Professor Noriaki Kano and supported by Nilsson and Fundin (2005) which the method focuses on how customers evolve, evaluate and perceive quality attributes. It studies the more important attributes considered by the customers, and then improves them. These quality attributes are dynamic, meaning that it changes over time. It may go from satisfactory to unsatisfactory, and vice versa. Kano method also states that satisfaction and dissatisfaction should be considered separately as they are two different concepts. There are five categories of features, quality as categorized by Noriaki Kano as depicted in Table 1, based on these qualities, we can determine in which group the product feature belongs to by using its bipolar (calculation of the extent of satisfaction and dissatisfaction) as mathematical approach.

Kansei engineering will be our main approach in this study supported by Kano Method. Different people will have different ways of expressing their emotions and feelings, thus requires us to have a high understanding of language semantics

IOP Conf. Series: Materials Science and Engineering **344** (2018) 012024 doi:10.1088/1757-899X/344/1/012024

Table 1. Kano Quality Atrributes

Quality		Description
I.	**Must-be (M)**	When this quality is fulfilled, it will not increase the level of satisfaction, however, when it is not fulfilled it will cause major dissatisfaction.
II.	**One-dimensional (O)**	When this quality is fulfilled, it will increase satisfaction, and when it does not fulfill it will decrease the satisfaction. This is basically what makes the product different from its other competitors. This quality also builds up customer loyalty.
III.	**Attractive (A)**	When this quality is fulfilled, it will increase the level of satisfaction, when it does not fulfil it will not cause dissatisfaction. These qualities are not expected when the customer buys the product. .
IV.	**Indifferent (I)**	Whether this quality is fulfilled or not, it makes no difference to the level of satisfaction and dissatisfaction of customers. It gives a minimal impact to the satisfaction level of customers. To save costs and design considerations, this attribute should be excluded if possible.
V.	**Reverse (R)**	This element gives dissatisfaction to the customer. Most customers hate this quality factor. In designing a product, this quality should be avoided at all costs.

3. Methodology

Kansei Engineering's purpose is to improve quality. Product development can be thought as the process to transform market demands and opportunity into a product for sale. It requires knowledge not only from engineering but also from economics, art, and others. The whole process can be roughly divided into investigation of the market, development of product design specification, concept design, detailed design, manufacture and selling. The main of focus of this study is on the 'investigation of the market' and 'development of product design specifications'. On the scope of studies, a headphone product (Figure 1) is chosen (popular among students) as our product domain in order to apply a qualitative measurement in the real situation. Fourteen, 14 types of headphones have been chosen from the market existing design based on their designs and the current trend. These headphones are broken into a few parts (Table 2), headphone models based on these chosen combinations will be developed using 3D Solidwork software (Nine, 9 models, Table 3) and will be used in the final survey.

Figure 1. Headphone parts

Table 2. Headphone features

Features	Parts / designs
Headband	1) Solid headband 2) Framed headband 3) Cushioned / leather headband
Ear cups	1) Over-ear 2) On-ear
Ear piece shape	1) Square 2) Round 3) Triangle
Folding mechanism	1) Foldable 2) Not foldable

Table 3. Headphone features

DESIGN	PARTS
Design 1	leather headband, over ear pad, round shape of earpiece, foldable
Design 2	leather headband, on ear pad, round shape of earpiece, foldable
Design 3	full solid headband, on ear pad, round shape of earpiece, foldable
Design 4	leather headband, over ear pad, round shape of earpiece, no foldable
Design 5	full solid headband, over ear pad, square shape of earpiece, no foldable
Design 6	full solid headband, over ear pad, round shape of earpiece, no foldable
Design 7	frame headband, over ear pad, round shape of earpiece, foldable
Design 8	frame headband, on ear pad, round shape of earpiece, foldable
Design 9	leather headband, over ear pad, round shape of earpiece, foldable

The focus on surveys and studies to find out the demands of the customers and will be rated by using the Kansei emotional feeling of words collection (Table 4). Table 4 contains of the Kansei words used with the semantic differential to articulate the design characteristics in terms of emotional or quality affective ((D/F, P/I, C/M, A/U, M/O, U/C, P/I, D/P). Furthermore, in the Kano designing process, the final design from 'development of product design specifications' will be merged with emotional words (Kansei). Then, both functional and dysfunctional questions are formed with the combinations of the features and the Kansei words (Table 5). For each pair of questions, it needs to be constructed so that respondents can answer them in five different ways in Likert style (Kano, 1984).

Table 4. Headphone features with its related Kansei words

HEADPHONE FEATURES	KANSEI WORDS	SEMANTIC DIFFRENTIAL SCALE
Headband (leather)	Durable- Fragile	(D/F)
Headband (solid)	Portable - Inconvenient	(P/I)
Headband (framed)	Comfortable- Miserable	(C/M)
Round Earpiece	Attractive - Ugly	(A/U)
Square Earpiece	Modern - Old-fashioned	(M/O)
Triangle Earpiece	Unique - Common	(U/C)
On-ear pads	Portable - Inconvenient	(P/I)
Over-ear pads	Durable - Portable	(D/P)

Table 5. Kano functional and Dysfunctional question

	Functional Questions
1.	A headphone with full solid headbands makes me feel *durable* and secure when using it.
2.	A headphone with framed headband is more *portable* when I'm listening music on the go.
3.	A headphone with leather headband feels more *comfortable* when wearing it.
4.	A headphone with over-ear cups feels *durable* when wearing the headphone.
5.	A headphone with on-ear cups is more *comfortable* when wearing the headphone.
6.	A headphone with a round shaped earpiece looks more *attractive* to me.
7.	A headphone with a square shaped earpiece is more *modern* and futuristic looking.
8.	A headphone with triangle shaped earpiece is *unique* and seamless looking.
9.	A headphone that can be folded is *portable* and easy to bring to anywhere.
10	A headphone without folding mechanism is more *durable* and lasts longer.
	Dysfunctional Questions
1.	A headphone with full solid headbands makes me feels *fragile* and loose when using it.
2.	A headphone with framed headband is more *inconvenient* when I'm listening music on the go.
3.	A headphone with leather headband feels more *miserable* when wearing it.
4.	A headphone with over-ear cups feels *inconvenient* when wearing the headphone.
5.	A headphone with on-ear cups feels *miserable* when wearing the headphone.
6.	A headphone with a round shaped earpiece looks *ugly* to me.
7.	A headphone with a square shaped earpiece is more *old-fashioned* and worn looking.
8.	A headphone with triangle shaped earpiece is *common* and sluggish looking.
9.	A headphone that can be folded is *inconvenient* and to bring to anywhere.
10	A headphone without folding mechanism is more *fragile* and lasts shorter.

In addition, the functional Kano question is "how the respondent would react with the Kansei words", while the dysfunctional Kano question is "how the respondent would react with the antonyms of the Kansei words". The "voice of the customer" is the prime importance in creating Kano questions (Hauser&Clausing, 1998). Consequently, the juxtaposition idea of KE and KM is merged, indirectly it gives the translating of emotional feeling while using the headphone.

Therefore, the Kano evaluation table (Table 6) is used as our decision maker, the mathematical approach in order to determine in which categories the feature of the design is belongs to and its level of satisfaction.

Table 6. Kano evaluation table

Functional Answers	**Dysfunctional Answers**				
	Like	**Must-be**	**Neutral**	**Live-with**	**Dislike**
Like	Q	A	A	A	O
Must-be	R	I	I	I	M
Neutral	R	I	I	I	M
Live-with	R	I	I	I	M
Dislike	R	R	R	R	Q

4. Results and Discussion

The total 75 final year student (mechanical engineering) in one of the higher institution is chosen as a sample size of the population, the majority of respondents are male with percentage of 65%. Results show that 69% of respondents use their headphones mostly to listen to music, next to gaming usage with 25%. The rest uses it as phone accessories. Most users only replace their headphones when their current one is broken. It can say that they would prefer their headphones be durable and long lasting. Most people prefer to buy the headphones with affordable prices. They would look into the specifications of the headphones before buying it. Respondents would prefer to use earphones as an alternative if headphones if not available. Lastly, respondents prefer to keep their headphones in their own bags instead of using custom cases. This is due to the ease of access. This means that we need to make sure the headphones are durable, so it will not break when being placed in a bag.

4.1 KM Vs KE Result

Using Kano evaluation table, we need to identify the Kano quality based on the answers from both Kano functional and Kano dysfunctional questions. Kano evaluation table converts both positive (functional) and negative (dysfunctional) answers to get our desired Kano Quality. In order to get the total percentage for each Kano quality acquired, the calculation towards the extent of satisfaction and dissatisfaction using the formulas below.

Extent of Satisfaction, CS = $$\frac{A+O}{A+O+M+I} \quad (1)$$

Extent of Dissatisfaction, DS = $$\frac{O+M}{-(A+O+M+I)} \quad (2)$$

The values of CS and DS (sample of calculation, Table 7) towards Kano quality are tabulated in Table 8. The lower the number means the higher the importance of the feature.

Table 7. Kano& Kansei Results

Product features	**Percentage of replies**							
	A	**M**	**I**	**O**	**R**	**Q**	**Total**	**Category**
Headbands (leather)	**8**	**41.33**	**21.33**	**29.33**	**0**	**0**	**100**	**Must-be (M**

Example of Must be (M)	**CS: 0.37**	**DS: -0.71**

Four product features are recognize as a **Must-be (M)** requirement are leather headband, solid headband, and both of the folding options. Here we can deduce that headband comes with a leather/ solid set of design and the folding/no-foldable options is **MUST** in designing a headphone. The other six features are recognized as **INDIFFERENT (I)** , which means the features play a minimal role in increasing customer's satisfaction towards the product. In the ranking analysis, headband (leather) is the top in the rank, further approves that headband (leather) is an important feature, **Must-be (M)** in Kano analysis, and the feeling of **durability** is an important factors need to be considered in designing process in order to give full satisfaction.

Table 8. Kano& Kansei Results

Headphone features	Kano Quality	CS-DS	Rank	Kansei Words
headband (leather)	**Must-Be (M)**	**-0.34**	**1**	**durable**
ear pad (on-ear)	**Indifferent (I)**	**-0.31**	**2**	**comfort**
headband (framed)	Indifferent (I)	-0.24	3	portability
headband (solid)	**Must-Be (M)**	-0.24	4	**comfort**
ear pad (over-ear)	Indifferent (I)	-0.21	5	durable
shape (square)	Indifferent (I)	-0.14	6	modern
shape (round)	Indifferent (I)	-0.13	7	attractive
shape (triangle)	Indifferent (I)	0.11	8	unique
No foldable	**Must-Be (M)**	0.24	9	**durable**
folding	**Must-Be (M)**	0.24	10	**portability**

As a most design preference, the evaluation towards all nine models (Table 3) is done by using the Expert Choice software. The software applies the use of analytical hierarchy, which helps us to make decisions with complex problem involving multiple conflicting and subjective criteria (Zahedi, 1986) as depicted in Fig. 2.

Figure 2. Expert Choice results

The top design chosen by the respondents is **design 2**, which made up of leather headband, round earpiece, on-ear pads, and it is foldable (Fig. 3). In fact, leather headband and on ear pad is in all of the top 2 of design (Table 7). We can strongly agree that **leather headband** and **on ear pad** are the most preferred as the highest attractiveness, which is giving them the feeling of durable and comfortable (Kansei emotional feeling) while they wear it (Kano-functional) and become miserable if this element did not exist in design.

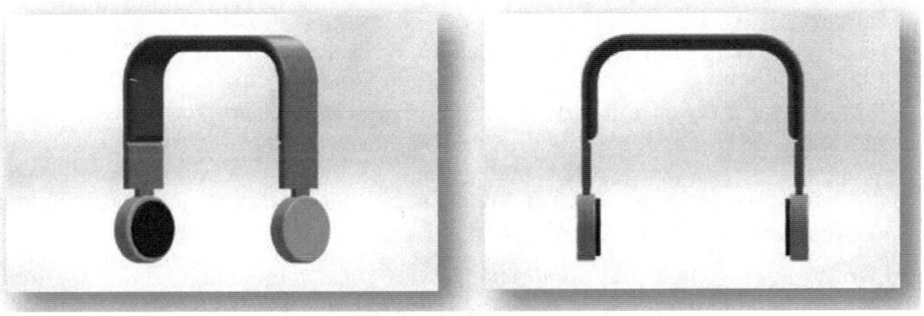

Figure 3. Design 2

5. Conclusion

The qualitative measurement through the collection of the Kansei words is achieved through the words represent of full customer's satisfaction values (KE) and how important the design part of designing process through KM. Based on this study towards the headphone design, we found as follows:

1. Leather headband is the top feature desired by respondents (based on ranking analysis and Kano quality, Must-be (M)). It is chosen for its **durability** (D/F) (emotional feeling) that increases customers' satisfaction towards the headphones. In addition, which is giving them the feeling of durable (Kansei emotional feeling) while they wear it (Kano-functional) and become miserable if this element did not exist in design, Must-Be category (Table 1) in Kano analysis.

2. Other than leather, full solid headband is also favored. Customers also would prefer that their headphone is portable with the folding feature. On the other hand, the shape of the earpiece is not an important part.

3. It is recommended on focusing on the development of headbands especially the **leather** and **the solid headbands**. Developers need to increase its **durability** and **comforts**. These qualities will increase customer's satisfaction while using the new improved product. Customers also prefer to have the option to choose between the foldable and non-foldable

headphones. With the folding feature, it will increase the portability and without the folding feature, it will increase durability.

Since the study carried out is against the engineering students, any further study is required, especially when it is extended to the using of Analytical Hierarchy Process (AHP) based on Semantic Differential towards Kansei Engineering. Therefore, further study can be done towards different product (product development) as a case study. Since this scope of study towards an emotional feeling of design (existing part/appearance) in Kano categories, further requirement can be done in the basic function of product in technical requirement such as sound.

REFERENCES

[1] Octavian A 2012 Kano Model, *Economy Trandisciplinarity Cognition* Vol. 15.
[2] Bearden O, William, Jess E Teel 1983 Selected Determinants of Consumer Satisfaction and Complaint Reports, *Journal of Marketing Research* Vol.20 pp.21-28.
[3] Churchill A, Gilbert Jr and Carol S 1982 An Investigation into the Determinants of Customer Satisfaction *Journal of Marketing Research* Vol. 19 pp.491-504.
[4] Norman D 1988 the design of everyday things *New York: Doubleday.*
[5] Hauser J R, and Clasuing D 1988 the House of Quality *Harvard Business Review* pp. 164-174.
[6] Hartono M and Chuan T K 2011 A Proposed Integrative Framework of Kansei Engineering and Kano Model Applied to Services *the 2nd International Research Symposium in Service Management Yogyakarta* pp.484-492.
[7] Swan E, John and Linda J C 1976 Product Performance and Consumer Satisfaction: A New Concept *Journal of Marketing* Vol. 40 pp.25-33.
[8] Zahed F 1986 The analytic hierarchy process-a survey of the method and its applications. *Interfaces* Vol. 16 pp. 96-108.
[9] Nilsson-Witell L and Fundin A 2005 Dynamics of service attributes: a test of Kano's theory of attractive quality *International Journal of Service Industry Management* pp. 152-168.
[9] Nagamachi M 1995 Kansei engineering: A new ergonomic consumer-oriented technology for product development, *International Journal of Industrial Ergonomics* Vol. 15 pp.3–11.
[10] Nagamachi M 1997 Kansei Engineering: The Framework and Methods Kansei Engineering, M. Nagamachi *Kure, Kaibundo Publishing co.,LTD* pp. 1-9.
[11] Oliver LR and Wayne S 1988 Response Determinants in Satisfaction Judgments Journal *of Consumer Research* Vol. 14 pp.495-507.
[12] Oliver L R 1980 A Cognitive Model of the Antecedents and Consequences of Satisfaction Decisions *Journal of Marketing Research* Vol. 17 pp.460-469.
[13] Oliver L R 1997 Satisfaction: A Behavioral Perspective on the Consumer *New York: The McGraw-Hill Companies, Inc.*
[14] Schutte S 2002 Designing Feelings into Products: Intergrating Kansei Engineering Methodology in Product Development *Institute of Technology Linköpings University.*
[15] Syaifoelida F, Hamdan M.A.M, Murrad M 2016 The Hierarchy of Need, Fev Matrix in the Concept Generation towards the Existing Product Design in Product Development: A Watch Product *4TH International Conference On Engineering Management (ICEM)* pp.26-27.
[16] Watson W B 2011 towards an Understanding of Human Behaviour for Design Action, Loughborough University.

Experimental Equipment Validation for Methane (CH_4) and Carbon Dioxide (CO_2) Hydrates

Muhammad Saad Khan[a]*, Sana Yaqub, Naathiya Manner, Nur Ani Karthwathi, Ali Qasim, Nurhayati Binti Mellon, Bhajan Lal[b]

Chemical Engineering Department, Universiti Teknologi PETRONAS, Bandar Seri Iskandar, 32610 Tronoh, Perak, Malaysia.

*Corresponding author. Tel.: +60-17-9486804; E-mail address: [a]khansaaad@gmail.com; [b]bhajan.lal@utp.edu.my

Abstract. Clathrate hydrates are eminent structures regard as a threat to the gas and oil industry in light of their irritating propensity to subsea pipelines. For natural gas transmission and processing, the formation of gas hydrate is one of the main flow assurance delinquent has led researchers toward conducting fresh and meticulous studies on various aspects of gas hydrates. This paper highlighted the thermodynamic analysis on pure CH_4 and CO_2 gas hydrates on the custom fabricated equipment (Sapphire cell hydrate reactor) for experimental validation. CO_2 gas hydrate formed at lower pressure (41 bar) as compared to CH_4 gas hydrate (70 bar) while comparison of thermodynamic properties between CH_4 and CO_2 also presented in this study. This preliminary study could provide pathways for the quest of potent hydrate inhibitors.

Keywords: CH_4 Hydrate; Compressibility Factor; CO_2 Hydrate; HLwVE; Thermodynamic Analysis.

1. Introduction

Gas hydrates are an ice-like crystalline structure shaped by water molecules through hydrogen bonding with suitable gas particles grasping inside polyhedral cavities. The key constituents, such as water particles (hosts), gas molecule (visitors), and thermodynamically appropriate environments including high pressure/low temperature are essential prerequisites for the formation of gas hydrate [1, 2]. The studies on the gas hydrates are ongoing in two fields; gas hydrate inhibitors for decreasing hydrate formation and gas hydrate promoters to enhance hydrate formation. Hydrates removals are industrially vital because the existence of gas hydrates in off-shore pipelines presents extraordinary financial losses with increased ecological and safety hazards. On the other hand, hydrates also have industrial significance in separation and storage processes of gases [3, 4]. Methane hydrates are also seen as a prospective fuel source in a future, for the reason that vast deposits of methane hydrate are found globally in continental margins and permafrost regions [5]. Other progressive applications of Gas hydrate incorporate; storage and transportation of gases in hydrate form (solid), separation of various gases via hydrates and carbon dioxide sequestration [6, 7]. At last, the utilization of their separation vitality can be applied in refrigeration procedures and cool stockpiling.

In petroleum industry, one of topmost obstruction confronted in flow assurance engineering is the formation of gas hydrate in pipelines, which are the root cause of the impasse of the oil and gas production, transportation and processing [7–9]. Gas hydrate formation may possibly be a reason to appalling ecological and economic risks. Hydrate formations problem also arise during the offshore drilling operations when drilling fluids prone to hydrate formation could result in severe dangers towards the operational safety. Hydrates problems are regularly encountered in deep waters like Gulf of Mexico, North Sea and Caspian Sea alongside permafrost region like Alaska. Hydrate problem alone accounted the oil and gas industry billions of dollars annually to control hydrate formation with no perpetual solution in focus. Initially, Hammer Schmidt (1934) has found the hydrate blockage problem in the gas pipeline, and since then extensive investigations are being conducted to discover efficient and better mitigation solutions especially chemical inhibitors [2, 10].

Incrassating attention has been found of late in the gas hydrate of CO_2 and CH_4 due to the exploration of higher CO_2 content gas fields. One of the most significant prone for CO2 rich gas is the possibility of hydrate formation during production, transportation and processing stages [8]. Subsequently, CO_2 is more susceptible to form hydrate than CH_4, therefore increase quantities of CO_2 up to 70 mole % is additionally inclined to form gas hydrate mostly during the separation and transportation stages [11].

There are four conventional methods employed in oil and gas industry for preventing gas hydrate, which are (i) Dehydration (ii) Depressurization (iii) Heating and (iv) Chemical inhibitions. Dehydration refers as a preventive method for avoiding gas hydrate plug formation. Dehydration from offshore sources may not be liable for all the operations owing to constraints of physical footprints in the offshore production facilities. Depressurization also considered as a remedial method for hydrate plug removal; depressurization should be done in the cautious manner since sudden pressure drop caused a rapid increase in hydrate plug velocity can lead to catastrophic damages to plate form or engineering machinery. Similarly, thermal Heating involves insulation of entire pipeline which caused an enormous amount of economic losses since pipelines nowadays last for hundreds of miles, and capital cost of 1 million US$/mile contain it to the remedial method, once gas hydrate formed inside the pipeline. Therefore, the most efficient method for removal and prevention of gas hydrate is the insertion of chemicals [4], [12]. Chemical inhibition involves the addition of certain chemicals which make a hydrogen bond with the water molecule and delay /avoid gas hydrate plug formation. For that reason, under most of the circumstances, the application of gas hydrate inhibitors is the solitary practicable choice [6], [13]. From the knowledge of the chemical composition requirements and the thermodynamics of hydrates formation, the industry has tried to prevent the hydrate formation by using conventional hydrate prevention methods which are either ineffective or required an enormous quantity of chemicals resulting in extraordinary operational cost together with adverse environmental influence of operating gas and oil facilities. Thus, in numerous cases, the only viable solution for hydrate inhibition is the addition of chemical inhibitors in the pipelines [14]–[18]. An illustration of a gas hydrate plug is shown in Figure 1 [19].

Thermodynamic and kinetic inhibitors usually applied in the off-shore flow assurance strategies to avoid hydrate formation in the pipeline [20]. The purpose of these inhibitors to make sure the operational conditions in the transmission pipeline are set well beyond the hydrate formation region. Owing to this motive, accurate hydrate equilibrium data measurement and thermodynamic properties of the gas hydrates are essential [11] and measured in this initial study.

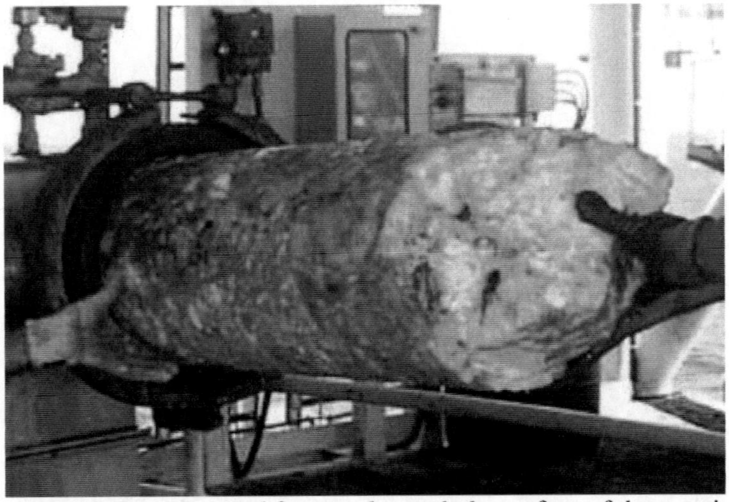

Figure 1. Gas hydrate plug recovered from underneath the surface of the sea pipeline [19].

2. Methodology

In this experiments, CH_4 (≥99.995%), CO_2 (≥99.995%) gases were bought from Gas Walker Sdn Bhd. Malaysia. All gases and de-ionized water are used without any further purification in all experiments as listed in Table 1.

Table 1. Material used for gas hydrate mitigation study

No	Chemical Name	Symbol	Purity	Chemical Structure
1	Methane	CH_4	99.95 mole %	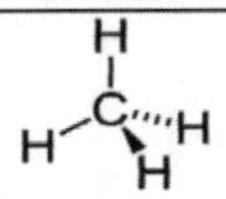
2	Carbon Dioxide	CO_2	99.99 mole %	O=C=O
3	Water	H_2O	Deionized	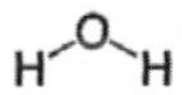

2.1 Equipment

In this work, a specifically designed custom-made sapphire cell hydrate reactor was used. A PVT cell apparatus can perform formation/dissociation studies for the gas hydrates. Fig. 1 shows the complete schematic diagram of the experimental setup and Fig 2 provide the pictorial view of the sapphire cell. The sapphire apparatus consists of equilibrium cell with a volumetric capacity of 35 cm^3, with operatable temperature and pressure range from (253– 523) K 1-30 MPa respectively. Experimental conditions like pressure and temperature are recorded for every subsequent second with an accuracy of ±0.01 MPa and ±0.1K respectively. Furthermore, the equipment is fitted with a magnetic rod stirrer to provide adequate mixing of the sample in test conditions.

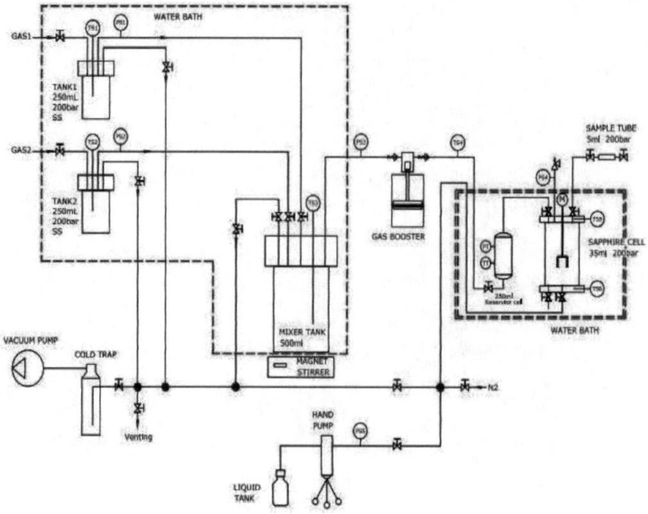

Figure 1. Schematic diagram of Sapphire Autoclave Cell

Figure 2. Picture of the experimental setup used

A well renowned T-cycle technique by isochoric step heating method was employed in this study [21]. Before experiments, the sapphire cell was thoroughly rinsed with water then vacuumed it to confirm there is no traces of the air remains in the cell. Now the cell was charged initially with 19 ml of deionized water into the cell, followed by gas sample. When temperature-pressure conditions stabilized, the mechanical stirrer was set at 300 rpm. Then the temperature of the cell was gradually condensed step-wise for a rate of 0.1 K/min up to hydrate formation regions. The span to each step usually differs from 2 to 6 hours. Hydrate formation was confirmed via both visibly from the sapphire frame along with sudden pressure drops seen in the data. As soon as the hydrate is entirely formed with no more pressure drip observed, then the cell was heated gradually for complete dissociation of gas hydrate. Figure 3 represents the temperature-pressure profile and plotted by Bavoh et al. [22]. The gas hydrate equilibrium point was reserved as the initial point, where cooling line encounters with the heating line.

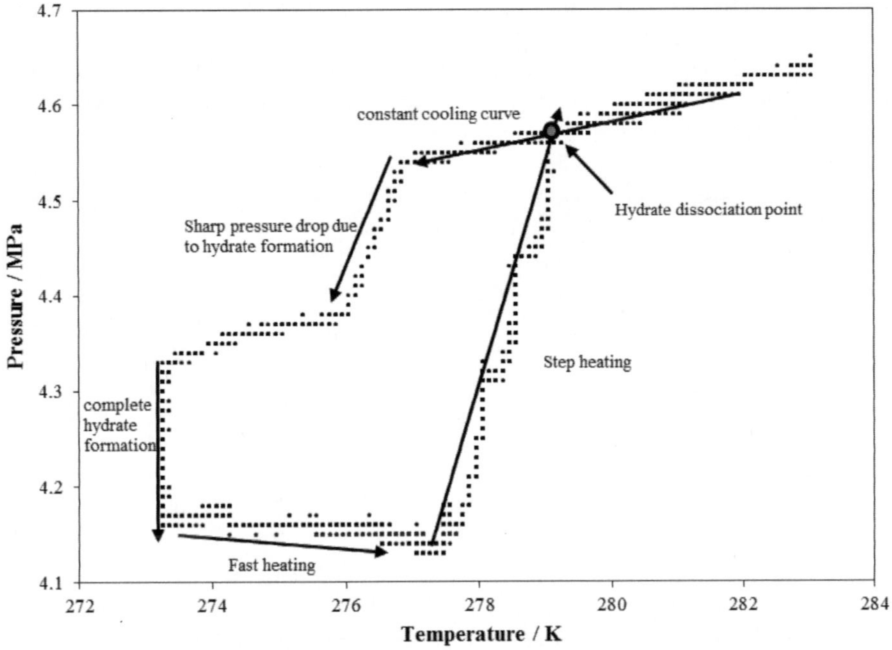

Figure 3. Temperature-pressure profile measurement for hydrate dissociation point [22]

2.2. Calculation

Pitzer correlation of gases [23] equation was employed for the determination of compressibility factor Z shown in equation 1.

$Z=Z_\circ+\omega Z1$ (1)

$Z_\circ=1+ B^\circ(Pr/Tr)$ (2)

$Z1=B1\ (Pr/Tr)$ (3)

$B^\circ=0.083-0.422/[Tr]1.6$ (4)

$B1=0.139-0.172/[Tr]4.2$ (5)

Where; Z is the compressibility factor, Z○ and Z1 are functions of reduced Pressure (Pr) and reduced temperature (Tr). Z○ is a contribution for of simple fluid to compressibility factor, and Z1 is the generalized term for molecules with behaviour different from the simple fluid; ω is known as acentric factor and the values for CH_4 and CO_2 are 0.0115 and 0.228 respectively [24]. B○ and B1 are function of reduced temperatures. Then, some moles of methane were calculated by the general gas equation:

$PV=nZRT$ (6)

$n=PV/ZRT$ (7)

Where; n is the number moles of methane, P is the pressure, T is the temperature, Z is the calculated compressibility factor, V is the volume of the vessel (gas phase) = 20 cm^3 and R is the universal gas constant = 83.147.

3. Results and Discussion

3.1. H-Lw-V Equilibrium Curve

This study is for preliminary experimental validation of newly design apparatus (sapphire cell reactor) for gas hydrate thermodynamic and kinetic studies. Equilibrium data collected and compared with the existing literature data for CH_4 hydrate. Figure 4 shows the calibration data which revealed that the values remain within the range of hydrate.

For instance, it was stated earlier that Figure 4 refers to the numerous chronological work done on CH_4 hydrates. Besides this data also approaching in line with the preceding efforts, therefore, it can be said that the apparatus is ready for further experiments.

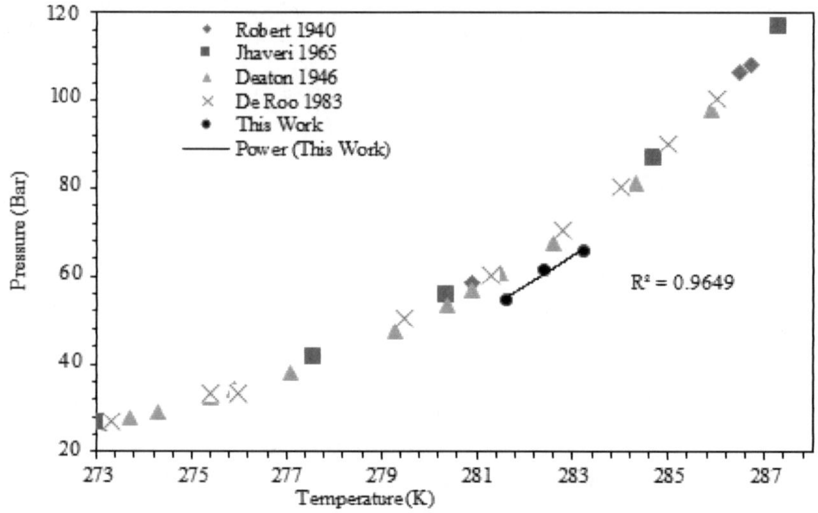

Figure 4. Schematic diagram of Sapphire Autoclave Cell

3.2 Thermodynamic Analysis of CH_4 and CO_2

In thermodynamic testing of methane hydrate equilibrium data was achieved and plot the pressure against temperature relation. HLwVE data shown in Figure 5 reflect that cooling line initially meeting with the heating line which also considered as HLwV equilibrium point at point E (285.1 K and 70.8 bar pressure) for CH_4 hydrate. Figure 6 further highlighted the different stages exists during the various parts of the HLwVE curve for CH_4 hydrate. Initially, at point A; only gas vapours and liquid water phases existed while cooling the system continues, the system then enters into B state where hydrate nucleation is started and considered as a metastable zone where conditions is favourable to hydrate formation. Sudden pressure drop observed (from 281 K at 41.4 Bar) reflect the starting of hydrate nucleation along with complete hydrate formation at point C. Point C and D possess three phases vapour, Hydrate, and liquid water which make ensure complete hydrate formation. Subsequently, step heating was initiated to dissociate hydrate by slowly increasing in pressure shown the dissociation curve until point E where no further increase in pressure was observed. Point E also possessed vapour phase, liquid and nucleation states. Finally, point F achieved after further heating for ensuring complete dissociation of hydrate. The temperature difference among point E to D considered as a degree of subcooling was around 11K due to complete hydrate formation around 274K.

Hydrate Equilibrium data of CO_2 hydrate was achieved and plotted as pressure and temperature profile. HLwVE data for CO_2 presented in Figure 6, reflects the meeting of cooling line with the heating line at 282.7 K and 41.1 bar pressure respectively. Furthermore, the degree of sub-cooling also deduced from this graph which almost is around 9.75 K due to complete hydrate formation around 272.95 K.

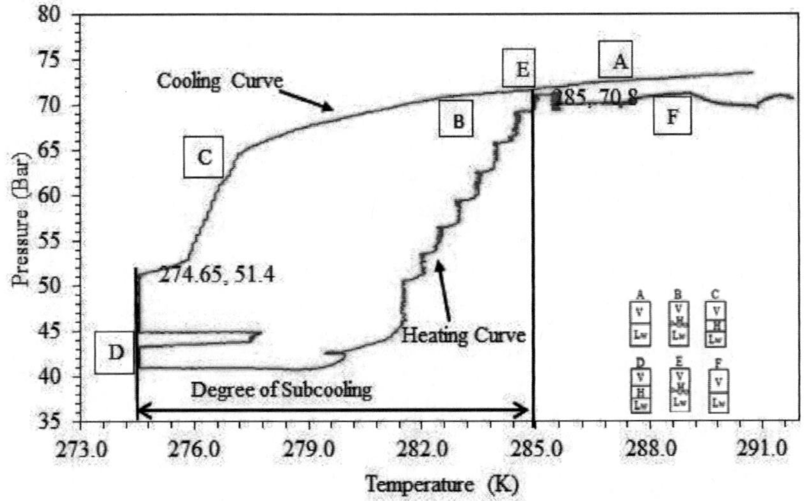

Figure 5. HLwVE Curve for Pure Methane Hydrate

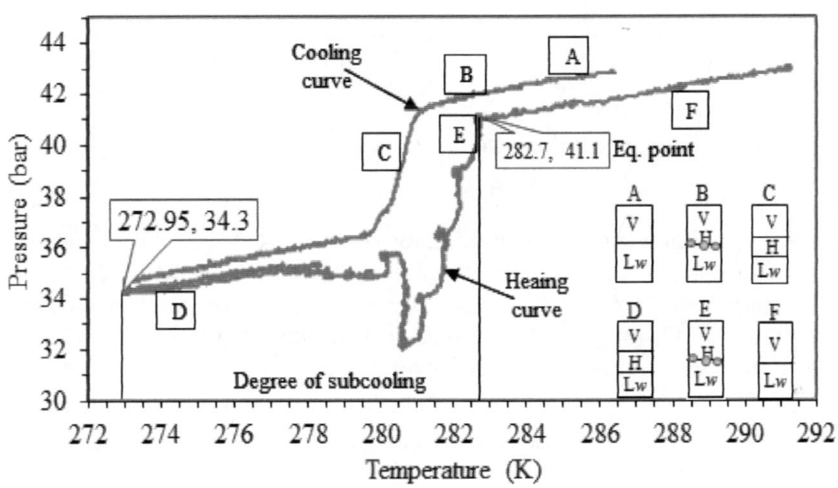

Figure 6. HLwVE curve for CO_2 Hydrate

From figure 6 further detailed about the different existing states during the various parts of the HLwVE curve for CO_2 hydrates. Initially, at A position only gas vapours and liquid water exists while cooling the system continues, the system enters into B state where hydrate nucleation is started and considered as a metastable zone where conditions are favourable to hydrate formation. Sudden pressure drop observed (from 281 K at 41.4 MPa) reflect the starting of hydrate nucleation along with complete hydrate formation at point C. Point C and D possess three phases vapour, Hydrate, and liquid water which make certain complete hydrate formation. After that again step heating is performed to dissociate hydrate slowly, increase in pressure shown the dissociation curve until point E where no further increase in pressure was observed. Point E also has vapour, liquid and nucleation. As a final point, point F achieved after further heating for ensuring the complete dissociation of hydrate. The temperature differences among point E to D considered as a degree of subcooling; as subcooling temperature higher there is more chance of hydrate formation. In some cases; CO_2 not able to attain initial pressure due to the dissolution of CO_2 in the aqueous phase as evident above (Figure 6).

3.3. Determination of Compressibility vs Temperature for CH_4 and CO_2 Hydrate

A Compressibility factor describes the deviation of gas from the ideal state. Previously Qazi et al., [11] calculated the compressibility factors and enthalpies of CH_4 and CO_2 gas hydrate. Figure 7 describes the relationship of compressibility factor of CH_4 on temperature. From Figure 7, it is indicated that the value of Z increased when hydrate formation started while Z values depressed in the dissociation of hydrates. The values of Z for CH_4 alleged in between 0.85 to 0.91 which included in CH_4 boundary region.

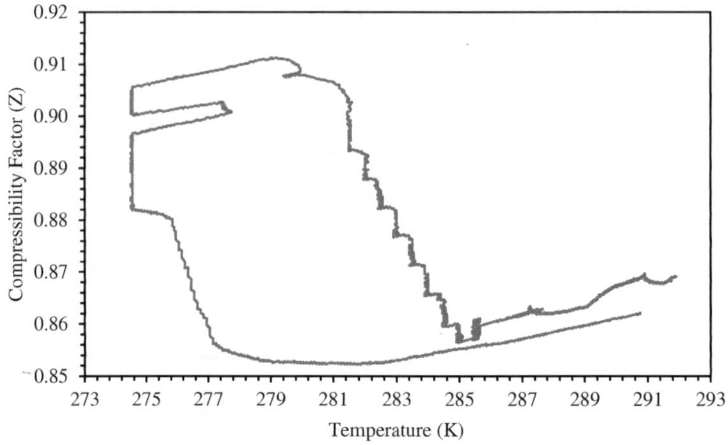

Figure 7. Compressibility Factor (Z) vs Temperature for CH_4 hydrate

Figure 8 discussed the relationship between compressibility factors of CO_2 gas on temperature. As mentioned earlier, figure 8 also provide the agreement that compressibility factor Z increased when hydrate formation started while Z values were depleted in the dissociation of CO_2 hydrate as well. However the Z values for CO_2 ranges between 0.73-0.79 which were lower than CH_4 compressibility values due to lesser pressure requirement for CO_2 hydrates.

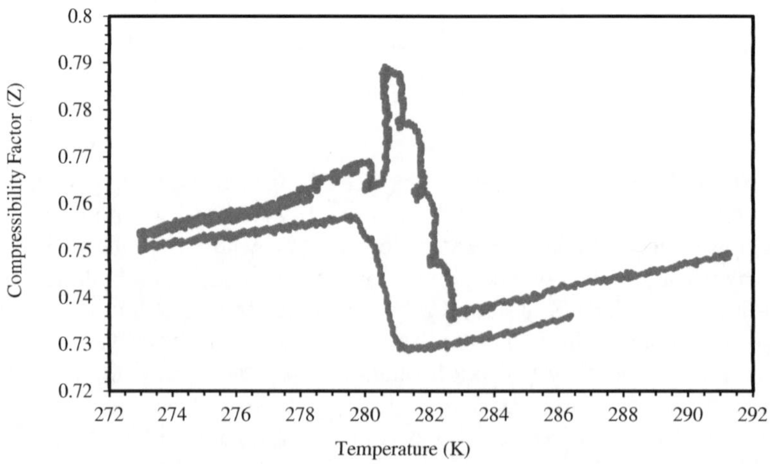

Figure 8. Compressibility Factor (Z) vs Temperature for CO_2 hydrate

4. Conclusion

In this preliminary work, H-Lw-V equilibrium data of CH_4 and CO_2 hydrates were measured along with the thermodynamic characterization. Obtained result were compared to existing literature data for CH_4 hydrate and found acceptable agreement for H-Lw-VE data. From thermodynamic results, it can clearly observe that CO_2 hydrates formed relatively low pressure compared to CH_4 hydrate at comparatively similar temperature conditions. Initial results revealed that the apparatus could route for extensive gas hydrate studies.

Acknowledgments

This study is supported by PETRONAS Research Sdn Bhd (PRSB) under TD Project (MRA) Grant No. 053C1-024. The authors would like to thanks, Chemical Engineering Department, Universiti Teknologi PETRONAS for providing facilities throughout the studies.

References

[1] L. Xingcai, H. Zhen, Z. Wugao, and L. Degang, "the Influence of Ethanol Additives on the Performance and Combustion Characteristics of Diesel Engines," Combust. Sci. Technol., vol. 176, no. 8, pp. 1309–1329, Aug. 2004.

[2] M. R. Riazi, Characterization and Properties of Petroleum Fractions, First. West Conshohocken, USA: ASTM International, 2005.

[3] G. Nagarajan, A. N. Rao, and S. Renganarayanan, "Emission and performance characteristics of neat ethanol fuelled DI diesel engine," Int. J. Ambient Energy, vol. 23, no. 3, pp. 149–158, 2002.

[4] M. L. Sattler, "Technologies for Reducing NO x Emissions from Non-Road Diesel Vehicles : An Overview," Environmental Protection, 2002.

[5] M. S. Khan, I. Ahmed, A. Mutalib, and A. Bustum, "Characterization Of Diesel-Hydrogen Peroxide Fuel Blend," J. Energy Technol. Policy, vol. 3, no. 12, pp. 273–279, 2013.

[6] C. S. Cheung, R. Zhu, and Z. Huang, "Investigation on the gaseous and particulate emissions of a compression ignition engine fueled with diesel – dimethyl carbonate blends," Sci. Total Environ., vol. 409, no. 3, pp. 523–529, 2011.

[7] B. B. Sahoo, N. Sahoo, and U. K. Saha, "Effect of engine parameters and type of gaseous fuel on the performance of dual-fuel gas diesel engines — A critical review," Renew. Sustain. Energy Rev. 13, vol. 13, pp. 1151–1184, 2009.

[8] A. Henham and M.. Makkar, "Combustion of simulated biogas in a dual-fuel diesel engine," Energy Convers. Manag., vol. 39, no. 16–18, pp. 2001–2009, Nov. 1998.

[9] Muhammad Saad Khan, Z. Bin Hassan, and A. Iqbal, "Influence Of Hydrogen Peroxide on Liquefied Petroleum Gas (LPG) Performance," in National Conference on Postgraduate Research (NCON-PGR) 2009, 2009, no. October, pp. 107–113.

[10] M. G. Shirk, T. P. Mcguire, G. L. Neal, and D. C. Haworth, "Investigation of a hydrogen-assisted combustion system for a light-duty diesel vehicle," Int. J. Hydrogen Energy, vol. 33, no. 23, pp. 7237–7244, 2008.

[11] R. G. Papagiannakis, P. N. Kotsiopoulos, T. C. Zannis, E. A. Yfantis, D. T. Hountalas, and C. D. Rakopoulos, "Theoretical study of the effects of engine parameters on performance and emissions of a pilot ignited natural gas diesel engine," Energy, vol. 35, no. 2, pp. 1129–1138, 2010.

[12] J. Wang, F. Wu, J. Xiao, and S. Shuai, "Oxygenated blend design and its effects on reducing diesel particulate emissions," Fuel, vol. 88, no. 10, pp. 2037–2045, 2009.

[13] H. Chen, J. Wang, S. Shuai, and W. Chen, "Study of oxygenated biomass fuel blends on a diesel

engine," Fuel, vol. 87, no. 15–16, pp. 3462–3468, 2008.

[14] Y. Ren, Z. Huang, H. Miao, Y. Di, D. Jiang, K. Zeng, B. Liu, and X. Wang, "Combustion and emissions of a DI diesel engine fuelled with diesel-oxygenate blends," Fuel, vol. 87, pp. 2691–2697, 2008.

[15] M. S. Khan, A. Iqbal, I. bin A. Mutalib, and Azmi Bostum, "Role of Oxygenated Additives for Diesel Fuel Blend (A Short Review)," J. Appl. Sci., vol. 15, pp. 1–7, 2015.

[16] M. P. Ashok and C. G. Saravanan, "Performance and Emission of the Emulsified Fuel in a DI Diesel Engine Using Oxygenated Additive Diethyl Ether with Surfactant of Span−80," energy & fuels, vol. 21, no. 4, pp. 1878–1882, 2007.

[17] Z. A. M. Yusof, I. Ahmed, M. S. Khan, S. A. Hussain, A. Hussain, I. bin A. Mutalib, K. S. Balkhair, and M. H. Albeirutty, "Thermal Evaluation of Diesel/Hydrogen Peroxide Fuel Blend," Chem. Eng. Technol., vol. 38, no. 12, pp. 2170–2180, 2015.

[18] D. Li, H. Zhen, L. Xingcai, Z. Wu-gao, and Y. Jian-guang, "Physico-chemical properties of ethanol–diesel blend fuel and its effect on performance and emissions of diesel engines," Renew. Energy, vol. 30, no. 6, pp. 967–976, May 2005.

[19] V. Í. Pirozfar, A. Z. Moghadam, M. R. Omidkhah, and A. Ameri, "Effect of additive-ethanol-diesel blend fuel on Physico- chemical properties and emission," pp. 1–6, 2001.

[20] G. Labeckas and S. Slavinskas, "Comparative performance of direct injection diesel engine operating on ethanol, petrol and rapeseed oil blends," Energy Convers. Manag., vol. 50, no. 3, pp. 792–801, 2009.

[21] C. D. Ã. Rakopoulos, K. A. Antonopoulos, and D. C. Rakopoulos, "Experimental heat release analysis and emissions of a HSDI diesel engine fueled with ethanol – diesel fuel blends," Energy, vol. 32, pp. 1791–1808, 2007.

[22] M. A. Ak, A. Ulas, B. Sümer, B. Yazıcı, C. Yıldırım, L. O. Gönc, and F. E. Orhan, "An experimental study on the hypergolic ignition of hydrogen peroxide and ethanolamine," Fuel, pp. 8–11, 2010.

[23] P. Kwanchareon, A. Luengnaruemitchai, and S. Jai-In, "Solubility of a diesel–biodiesel–ethanol blend, its fuel properties, and its emission characteristics from diesel engine," Fuel, vol. 86, no. 7–8, pp. 1053–1061, May 2007.

[24] M. Ribeiro, A. C. Pinto, C. M. Quintella, G. O. Rocha, L. S. G. Teixeira, L. N. Guarieiro, C. Rangel, C. C. Veloso, M. J. C. Rezende, R. Serpa, A. M. De Oliveira, E. A. Torres, and J. B. De Andrade, "The Role of Additives for Diesel and Diesel Blended (Ethanol or Biodiesel) Fuels : A Review Nu," no. 1, pp. 2433–2445, 2007.

IOP Conf. Series: Materials Science and Engineering **344** (2018) 012026 doi:10.1088/1757-899X/344/1/012026

Physicochemical and FTIR Study of Diesel-Hydrogen Peroxide Fuel Blend

Muhammad Saad Khan[*1], Iqbal Ahmed[2,3], Bhajan Lal[*1], Al-Amin Idris[1], Muhammad H. Albeirutty[2], Muhammad Ayoub[1], Suriati binti Sufian[1]

[1]Chemical Engineering Department, Universiti Teknologi PETRONAS, Bandar Seri Iskandar, 32610 Tronoh, Perak, Malaysia.

[2]Center of Excellence in Desalination Technology, King Abdulaziz University, P.O. Box 80200, Jeddah 21589, Saudi Arabia.

[3]Faculty of Engineering - Department of Mechanical Engineering, King Abdulaziz University, P.O. Box 80200, Jeddah 21589, Saudi Arabia.

*Corresponding author. Tel.: +0-060-17-9486804; E-mail address: khansaaad@gmail.com; bhajan.lal@utp.edu.my

Abstract. Physicochemical properties of combustion fuels play a key role in determining the qualitative and quantitative characteristics, reliability and health effects associated with emissions. This paper reports the preparation of polysaccharide (PS) based emulsifier for stable blending of petroleum diesel-hydrogen peroxide (H_2O_2) and investigated the influence of H_2O_2 as diesel fuel blends on the physicochemical properties and characteristics. The quantity of PS-emulsifier was kept at 5 volume % (vol. %) and the volume ratio of H_2O_2 were varied 5-15 vol. % to reference diesel (RD), respectively. The blended diesel/H_2O_2 fuel were prepared under inert oxygen (O_2) gas closed heating system; afterthought, physiochemical properties of diesel/H_2O_2 blend were evaluated at standard ASTM D-975 testing method. The kinetic properties show the interaction of RD and H_2O_2 blend at presence of PS emulsifier which exhibit the phenomenon to diminish the interfacial tension among the two different phases to form a homogenized stable solution. Results revealed that H_2O_2 is capable of enhancing the diesel fuel properties and showed that the addition of H_2O_2 in a diesel fuel blend are lied within the ranges of standard ASTM D-975. Due to further oxygen atom present in H_2O_2, it can facilitate the combustion process which ultimately effect on exhaust emission.

Keywords: ASTM test, diesel, FTIR, H_2O_2, physicochemical properties and PS emulsifier

1. Introduction

After the foremost breakthrough invention in 19th century by Rudolf Diesel made an invention in the field of internal-combustion engines (ICE). Ever since this invention, petro- diesels have been a vital fuel in energy generation and power of engineering equipment all around the world [1]. Diesel derived engines were turned out to be extremely productive in terms of supplying the required energy levels for their utilization use at very low working (operational) and maintenance costs when contrasted to other power generation machines. In reality the flow has dramatically switching from a gasoline driven world economy towards the diesel based world economy. Generally, petroleum derived diesel includes around 75% of saturated hydrocarbons (primarily paraffin's including n, iso, and cycloparaffins), along with 25% aromatic constituents (including alkylbenzenes and naphthalenes). The approximate range of hydrocarbon in diesel fuel are from $C_{10}H_{20}$ to $C_{15}H_{28}$ with average chemical formula of common diesel fuel is $C_{12}H_{23}$, [2].

However, crude diesel fuel produces major hazard of a variety of types of air pollutant emission with adverse environmental impact especially on the urban air quality in the form of harmful oxides of nitrogen (NOx), carbon monoxide (CO), Particulate matter (PM), and other carcinogenic compounds [3]. Even though ULSD (blended diesel) diesel engine exhaust produces crowd of destructive materials including airborne carbon soot, PM, polycyclic aromatic hydrocarbons, toxic metals, carbon dioxide (CO_2), CO, NOx which persuade ozone formation along with volatile organic compounds includes acroleine and formaldehyde [4]. Concerns over the availability, price and carbon emissions from petroleum sources are among major concerns which provide motivation for the research on alternative sources, especially for automobiles fuels [5].

Various types of fuels had investigated in recent years for improving the quality and performance of diesel fuel. Principally for the decrease of released pollutants; researchers have engrossed their importance on the fuel modification techniques. For instance, the use of alternative fuels; frequently in fumigated form or gaseous fuels of renewable nature that are considered as environmentally friendly in nature [6]–[11] or oxygenated compounds as fuels, that display the capability to diminish hazardous pollutants, specially particulate emissions [12]–[17]. Lot of attention has given on alternative fuel with superior physiochemical properties for protecting environmental and enhancing the fuel efficiency aspect, especially alcoholic fuels. In alcohols, principal attention has given to ethanol fuel because of its admirable properties [18]–[24]. Methanol (M100) had looked into as an alternative diesel fuel for used in heavy-duty vehicles [25]–[28]. Blends of methanol and ethanol used in gasoline, with a focus on blends contained 85% alcohol (M85 and E85) [29] were evaluated as alternative light-duty vehicle fuels. To attain substantial reductions in emissions, it is believed that, reformulation of diesel fuel properties has play a very significant role. The reformulation of diesel fuels embrace; lowering of the aromatic and sulfur contents, or addition of the potential oxygen carriers within the fuel. It has been revealed that, numerous oxygenates are quite effective in dropping particulate emissions from diesel engines sources. For that reason, considerable research has attentive on selection of oxygenative fuel additives including alcohols, ethers and esters.

Hydrogen peroxide (H_2O_2) is light blue liquid, somewhat more viscous than water, which seems colorless in dilute aqueous solution. It's weak acid which possess strong oxidizing and bleaching properties. Recently few of the researches explore the oxidative nature of H_2O_2 as oxidizer which could provide additional oxygen during fuel combustion process in the internal combustion (IC) engines. H_2O_2 also possesses higher cetane value (125) as reported by Ashok and Saravanan [16], [30]. This agreement has also proven by the recent study from Yusuf et al.,[17] in their thermal evaluation of diesel-H_2O_2 fuel blend.

Chemical Equation

H_2O_2 spontaneously decomposes (disproportionate) exothermically into oxygen and water:

$$H_2O_2 \rightarrow \tfrac{1}{2} O_2 + H_2O \text{ (steam)} + \text{heat} \quad (1)$$

This is thermodynamically favorable process has a ΔG_o of −119.2 kJ/mol−1 with ΔH_o of −98.2 kJ/mol−1 along with ΔS value of 70.5 J/mol−1K−1. The rate of H_2O_2 decomposition are depends upon certain things which includes temperature, concentration of the H_2O_2, pH and along with presence of stabilizers and impurities.

H_2O_2 decomposes in presence of catalyst at ordinary temperature produce water in the form of steam and gives free half mole of oxygen with enormous amount of heat that could enhance the efficiency of the diesel engine. This excess oxygen will also decrease the emission of carbon monoxide (CO) because it converts CO into carbon dioxide (CO_2). For Diesel, Equation becomes:

$$2C_{12}H_{26} + 37O_2 \rightarrow 24CO_2 + 26H_2O \text{ (steam)} + \text{heat} \quad (2)$$

This reaction require air fuel ratio of 20:1. It means for every mole of diesel requires 20 moles of air. Appropriate air fuel ratio and composition of H_2O_2, would increase the lean operation limit, improve the lean burn ability and it will decrease the exhaust emission of compression ignition engine. Thus, based on above hypothesis in this project, introduction of H_2O_2 as blend for diesel were carried out experimentally. Various compositions of H_2O_2 were tested in order to enhance the properties of diesel, starting from lean crude diesel until obtaining optimum condition. Polysaccride based emulsifier reduces the surface tension between the diesel and H_2O_2 and stabilizes the blend for the longer period.

2. Experimental Design & Set-Up

2.1. Material

Reference diesel fuel (commercial diesel) was obtained from PETRONAS Sdn Bhd and Table 1 shows the physical properties of commercial diesel fuel that was used in this study. Commercial grade H_2O_2 (30%) was obtained from Chemical Company of Malaysia Berhad (CCM). Commercial grade acetone (98.8%) was purchased from university registered supplier. Polysaccharide (PS) bio-polymer (contain hydrophilic and lipophilic properties) were purchased from Merck. Laboratory characterization equipment have utilized to complete characterization for the physical and chemical properties of each material (except PS) at room temperature (25±1oC); and the comparisons were made with their corresponding literature values. The comparisons results are summarized in Table 1. Prior to the measurements, all liquids were partially degassed under vacuum before experiments. Then, the densities of each material and blend fuels all the mass measurements were performed by using an electrical balance (Dhona 200 D, India) accurate to 0.0001mg. All apparent masses were corrected for buoyancy. The densities of each liquid (RD, H2O2, and acetone) were measured with a bi-capillary pycnometer which gave an accuracy of 5 parts in 105.

2.2. Preparation of emulsifier and diesel/H_2O_2 fuel blend using closed heating technique

The emulsifier was prepared beforehand mixing of reference diesel (RD) and H_2O_2 diesel fuel blend which briefly described in our earlier work [17].

2.3. Properties of Diesel Fuel Blends

Laboratory experiments were performed to more completely describe the physical and chemical properties of stable diesel/H_2O_2 blend emulsion in presence of PS emulsifier. ASTM D-975testing method was adopted using diesel test standard to determine the following properties; density (ASTM D-1298), viscosity and kinematic viscosity (ASTM D-445) and FTIR characterization. Table 1; describe the literature and experimental values of components of the fuel in accordance with ASTM D-975 diesel fuel testing method.

3. Results and Discussion

3.1. Influence of H_2O_2 on density of the fuel blend

The density of diesel fuel varies between 0.81 and 0.89 g/cm^3. In this work, the densities of prepared blend fuel were laid between these values. Since, density was much easier to model over wide temperature ranges for most liquids including fuels. PS emulsifier possess the lowermost density of 0.6123g/cm^3 amid the different component, while H_2O_2 density is 1.130 g/cm^3 which having a higher value than pure diesel fuel (see Table 1). The density of the RD and each diesel/H_2O_2 fuel blends at different temperatures shown in Figure 1. It can be observed that as the vol. % of H_2O_2 in the diesel fuel blends was increases, the density also increases. However, the density difference of each blend was about 0.89-1.3% higher than RD. Yet, the difference of each fuel blend density was lying within the ASTM D-1298 standard and the difference of blend diesel was not too high as what we expect due to higher water concentration in H_2O_2, which was almost 1.3 times higher than RD. Consequently, the density of the higher concentration of H_2O_2 in diesel blend increases a bit higher particularly at 13 vol.%- 15 vol.% of H_2O_2. However, for the concentration of 5 vol. % to 8 vol. %. Results from Figure 1 shown that the density decreased less as compared to higher concentration of H_2O_2 in diesel.

The obtained results suggested that the presence of PS based emulsifier addition had the important impact on diesel blend density and stability due to its hydrophilic and hydrophobic balance which is significantly participate to diminish the interfacial tension among polar and no-polar to formulate homogenized solution. Nevertheless higher density of diesel/H_2O_2 fuel blend proven as a good agreement of higher cetane number several researchers were also reported same tendency [5], [31]. Figure 1 were also established that the density of the fuel blends dependent on temperature variations. With temperature increase, the densities of the fuel blends decreasing. Similar results were seen for conventional diesel and neat biodiesel [32].

3.2 Properties of Diesel Fuel Blends

Usually, the viscosity of diesel fuel usually varies 2-5 cP. In this experimental work, the viscosities of blend fuel have laid well between these values. The value of viscosity of R.D is 4 cP. PS Emulsifiers possess the maximum kinematic viscosity value of 17 cP among the individual components [5]. H_2O_2 having a viscosity value of 1.2 cP, which was minimum value of all the constituents of the blend fuel. The experimental values of viscosities of R.D and diesel fuel blends at various temperatures presented in Figure 2.

Table 1: Comparison of physical properties between experimental and literature of each materials used in this study

Physical Properties	PETRONAS Diesel (R.D)	H_2O_2 (30% conc.)	Acetone	PETRONAS Diesel	H_2O_2 (30% conc.)	Acetone
	Experimental Values			Literature Values		
Chemical Formula	$C_{12}H_{23}$	H_2O_2	C_3H_6O	–	–	–
Density (kg/m^3)	827.98 ± 0.2	1128.87 ± 0.2	790.95 ± 0.2	830	1130	791
API gravity	37.257	–	–	0.83	1.113	0.791
Viscosity (centipoises) at 27°C	3.78 ± 0.2	1.223 ± 0.2	0.331 ± 0.2	3.9	1.245	0.322
pH	4.000	3.20	7.2	–	3.22	5 - 6
Boiling Point (°C)	287 ± 0.5	148 ± 0.5	56.5 ± 0.5	180-330	150.2	56-57
Flash Point (°C)	88	–	−17.5 ± 0.5	52-96	–	-17
Fire point (°C)	97.5	–	-13.015	68-116	-	-12
Cetane Number	49 ± 0.5	120 ± 0.5	7 ± 0.5	50	>125	–
Refractive Index at 22°C	1.476	1.348	-	1.4798	1.355	1.358
Surface tension (cm/dyne) at 22°C	23.810	73.897	26.062	23.721	74.218	25.23
Self-ignition Temperature (°C)	–	–	–	200-420	160	465
Stoichiometric air/fuel ratio (wt. /wt.)	–	–	–	14.6	11.1	–
Lower heating value (MJ/kg)	44	33.9	29.98 ± 0.5	–	–	30.86
Specific heat capacity (J/kg°C)	1851.22 ± 0.5	2335 ± 0.5	2089 ± 0.5	1850	2619	2175

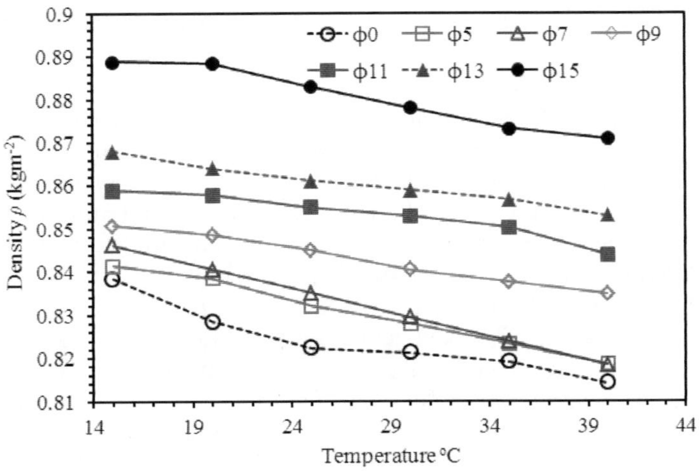

Figure 1: Density (g/cm^3) of reference diesel and diesel/H_2O_2 vol. %, blend fuel at various temperatures

The effects of temperature on diesel/H_2O_2 density were presented over the temperature range of 15oC to 60oC. The viscosity of each fuel blends are reduces with increased the amount of H_2O_2 in the blend due to higher contents of water of H_2O_2 along with lesser viscosity. However, all the viscosities of the fuel blends showed very close values to each other, similar trend also observed at various temperature levels. Hypothetically, the viscosity of the blends fuel approach to be decreased, in comparison with the diesel fuels as the H_2O_2 concentration increases in the fuel blend [5], [33], [34].

3.3. Influence of H_2O_2 on density of the fuel blend

Kinematic viscosities are a very important role for determination of kinetic properties of the blend like excess energy of activation and excess viscosities of fuel blend [17]. Usually, the kinematic viscosity of diesel fuel usually varies 2-6 cSt (mm^2/s). In this work, the kinematic viscosities of blend fuel have laid well between these values. The value of kinematic viscosity of RD is 5.5 cSt at ambient temperature. PS emulsifier possess the maximum kinematic viscosity value of 17 mm^2/s among the individual components while, H_2O_2 having a kinematic viscosity value of 1.10 mm^2/s, which was minimum value of all the constituents of the blend. The experimental values of kinematic viscosities of RD and fuel blends have presented in Figure 3. Results revealed that, as the quantity of H_2O_2 in the blends upsurges, oxygen contents improved due to the peroxiding nature of H_2O_2, although viscosity, density and high heat value was decreased slightly. Generally, it already known that higher density and lower viscosity leads to higher flow, thus, these results suggested that the lower viscosity of diesel/H_2O_2 blend fuel could succeed to lower fuel injection with an early ignition time. Moreover, the drop of viscosity of each stable blend as compared to RD and lower concentration of H_2O_2 in fuel blends observed due to the higher molar ratio of peroxiding group. It also suggested that the 70% water content of H_2O_2 were formed water droplet inside the diesel and these droplet were mixed well due to PS emulsifier, thus the viscosity of blend fuels were lower than RD. In addition, all the diesel-H_2O_2 blends possess densities and viscosities data which are quite acceptable for the standard limit for diesel fuels. These results indicate the same trend as those of prior efforts [5], [15], [35]. Additionally, the effects of temperature on diesel/H_2O_2 density were displayed in Figure 2, over the temperature range of 15 °C to 60 °C. The kinematic viscosity of each fuel blends are decreases with increased the amount of H_2O_2 in the blend due to higher contents of water of H_2O_2 value 1.110 (Pa.s). However, the kinematic viscosities of the blends showed very close values to each other. Similar trend were also observed at various temperature level. Hypothetically, the viscosity of the blends approach to be decreased those of the diesel fuels as the H_2O_2 concentration increases in the fuel blend.

3.4. Influence of H_2O_2 on density of the fuel blend

The significance of absorption phenomenon on the radiation scattering of the fuel droplets in diesel devices is among an interesting fuel research area. This phenomenon indirectly affects fuel evaporation and heating capabilities, which directly affect the auto-ignition of fuel vapors. Fourier Transform Infrared (FTIR) Spectroscopic analyses were conducted to determine the presence of H_2O_2 in all fuel blends. Following Figure reflects the trend of RD compare with various percentages of H_2O_2 in diesel fuel blend such 0 vol.% H_2O_2, 5 vol.% H_2O_2, 10 vol.% H_2O_2, and 15 vol.% H_2O_2, respectively and combined results presented in Figure 4. From Figure 4 it was observed that the infrared rays reflected greater peaks level when increases the amount of H_2O_2 due to strong electro-negativity of H_2O_2 [4]. The experimental data of density and viscosity were showing a good agreement to explain the greater peaks levels of diesel blend fuels. In addition, H_2O_2 showed a definite trend in peak like in 1st region around 3850-3700 wavelength, where addition of H_2O_2 enhances the peak concentration due to higher OH molecules presents in the H_2O_2. In 2nd region that ranges between 2900- 2750 which has shown a increments in peaks as 15 % H_2O_2 have a very clear indication of peak shown the functional group of Saturated aliphatic (alkanes/alkyl) group frequencies which are the essential constituents of diesel fuel. 3rd region has belongs to a functional group of frequencies for thiols or thio-substituted molecules which have a range between 2650- 2500; mainly due to an occurrence of inorganic mixtures in the shape of amino and ammonium based compounds or salts of carboxylic acids in the form of Polysaccride [33], [34]. These peaks verify the presence of the emulsifier. However, FTIR result also proved that the H_2O_2 present in the diesel fuel blend causes the enhancement of the OH and oxygen contents for the diesel fuel blend; other regions also strongly indicate presence of H_2O_2 significantly enhances the peaks of hydrocarbons.

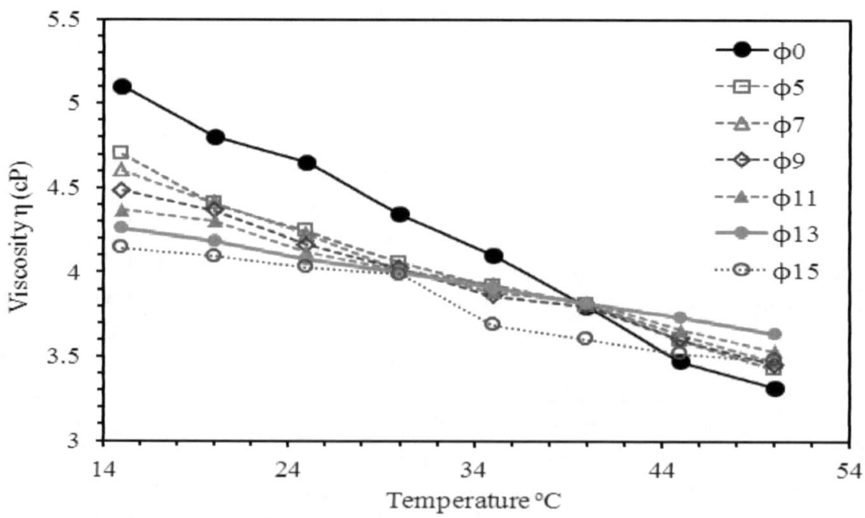

Figure 2: Viscosity (cP) of H_2O_2 vol. vol. %, in diesel blend at various temperatures

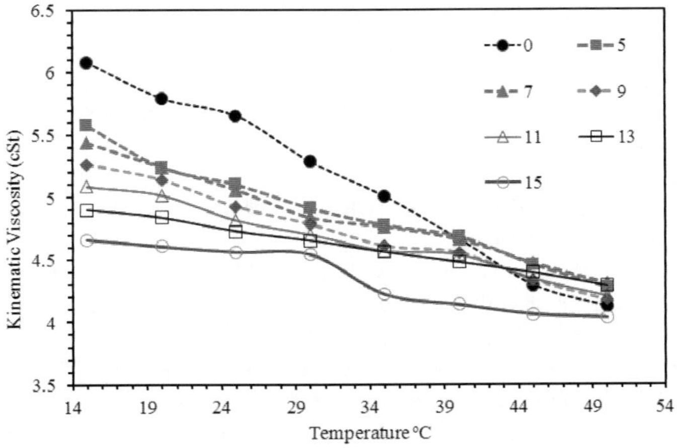

Figure 3: Kinematic viscosity (cSt) of various vol. % of H_2O_2 in diesel blend at different temperatures

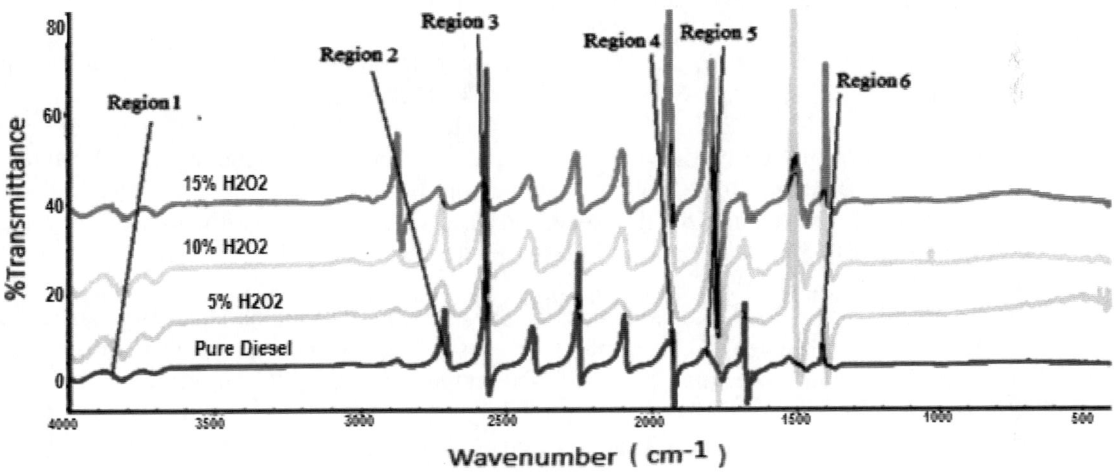

Figure 4: Comparative FTIR Spectrums of diesel and blend fuel

4. Conclusion

In this experimental study, influence of the addition of H_2O_2 with diesel fuel has been carried out using various compositions of diesel fuel blend. The results of physicochemical characterizations found in accordance with ASTM D-975 which showed that addition of H_2O_2 in a blended form lied within the ranges of standard diesel fuel (compared RD fuel). Fourier Transform and Infra-Red (FTIR) results were also illustrated the spectroscopic presence of H_2O_2, which proved the enhancement of the hydrogen and oxygen contents in diesel fuel blend.

Acknowledgments
The authors would like to acknowledge and appreciate the Universiti Teknologi PETRONAS for providing laboratory and financial facilities throughout the studies.

References

[1] L. Xingcai, H. Zhen, Z. Wugao, and L. Degang, "the Influence of Ethanol Additives on the Performance and Combustion Characteristics of Diesel Engines," Combust. Sci. Technol., vol. 176, no. 8, pp. 1309–1329, Aug. 2004.

[2] M. R. Riazi, Characterization and Properties of Petroleum Fractions, First. West Conshohocken, USA: ASTM International, 2005.

[3] G. Nagarajan, A. N. Rao, and S. Renganarayanan, "Emission and performance characteristics of neat ethanol fuelled DI diesel engine," Int. J. Ambient Energy, vol. 23, no. 3, pp. 149–158, 2002.

[4] M. L. Sattler, "Technologies for Reducing NO x Emissions from Non-Road Diesel Vehicles : An Overview," Environmental Protection, 2002.

[5] M. S. Khan, I. Ahmed, A. Mutalib, and A. Bustum, "Characterization Of Diesel-Hydrogen Peroxide Fuel Blend," J. Energy Technol. Policy, vol. 3, no. 12, pp. 273–279, 2013.

[6] C. S. Cheung, R. Zhu, and Z. Huang, "Investigation on the gaseous and particulate emissions of a compression ignition engine fueled with diesel – dimethyl carbonate blends," Sci. Total Environ., vol. 409, no. 3, pp. 523–529, 2011.

[7] B. B. Sahoo, N. Sahoo, and U. K. Saha, "Effect of engine parameters and type of gaseous fuel on the performance of dual-fuel gas diesel engines — A critical review," Renew. Sustain. Energy Rev. 13, vol. 13, pp. 1151–1184, 2009.

[8] A. Henham and M.. Makkar, "Combustion of simulated biogas in a dual-fuel diesel engine," Energy Convers. Manag., vol. 39, no. 16–18, pp. 2001–2009, Nov. 1998.

[9] Muhammad Saad Khan, Z. Bin Hassan, and A. Iqbal, "Influence Of Hydrogen Peroxide on Liquefied Petroleum Gas (LPG) Performance," in National Conference on Postgraduate Research (NCON-PGR) 2009, 2009, no. October, pp. 107–113.

[10] M. G. Shirk, T. P. Mcguire, G. L. Neal, and D. C. Haworth, "Investigation of a hydrogen-assisted combustion system for a light-duty diesel vehicle," Int. J. Hydrogen Energy, vol. 33, no. 23, pp. 7237–7244, 2008.

[11] R. G. Papagiannakis, P. N. Kotsiopoulos, T. C. Zannis, E. A. Yfantis, D. T. Hountalas, and C. D. Rakopoulos, "Theoretical study of the effects of engine parameters on performance and emissions of a pilot ignited natural gas diesel engine," Energy, vol. 35, no. 2, pp. 1129–1138, 2010.

[12] J. Wang, F. Wu, J. Xiao, and S. Shuai, "Oxygenated blend design and its effects on reducing diesel particulate emissions," Fuel, vol. 88, no. 10, pp. 2037–2045, 2009.

[13] H. Chen, J. Wang, S. Shuai, and W. Chen, "Study of oxygenated biomass fuel blends on a diesel engine," Fuel, vol. 87, no. 15–16, pp. 3462–3468, 2008.

[14] Y. Ren, Z. Huang, H. Miao, Y. Di, D. Jiang, K. Zeng, B. Liu, and X. Wang, "Combustion and emissions of a DI diesel engine fuelled with diesel-oxygenate blends," Fuel, vol. 87, pp. 2691–2697, 2008.

[15] M. S. Khan, A. Iqbal, I. bin A. Mutalib, and Azmi Bostum, "Role of Oxygenated Additives for Diesel Fuel Blend (A Short Review)," J. Appl. Sci., vol. 15, pp. 1–7, 2015.

[16] M. P. Ashok and C. G. Saravanan, "Performance and Emission of the Emulsified Fuel in a DI Diesel Engine Using Oxygenated Additive Diethyl Ether with Surfactant of Span−80," energy & fuels, vol. 21, no. 4, pp. 1878–1882, 2007.

[17] Z. A. M. Yusof, I. Ahmed, M. S. Khan, S. A. Hussain, A. Hussain, I. bin A. Mutalib, K. S. Balkhair, and M. H. Albeirutty, "Thermal Evaluation of Diesel/Hydrogen Peroxide Fuel Blend," Chem. Eng. Technol., vol. 38, no. 12, pp. 2170–2180, 2015.

[18] D. Li, H. Zhen, L. Xingcai, Z. Wu-gao, and Y. Jian-guang, "Physico-chemical properties of ethanol–diesel blend fuel and its effect on performance and emissions of diesel engines," Renew. Energy, vol. 30, no. 6, pp. 967–976, May 2005.

[19] V. Í. Pirozfar, A. Z. Moghadam, M. R. Omidkhah, and A. Ameri, "Effect of additive-ethanol-diesel blend fuel on Physico- chemical properties and emission," pp. 1–6, 2001.
[20] G. Labeckas and S. Slavinskas, "Comparative performance of direct injection diesel engine operating on ethanol, petrol and rapeseed oil blends," Energy Convers. Manag., vol. 50, no. 3, pp. 792–801, 2009.
[21] C. D. Ã. Rakopoulos, K. A. Antonopoulos, and D. C. Rakopoulos, "Experimental heat release analysis and emissions of a HSDI diesel engine fueled with ethanol – diesel fuel blends," Energy, vol. 32, pp. 1791–1808, 2007.
[22] M. A. Ak, A. Ulas, B. Sümer, B. Yazıcı, C. Yıldırım, L. O. Gönc, and F. E. Orhan, "An experimental study on the hypergolic ignition of hydrogen peroxide and ethanolamine," Fuel, pp. 8–11, 2010.
[23] P. Kwanchareon, A. Luengnaruemitchai, and S. Jai-In, "Solubility of a diesel–biodiesel–ethanol blend, its fuel properties, and its emission characteristics from diesel engine," Fuel, vol. 86, no. 7–8, pp. 1053–1061, May 2007.
[24] M. Ribeiro, A. C. Pinto, C. M. Quintella, G. O. Rocha, L. S. G. Teixeira, L. N. Guarieiro, C. Rangel, C. C. Veloso, M. J. C. Rezende, R. Serpa, A. M. De Oliveira, E. A. Torres, and J. B. De Andrade, "The Role of Additives for Diesel and Diesel Blended (Ethanol or Biodiesel) Fuels : A Review Nu," no. 1, pp. 2433–2445, 2007.
[25] K. Anand, R. P. Sharma, and P. S. Mehta, "Experimental investigations on combustion, performance and emissions characteristics of neat karanji biodiesel and its methanol blend in a diesel engine," Biomass and Bioenergy, vol. 35, no. 1, pp. 533–541, 2011.
[26] H. Bayraktar, "An experimental study on the performance parameters of an experimental CI engine fueled with diesel – methanol – dodecanol blends," Fuel, vol. 87, pp. 158–164, 2008.
[27] S. Cenk, N. O. Ahmet, and C. Mustafa, "The influence of operating parameters on the performance and emissions of a DI diesel engine usubg methanol-blended-diesel fuel," Fuel, vol. 89, pp. 1407–1414, 2010.
[28] C. Sayin, M. Ilhan, M. Canakci, and M. Gumus, "Effect of injection timing on the exhaust emissions of a diesel engine using diesel – methanol blends," Renew. Energy, vol. 34, no. 5, pp. 1261–1269, 2009.
[29] L. H. Ajba, Z. E. Ller, E. N. Agy, and J. H. Ancsok, "Properties of Diesel-Alcohol Blends," HUNGARIAN J. Ind. Chem. VESZPRÉM, vol. 39, no. 3, pp. 349–352, 2011.
[30] M. P. Ashok, "Role of Emulsified Fuel in the Present IC Engines – Need of Alodine EC Ethanol Corrosion Resistant Coating for Fuel Injection System," in Fuel Injection in Automotive Engineering, 1st ed., K. Lejda and P. Woś, Eds. Croatia: InTech Janeza Trdine 9, 51000 Rijeka, Croatia, 2012, p. 154.
[31] M. P. Ashok and C. G. Saravanan, "Role of Hydrogen Peroxide in a Selected Emulsified Fuel Ratio and Comparing It to Diesel Fuel," Energy & Fuels, vol. 22, no. 3, pp. 2099–2103, 2008.
[32] S. H. Yoon, S. H. Park, and C. S. Lee, "Experimental Investigation on the Fuel Properties of Biodiesel and Its Blends at Various Temperatures," Energy & Fuels, vol. 22, no. 1, pp. 652–656, Nov. 2007.
[33] I. Ahmed, A. Idris, A. Hussain, Z. A. M. Yusof, and M. Saad Khan, "Influence of Co-Solvent Concentration on the Properties of Dope Solution and Performance of Polyethersulfone Membranes," Chem. Eng. Technol., vol. 36, no. 10, pp. 1683–1690, 2013.
[34] I. Ahmed, A. Idris, M. S. Khan, S. Chowdhury, and J. Akhtar, "Effect of Acetone on Physical Properties of PES Membrane," Appl. Mech. Mater., vol. 625, pp. 545–548, 2014.
[35] M. S. Khan, I. Ahmed, M. I. bin A. Mutalib, S. Nadeem, and S. Ali, "Influence of H_2O_2 on LPG fuel performance evaluation," in 3rd International Conference on Fundamental and Applied Sciences (ICFAS 2014), 2014, no. JUNE, pp. 763–768.

IOP Conf. Series: Materials Science and Engineering **344** (2018) 012027 doi:10.1088/1757-899X/344/1/012027

Machinability Study on Milling Kenaf Fiber Reinforced Plastic Composite Materials using Design of Experiments

H. Azmi[1,2], C.H. C. Haron[1*], J. A. Ghani[1], M. Suhaily[3], A.R. Yuzairi[2]

[1]Department of Mechanical and Materials Engineering, Faculty of Engineering and Build environment, Universiti Kebangsaan Malaysia, Selangor, Malaysia.
[2]School of Manufacturing Engineering, Universiti Malaysia Perlis, Perlis, Malaysia.
[3]Department of Manufacturing and Materials Engineering, Faculty of Engineering, Universiti Islam Antarabangsa Malaysia, Selangol, Malaysia.

*chase@ukm.edu.my , chehase@gmail.com, chehase@yahoo.com

Abstract. The surface roughness (R_a) and delamination factor (F_d) of a milled kenaf reinforced plastic composite materials are depending on the milling parameters (spindle speed, feed rate and depth of cut). Therefore, a study was carried out to investigate the relationship between the milling parameters and their effects on a kenaf reinforced plastic composite materials. The composite panels were fabricated using vacuum assisted resin transfer moulding (VARTM) method. A full factorial design of experiments was use as an initial step to screen the significance of the parameters on the defects using Analysis of Variance (ANOVA). If the curvature of the collected data shows significant, Response Surface Methodology (RSM) is then applied for obtaining a quadratic modelling equation that has more reliable in expressing the optimization. Thus, the objective of this research is obtaining an optimum setting of milling parameters and modelling equations to minimize the surface roughness (R_a) and delamination factor (F_d) of milled kenaf reinforced plastic composite materials. The spindle speed and feed rate contributed the most in affecting the surface roughness and the delamination factor of the kenaf composite materials.

1. Introduction

In historical records, the use of natural fiber composite materials had already been practiced in Egypt about 3000 years ago from now. Nowadays, in automotive field, the manufacturers are advised to take the environmental impact into account by considering renewable materials as alternative materials in their production especially European Union End of Life Vehicles (ELV) regulations had been introduced [1]. The fuel efficiency of a vehicle can be greatly increased due to reduction of its weight by using lightweight natural fiber reinforced plastic in the making of bumper and the door panel [2]. The East German car 'Trabant' was the first car that employed natural fiber composites in its production. The car's body was made from a composite of cotton and polyester [3]. This had encouraged the rapid growth of implementation of natural fiber reinforced plastic into the manufacturing field.

Kenaf (*Hibiscus Cannabinus*) can found abundantly in Malaysia. Its potential usage should not be limited only in typical application such making rope or twine but should be concerned in manufacturing field by reinforcing with plastic. In this research, the epoxy was chosen as a matrix polymer because of the epoxy's less hazardous nature and its ease of handling. Natural fiber composite materials causes less damage to cutting tool and less hazard to health and environment due to its biodegradable properties. Besides, the cost of raw material of kenaf fiber is much lower than synthetic fiber since it is renewable source.

The milling was conduct by making slot on the kenaf composites. Slotting is one of the essential machining process which facilitates the assembly operations [4]. However, delamination and poor surface roughness can lead to a reduction in product quality if inappropriate milling parameters are used. Delamination can cause substantial reduction of load bearing capacity even the damages are not noticeable. Delamination is usually expressed by determining its factor. A rougher surface has a higher friction coefficient and wear off more quickly. The delamination factor (F_d) and surface roughness (R_a) are the two main criteria for predicting the mechanical performance and quality of a product. Hence, an optimum milling parameter and modelling equation for minimizing the milled kenaf composite are required.

Response surface methodology (RSM) is a set of statistical and mathematical modeling which helps in developing, enhancing and optimizing processes. The measured output is known as the response (surface roughness and delamination factor). The investigated parameters are known as the factors or variables (spindle speed, feed rate and depth of cut). In this research, RSM and analysis of variance (ANOVA) were used to investigate the relationships between measured responses and the input variables. The approach of Box-Behnken in RSM was selected to do the analysis on the response data. Before the RSM is applied to analyse the data, the full factorial design of experiments was use as an initial step to screen the significance of the parameters on the defects using Analysis of Variance (ANOVA). If the curvature of the collected data shows significant, Response Surface Methodology (RSM) is then applied for obtaining a quadratic modelling equation that has more reliable in expressing the optimization.

1.1. Kenaf Fiber Composite
The main constituents of kenaf are cellulose (45–57 wt. %), hemicelluloses (21.5 wt. %), lignin (8–13 wt. %) and pectin (3–5 wt. %) [3]. Wambua et al. claimed that poor mechanical properties may be resulted due to insufficient firm between the polymers (hydrophobic) and the natural fiber (hydrophilic) [5]. Kenaf fiber offers a better interfacial adhesion to matrix polymer than other natural fibers [6]. Matrix polymer can support the fibers, transfer the stresses to them to bear the most of the load, prevent direct physical damage to fibers and improve the ductility and toughness of the whole composite [7]. Epoxy gives no toxic gases and offers high temperature resistance which enables it performs at a high temperature [8]. Hafizah et al. studied tensile behavior of kenaf fiber reinforced several polymer composites and concluded that kenaf reinforced epoxy composite had the highest ultimate tensile strength and the Young's modulus [9]. There were various researches had been conducted and proved that machining parameters possess significant effects to performance and quality of natural fiber composite[4], [10].

1.2. Milling Kenaf Fiber Composite
Erkan et al. claimed that increased spindle speed worsen the damage factor and increase the plastic deformation rate on an end milled fiber reinforced plastic composite [11]. Davim et al. claimed that feed rate was expected to affect delamination rather than cutting speed on the machining of fiber reinforced plastic [12]. Ramulu et al. studied the machining of polymer composite. They stated that a better surface finish can be obtained with higher cutting speed [13]. The impact of depth of cut is not as important as cutting speed and feed rate in composite machining, however, it still gives a significant effect on machining process [14]. Besides, the milled surface is getting smoother as the flute of the cutting tool increases [11]. The end mill cutter can be made of either high speed steel or carbide insert and usually has a straight shank or tapered shank. Normally, the cutter rotates on a workpiece in a perpendicular axis but it still can be tilted in order to match the need of producing a machine tapered or curve surface. However, milled composites always tend to have the surface roughness and delamination problem which much depends on cutting parameter and the composite's characteristic [19]. Thus, an optimum setting of these parameters is needed to find out in order to control the quality

of a workpiece.The advantages of High Speed Steel (HSS) cutting tool over carbide cutting tool are higher strength to withstand cutting forces and lower cost. However, Carbide insert cutting tool is taking over HSS in metal cutting due to its great abrasion resistance and high temperature resistance which allows it to serve a machining process at a high speed without considering of over-heat situation. For machining of natural fiber composites, HSS cutting tool is sufficient to meet this research's purpose.

1.3. Surface Roughness and Delamination Factor

Surface roughness is one of the criteria in determining the quality of workpiece based on surface texture. If the deviation of actual surface is large from the ideal surface, the actual surface is considered as a rough surface. The smaller the deviation, the smoother the actual surface. R_a is the value of the Roughness Average of surface measured microscopic peaks and valleys. Sorrentino and Turchetta claimed that a reduction of surface roughness (R_a) can be obtained by increasing the feed rate [15]. The other researchers also expressed that feed rate contributes the highest statistical and physical influence on surface roughness in end milling [16]. Rao et al. investigated the effect of cutting parameters on the surface roughness of multi-walled carbon nanotubes (MWCNT) reinforced epoxy composite using CNC end milling and claimed that depth of cut affected the most on the surface roughness, followed by feed rate and spindle speed [17].

Delamination of a natural fiber composite can be described as the loss of adhesion between the layers that can leads critical damage of the reinforcement layers (separation).Hocheng et al. believed the delamination can be worsened by increasing the feed rate on milling parameter [18].The damage of the composite is greater if operated with high cutting speed and high feed rate [19]. Delamination of a natural fiber composite can be described as the loss of adhesion between the layers that can leads critical damage of the reinforcement layers (separation). Hocheng et al. believed the delamination can be worsened by increasing the feed rate and tool life can be shortened by increasing the cutting speed which causes higher torque [18]. Davim et al. studied the effects of cutting speed and feed rate on delamination of glass fiber reinforced plastic during machining it with two different matrices [26].The damage of the composite is greater if operated with high cutting speed and high feed rate[19].

1.4. Response Surface Methodology (RSM)

Response Surface Methodology (RSM) is used to solve the industrial problems such as mapping a response surface over a particular region of interest, optimization of the response and selection of operating conditions to achieve specifications or customer requirements [20]. The sequential nature of the Response Surface Methodology is as follows [20]:

Phase 0 - In this phase, amount of variables is reduced to minimum in order to reduce the number of experiments and make the experiments more efficient. This is called as screening experiment and identification of independent variables is carried out in this phase.

Phase 1 - In this phase, method of steepest ascent (descent) is used as optimization technique with considering the use of the first-order model.

Phase 2 - In this phase, the true response surface exhibits curvature near the optimum and a second-order or higher-order model should be implemented. The model can be analyzed later for determining the optimum conditions once an approximately accurate model has been made.

Palanikumar analyzed about the surface roughness in machining glass fiber reinforced plastic using response surface methodology. A second-order model is used in his research and he concluded that the surface roughness decreases with the increase of cutting speed and depth of cut but increases with the increase of feed rate [14]. Hussain et al. studied the optimization of surface roughness of glass fiber reinforced plastic composite in turning process had been evaluated using RSM. They suggested that RSM model can be interfaced with an effective genetic algorithm in determining optimum process parameters and this helps in enhancing the quality of the surface finish of workpiece [21]. Sorrentino and Turchetta analyzed the milling cutting force and surface roughness of carbon fiber reinforced plastics composite and used ANOVA to determine that the cutting force is depending on three parameters which is feed rate, radial and axial depth of cut [22].

2. Methodology

2.1 Materials Preparation

The kenaf reinforced epoxy composites materials were prepared using Vacuum Assisted Resin Transfer molding (VARTM). VARTM was used to transfer the resin to the natural fiber in vacuum condition for molding with assistance of vacuum pump. Type of kenaf fiber is mat form were cut in dimension of 300 mm x 300 mm x 5 mm (thickness). Two layer of kenaf fiber mat that used and mix with epoxy resin. The resin was prepared in ratio 4 (800g epoxy): 1 (200g hardener). VARTM manufacturer recommended this ratio for fabricating the kenaf fiber reinforce plastic composite materials. The VARTM is gaining the interest of manufacturer due to its ease of handling and low cost of preparation. Figure 2 shows the actual set up of VARTM. The composite panels were then cut into small workpiece which about in dimension 75 mm X 30 mm X (8 mm – 10 mm) for milling process.

(a) (b) (c)

Figure 1. VARTM process for kenaf fiber reinforce plastic composite workpiece. (a) VARTM process, (b) actual composite materials, (c) small workpiece for milling process.

2.2 Milling Parameter

The milling parameters of this research were based on Babu et al. research since they studied machining of natural fiber composite. The parameter setting in this research base on previous parameters used in research by Babu et al. [4]. Table 1 shows the parameters used in this research which all the units are converted to suit the use of computer numerical control (CNC) milling machine. There were 3 parameters used in this research spindle speed (rpm), feed rate (mm/min) and depth of cut (mm) with low and high level for each parameter. Thus, a 2-level factorial design of experiments was selected with 8 experiment run and aid of Design Expert 7.0 software. 4 center points were added to seek for the significance of the curvature and error testing. If the curvature of the results show not significant, a linear modeling equation is sufficient to express the optimization. If it shows significant, there is a chance for applying RSM. In RSM, the Box-Behnken Design approach was used to get the quadratic modeling equation for parameter optimization. RSM is used to develop a second order modeling equation which is better in accuracy of expressing the optimization. In Box-Behnken design approach, 12 experiment run was selected with 3 level of factor were low, middle and high. 5 center points were added in this design approach to get more accurate on quadratic modeling equation.

IOP Conf. Series: Materials Science and Engineering **344** (2018) 012027 doi:10.1088/1757-899X/344/1/012027

Table 1. Parameter setting for milling process.

Milling parameters	Low	High	Center point
Spindle speed (rpm)	500	1000	700
Feed rate (mm/min)	200	1200	750
Depth of cut (mm)	1	3	2

Table 2 shows the runs of the 2-level factorial design (2^3) of experiment for screening process base on parameter setting shown in Table 1. These parameters setting also used for Box-Behnken Design in RSM. If the results from 2-level factorial design shown, the curvature is significant, the Box-Behnken Design approach was chosen to estimate a quadratic equation that satisfies the optimization of the parameters (Table 3). For maximum efficiency, the axial points should be place over the original factor range. For example, the low level and high level of spindle speed are 500 rpm and 1000 rpm respectively, thus, the axial point should be extended to 326 rpm and 1174 rpm.

Table 2. The 2-level factorial design of experiment (2^3)

Run	A (Spindle speed, rpm)	B (Feed rate, mm/min)	C (Depth of cut, mm)
1	500	200	3
2	1000	1200	1
3	700	750	2
4	1000	200	1
5	700	750	2
6	500	1200	3
7	1000	1200	3
8	500	200	1
9	1000	200	3
10	700	750	2
11	700	750	2
12	500	1200	1

Table 3. The Box-Behnken Design for RSM

Run	A (Spindle speed, rpm)	B (Feed rate, mm/min)	C (Depth of cut, mm)
1	750	700	2
2	500	700	3
3	1000	700	3
4	750	1200	3
5	750	700	2
6	750	700	2
7	750	700	2
8	500	700	1

9	750	200	3
10	750	700	2
11	500	1200	2
12	1000	1200	2
13	750	1200	1
14	750	200	1
15	500	200	2
16	1000	700	1
17	1000	200	2

2.3 Measurement of Surface Roughness and Delamination

The results of surface roughness and factor of delamination were obtained after milling processes. The surface roughness of milled kenaf composites was measure using a surface roughness tester (Mitutoyo CS-3000 525-780E-1). For workpiece with dimension 75 mm X 30 mm X (8 mm – 10 mm), 4 area were measured on slotted kenaf composite (see figure 2) for surface roughness. Every area the measured area was divide into 4 points and average of them (R_a) were recorded in Table 4. The delamination factor (F_d) can be determined using (1).

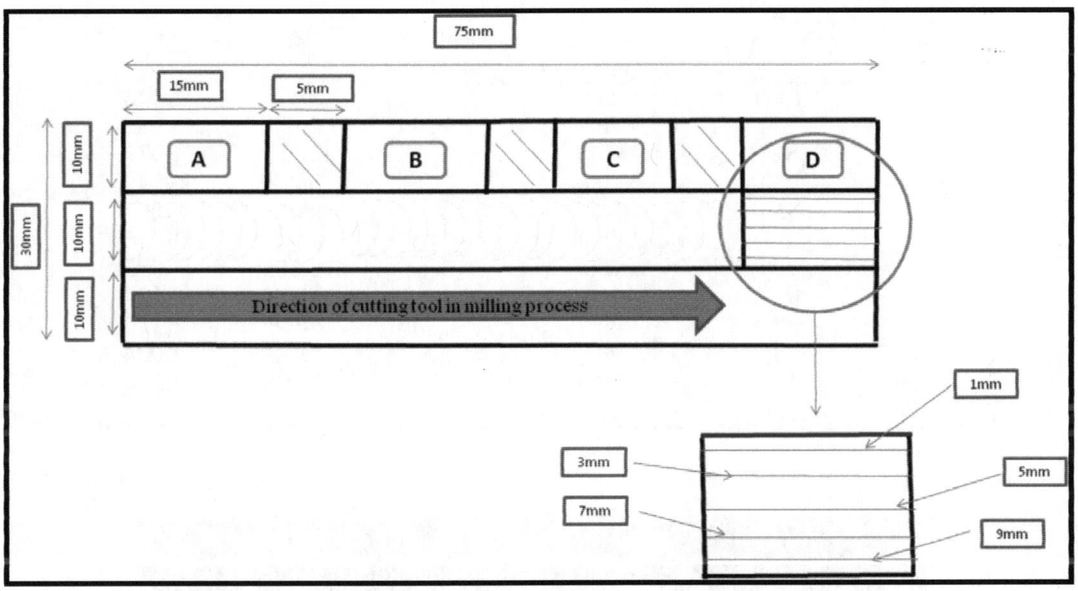

Figure 2. Schematic diagram for measurement of surface roughness (R_a).

$$F_d = W_{max} / W \quad (1)$$

Where,
F_d = Factor of delamination
W_{max}= maximum width of the delamination area
W = width of original cut

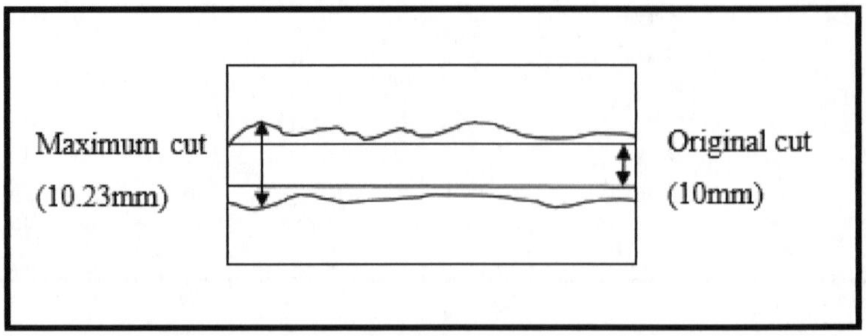

Figure 3. Example measurement of delamination factor (F_d)

3. Results and Discussion

Table 4. Result for surface roughness (R_a) and delamination factor (F_d) using screening parameter setting.

Run	A (Spindle speed, rpm)	B (Feed rate, mm/min)	C (Depth of cut, mm)	R_a	F_d
1	1000	1200	1	3.47	1.113
2	500	200	1	3.073	1.039
3	500	1200	1	6.725	1.077
4	750	700	2	3.28	1.042
5	750	700	2	2.952	1.067
6	1000	1200	3	4.743	1.016
7	750	700	2	4.02	1.062
8	1000	200	3	2.539	1.134
9	1000	200	1	2.248	1.124
10	750	700	2	4.133	1.034
11	500	1200	3	6.012	1.075
12	500	200	3	1.7	1.035

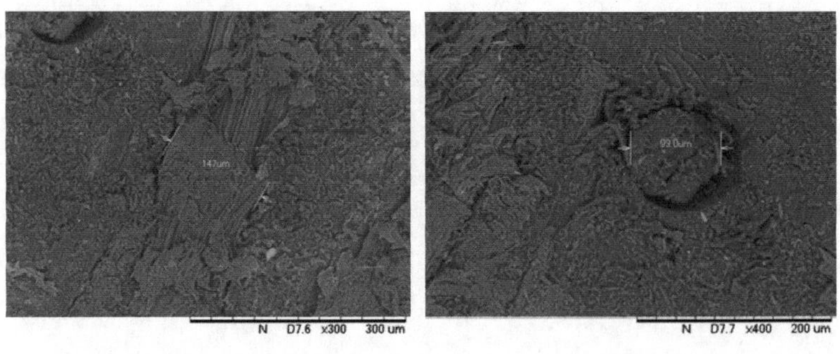

Figure 4. Example SEM for surface roughness

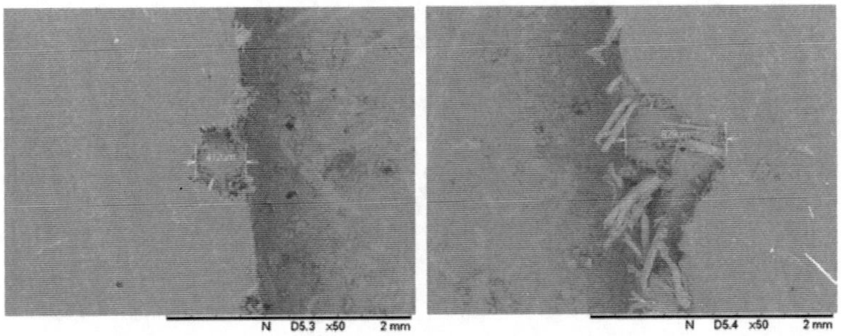

Figure 5. Example SEM for delamination

3.1 ANOVA analysis for screening process of surface roughness (R_a)

Table 5. ANOVA result of surface roughness (R_a)

Source	Sum of squares	Degree of freedom	Mean square	F-value	p-value Prob>F
Model	23.03	5	4.61	17.28	0.0036 significant
A-Spindle speed	2.54	1	2.54	9.54	0.0272
B-Feed rate	16.22	1	16.22	60.84	0.0006
C-Depth of cut	0.034	1	0.034	0.13	0.7353
AB	2.57	1	2.57	9.66	0.0266
AC	1.67	1	1.67	6.25	0.0545
Curvature	0.13	1	0.13	0.47	0.5221 Not significant
Residual	1.33	5	0.27	-	-
Lack of fit	0.35	2	0.17	0.53	0.6332 Not significant
Pure of error	0.98	3	0.33	-	-

According to Table 5, the feed rate was determined that made the largest contribution in affecting surface roughness (p-value 0.0006<0.05), followed by spindle speed (p-value 0.0272<0.05) whereas the depth of cut contributed the least (p-value 0.7353 > 0.05). The value of alpha α is 0.05. The depth of cut cannot be ignored since it was the main factor for interaction of AC (spindle speed and depth of cut). The interaction of AB and AC were determined as significant factors since the p-value is between 0.05 and 0.1. However, the curvature and 'lack of fit' show no significant, which means a linear equation is sufficient for optimizing the surface roughness using these factors. So, for surface roughness results, the analysis cannot be used in Box-Behnken Design analysis. For surface roughness, the mathematical modelling and the optimum parameter just used the ANOVA result in screening process.

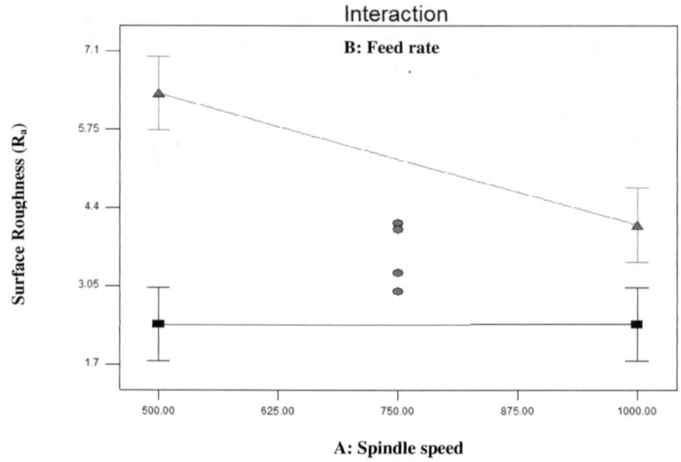

Figure 6. Effect of AB interaction for surface roughness.

Figure 6 shows the effect of AB interaction on surface roughness with low level (black line) and high level (red line) of feed rate (depth of cut remains 1.5 mm). The surface roughness decreases with a high spindle speed and low feed rate. To get a good surface roughness, the every milled segment of workpiece have to be milled repeatedly and evenly. With a high feed rate, the workpiece fail to do so and caused a rough surface. These findings were found similar to the research of Babu et al. which claimed that surface roughness decreases as the spindle speed increases whereas a low feed rate contributes the most in obtaining a good surface finish [4]. Although the depth of cut shows least significant, it still plays an essential role in affecting surface roughness but not in critical way. Palanikumar believed that high depth of cut leads to a total removal of fiber which results in having a good surface finish [14].

3.1.1 Optimum Parameters and modeling equation for Optimization of Surface Roughness

Design Expert software was used to predict the optimum milling parameters and to estimate the coefficient for the modelling equation. The suggested optimum parameters were 500 rpm spindle speed, 200 mm/min feed rate and 2 mm depth of cut. The prediction of surface roughness was 1.865 μm with the range between 0.13μm and 3.60 μm. This means the optimum parameters are valid if the response (R_a) is located at the range.The linear modelling equation for optimization of milling parameters on minimizing surface roughness is shown in (3).

$$R_a = 5.42561 + 4.46324 \times 10^{-3} A + 6.13008 \times 10^{-3} B - 2.86442 C - 4.36606 \times 10^{-6} A + 3.57843 \times 10^{-3} AC \quad (2)$$

Where A is spindle speed (rpm), B feed rate (mm/min), and C depth of cut (mm) in actual value.

Table 6. Optimum parameter and conformation test for surface roughness (R_a)

Spindle speed (rpm)	Feed rate (mm/min)	Depth of cut (mm)	Calculated R_a	Measured R_a	Percentage error (%)
500	200	2	2.267	2.287	0.88

IOP Conf. Series: Materials Science and Engineering **344** (2018) 012027 doi:10.1088/1757-899X/344/1/012027

Table 6 shows the percentage error between calculated (R_a) using the equation (3) and the measured R_a. The measured (R_a) was located at the prediction range (0.13µm-3.60 µm) and the percentage error was only 0.88%. These proved the both suggested optimum milling parameters and modelling equations were valid.

3.2 ANOVA analysis for screening process of delamination factor (F_d)

Table 7. ANOVA result of delamination factor (F_d)

Source	Sum of squares	Degree of freedom	Mean square	F-value	p-value Prob>F
Model	0.012	4	2.957E-003	20.75	0.0012 significant
A-spindle speed	2.319E-003	1	2.319E-003	16.27	0.0068
B-feed rate	2.622E-004	1	2.622E-004	1.84	0.2237
C-depth of cut	1.157E-003	1	1.157E-003	8.12	0.0292
AB	8.090E-003	1	8.090E-003	56.78	0.0003
Curvature	3.456E-003	1	3.456E-003	24.26	0.0026 significant
Residual	8.549E-004	6	1.425E-004	-	-
Lack of Fit	4.073E-004	3	1.358E-004	0.91	0.5300 Not significant
Pure Error	4.476E-004	3	1.492E-004	-	-

The curvature shows significant (see Table 7) and this indicates there is a chance for developing quadratic modeling equation using RSM with Box-Behnken approach. From the ANOVA analysis for delamination factor (F_d), the spindle speed was the most significant factor (p-value 0.0068) affecting the delamination. The depth of the cut was the second significant factor (p-value 0.0292) whereas the feed rate contributed the least (0.2237). In order to apply RSM, the Box-Behnken Design (see Table X) was added to further investigation of the parameters on the delamination by constructing a three order modeling equation.

3.3 Results of delamination factor (F_d) using Box-behnken Design

Table 8. Results of delamination factor (F_d) using Box-Behnken Design

Run	A (Spindle speed, rpm)	B (Feed rate, mm/min)	C (Depth of cut, mm)	F_d
1	750	700	2	1.045
2	500	700	3	1.054
3	1000	700	3	1.052
4	750	1200	3	1.042
5	750	700	2	1.047
6	750	700	2	1.038
7	750	700	2	1.05
8	500	700	1	1.042

9	750	200	3	1.048
10	750	700	2	1.043
11	500	1200	2	1.062
12	1000	1200	2	1.045
13	750	1200	1	1.064
14	750	200	1	1.032
15	500	200	2	1.041
16	1000	700	1	1.048
17	1000	200	2	1.055

3.3.1 ANOVA analysis using Box-Behnken Design of delamination factor (F_d)

Table 9. ANOVA analysis for delamination factor (F_d) using Box-Behnken Design

Source	Sum of squares	Degree of freedom	Mean square	F-value	p-value Prob>F
Model	7.850×10^{-4}	5	1.570×10^{-4}	5.97	0.0066 significant
A-Spindle speed	1.25×10^{-7}	1	1.250×10^{-7}	0.00475	0.9463
B-Feed rate	1.711×10^{-4}	1	1.711×10^{-4}	6.51	0.0269
C-Depth of cut	1.25×10^{-5}	1	1.250×10^{-5}	0.48	0.5048
AB	2.40×10^{-4}	1	2.403×10^{-4}	9.14	0.0116
BC	3.610×10^{-4}	1	3.610×10^{-4}	13.73	0.0035
Residual	2.892×10^{-4}	11	2.629×10^{-5}	-	-
Lack of fit	2.080×10^{-5}	7	2.972×10^{-5}	1.46	0.3743 Not significant
Pure of error	8.120×10^{-3}	4	2.030×10^{-5}	-	-

After obtaining the experimental result from Box-Behnken Design of experiments, a ANOVA had been constructed again to seek the significance of the three order factors and the significance of 'Lack of Fit' of the quadratic modeling equation as shown in Table 9. From Table 9, the main factor that significant to delamination factor (F_d) is Feed rate with F-value 5.97 (0.6%), interaction AB with F-value 9.14 (1.16%) and interaction BC with F-value 13.73 (0.3%). BC (feed rate and depth of cut) interaction made the largest contribution in affecting kenaf composite to laminate. Followed by AB (spindle speed and feed rate) and B (feed rate). The main factors (A and C) show non-significance values. However, these 2 factors cannot be eliminated due to the existence of AB and BC which include the factor A and C. The graph of delamination factor versus BC and AB interaction had been plotted in figure 5 and figure 6.

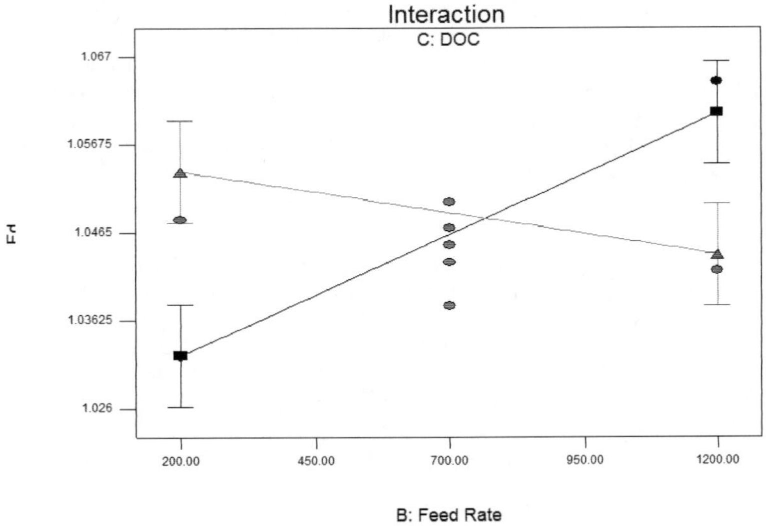

Figure 7. Effect of BC interaction for delamination factor

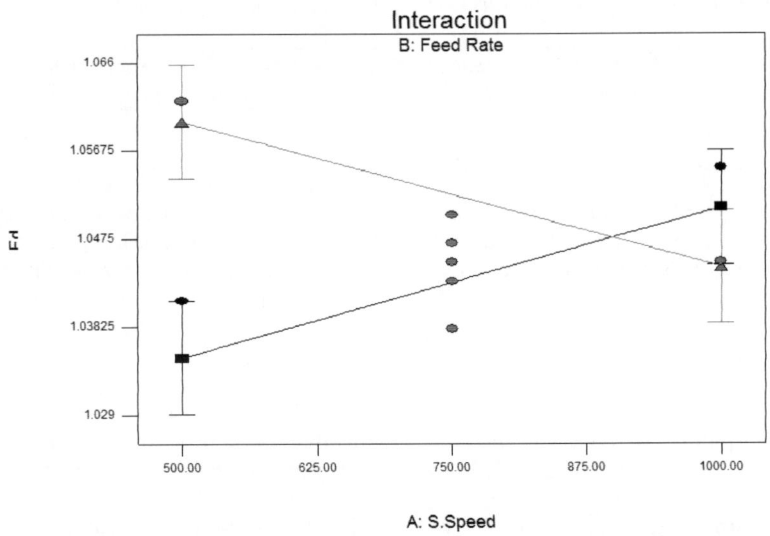

Figure 8. Effect of AB interaction for delamination factor

In figure 7, the spindle speed maintains at 750 rpm whereas the red line indicates the high level of depth of cut (3 mm) and the black line indicates the low level of depth of cut (1 mm). It can be clearly observed that the higher the feed rate (1200 min/mm), the worse the delamination. For BC interaction, the worse delamination factor (F_d) will course with high level of feed rate and low level of depth of cut In figure 8, the interaction of AB, the depth of cut remain at 2 mm. The red line indicates the high level of feed rate (1200 mm/min) and the black line indicates the low level of feed rate (200 mm/min). From the AB graph, the worse delamination factor (F_d) will accurse with high spindle speed and low feed rate. For both BC and AB graph, the factor B (feed rate) is most contributed to the delamination factor (F_d). Davim et al. had the same viewpoint that high cutting (spindle) speed caused great damage on glass fiber reinforced plastic composites during machining [19]. The non-linear relationship between the feed rate and the delamination might be caused by opposite milling to orientation of kenaf fiber. The obtained results might be less accurate if not all the specimens are milled in direction of orientation of fiber. In addition, there were previous researchers concluded that low feed rate favors small damage on composites [4], [19].

3.3.2 Optimum parameter and modeling equation for optimization of delamination factor

The three order modelling equation for optimization of milling parameters on minimizing delamination factor (F_d) is shown in (4).

$$Fd = 0.97903 + 4.390 \times 10^{-5} A + 9.375 \times 10^{-5} B + 0.01455 C - 6.2 \times 10^{-8} AB + 1.9 \times 10^{-5} BC \quad (3)$$

Where
A= spindle speed (rpm)
B= feed rate (mm/min)
C= depth of cut (mm) in terms of actual value.

Table 10. Optimum parameter and conformation test for delamination factor (F_d)

Spindle speed (rpm)	Feed rate (mm/min)	Depth of cut (mm)	Calculated F_d	Measured F_d	Percentage error (%)
500	200	2	1.027	1.032	0.48

Table 10 shows the percentage error between calculated (F_d) using the equation (4) and the measured F_d. The measured (F_d) was located at the prediction range (0.01- 0.04) and the percentage error was only 0.48% which proved the modelling equation and the suggested optimum parameters can be accepted.

4. Conclusion

Both suggested optimum milling parameters in this research for minimization of surface roughness (R_a) and delamination factor (F_d) were the same. 500 rpm spindle speed, 200 mm/min feed rate and 2 mm depth of cut gave the minimum surface roughness (R_a) and delamination factor (F_d).

For optimization of surface roughness, the ANOVA of the data showed no significance of curvature. This indicated a linear modelling equation was fit enough to express the optimization. RSM requires a significant curvature to prove the possibility of developing a second order equation to improve the reliability of the optimization. In this study, the feed rate affected the surface roughness the most. A low feed rate promised a minimum surface roughness. The spindle speed did not affect the surface roughness significantly. However, a high spindle speed tended to favor a good surface finishing. The effect of depth of cut on surface roughness was not displayed clearly. This might due to a larger range of depth of cut is required in order to amplify its impact on surface roughness. Nevertheless, a greater depth of cut is believed in producing product with a good surface finish.

For the optimization of delamination, the curvature shown in ANOVA after the 2^3 factorial design of experiments was significant. This enabled the proceeding of RSM with Box-Behnken Design approach that requires more experiment runs to collect data for constructing a second order modelling equation. A central composite design was selected and 8 more runs of experiments were added. A low spindle speed and a low feed rate caused less delamination on milled kenaf composites. In addition, with a deeper depth of cut, the possibility of delamination can be reduced.

Acknowledgments
Machine shop, laboratory facilities and their staff at the Faculty of Engineering and Built Environment of Universiti Kebangsaan Malaysia are gratefully acknowledged. Thank you Ministry of Higher Education Malaysia for supporting this project under grant LRGS/TD/2012/USM-UKM/PT/05.

References

[1] M. M. Davoodi, S. M. Sapuan, D. Ahmad, A. Ali, A. Khalina, and M. Jonoobi, "Mechanical properties of hybrid kenaf / glass reinforced epoxy composite for passenger car bumper beam," *Mater. Des.*, vol. 31, no. 10, pp. 4927–4932, 2010.
[2] S. Jeyanthi and J. J. Rani, "Improving Mechanical Properties by KENAF Natural Long Fiber Reinforced Composite for Automotive Structures," vol. 15, no. 3, pp. 275–280, 2012.
[3] R. M. Rowell, "A new generation of composite materials from agro-based fiber," in *Proceedings of the 3rd International Conference on frontiers of polymers and advanced materials*, 1995, pp. 659–665.
[4] G. D. Babu, K. S. Babu, and B. U. M. Gowd, "Effect of machining parameters on milled natural fiber-reinforced plastic composites," *J. Adv. Mech. Eng.*, pp. 1–12, 2013.
[5] P. Wambua, J. Ivens, and I. Verpoest, "Natural fibres : can they replace glass in fibre reinforced plastics ?," *Compos. Sci. Technol.*, vol. 63, pp. 1259–1264, 2003.
[6] B. F. Yousif, A. Shalwan, C. W. Chin, and K. C. Ming, "Flexural properties of treated and untreated kenaf / epoxy composites," vol. 40, pp. 378–385, 2012.
[7] S. R. Schmid and S. Kalpakjian, *Manufacturing Engineering and Technology*, 6th ed. New Jersey: Prentice Hall, 2009, pp. 172, 216–226, 662, 662–668.
[8] K. K. Chawla, *Composite Materials*. New York, NY: Springer New York, 2012, pp. 81–83.
[9] N. A. K. Hafizah, M. W. Hussin, M. Y. Jamaludin, M. A. R. Bhutta, M. Ismail, and M. Azman, "Tensile behaviour of kenaf fiber reinforced polymer composites," *J. Teknol.*, vol. 3, pp. 11–15, 2014.
[10] N. Abilash and M. Sivapragash, "Optimizing the delamination failure in bamboo fiber reinforced polyester composite," 2013.
[11] Ö. Erkan, I. Birhan, A. Çiçek, and F. Kara, "Prediction of Damage Factor in end Milling of Glass Fibre Reinforced Plastic Composites Using Artificial Neural Network," 2012.
[12] J. Davim, J. Rubio, and A. M. Abrao, "A novel approach based on digital image analysis to evaluate the delamination factor after drilling composite laminates," *Compos. Sci. Technol.*, vol. 67, no. 9, pp. 1939–1945, Jul. 2007.
[13] M. Ramulu, D. Arola, and K. Colligan, "Preliminary investigation on the surface Integrity of fiber reinforced plastics," *Am. Soc. Mech. Eng. Pet. Div.*, vol. 64, no. 2, pp. 93–101, 1994.
[14] K. Palanikumar, "Modeling and analysis for surface roughness in machining glass fibre reinforced plastics using response surface methodology," *Mater. Des.*, vol. 28, no. 10, pp. 2611–2618, Jan. 2007.
[15] L. Sorrentino and S. Turchetta, "Cutting Forces in Milling of Carbon Fibre Reinforced Plastics," *Int. J. Manuf. Eng.*, vol. 2014, pp. 1–8, 2014.
[16] J. P. Davim and P. Reis, "Damage and dimensional precision on milling carbon fiber-reinforced plastics using design experiments," *J. Mater. Process. Technol.*, vol. 160, no. 2, pp. 160–167, Mar. 2005.
[17] G. H. K. Rao, M. N. M. Ansari, and S. Begum, "Effect of cutting parameters on the surface roughness of MWCNT reinforced epoxy composite using CNC end-milling process," *Int. J. Sci. Res.*, vol. 2, no. 11, pp. 51–55, 2013.

[18] H. Hocheng, H. Y. Puw, and K. C. Yao, "Experimental aspects of drilling of some fiber-reinforced plastics," in *Proceedings of the Machining of Composites Materials Symposium*, 1992, pp. 127–138.
[19] J. P. Davim, P. Reis, and C. Conceição António, "A study on milling of glass fiber reinforced plastics manufactured by hand-lay up using statistical analysis (ANOVA)," *Compos. Struct.*, vol. 64, pp. 493–500, 2004.
[20] R. H. Myers, D. C. Montgomery, and C. M. A. Cook, *Response Surface Methodology: Process and Product Optimization Using Designed Experiments*, 3rd ed. New Jersey: John Wiley & Sons, Inc, 2009, pp. 1–11.
[21] S. A. Hussain, V. Pandurangadu, and K. P. Kumar, "Optimization of surface roughness in turning of GFRP composites using genetic algorithm," *Int. J. Eng. Sci. Technol.*, vol. 6, no. 1, pp. 49–57, 2014.
[22] L. Sorrentino and S. Turchetta, "Milling of carbon fiber-reinforced plastics: analysis of cutting forces and surface roughness," in *18th International Conference on Composite Materials*, 2014, pp. 1–6.
[23] J. Davim, J. Rubio, and A.M. Abrao, "A novel approach based on digital image analysis to evaluate the delamination factor after drilling composite laminates," Compos. Sci. Technol., vol. 67, no. 9, pp. 1939–1945, Jul. 2007.

IOP Conf. Series: Materials Science and Engineering **344** (2018) 012028 doi:10.1088/1757-899X/344/1/012028

Anticorrosion Coating using *Olea* sp. Leaves Extract

W M K W M Ikhmal[1], M Y N Yasmin[2], M F M Fazira[1], W A W Rafizah[2], W B Wan Nik[2] and M G M Sabri[1]*

[1]School of Fundamental Science, Universiti Malaysia Terengganu, 21030 Kuala Nerus, Terengganu, MALAYSIA
[2]School of Ocean Engineering, Universiti Malaysia Terengganu, 21030 Kuala Nerus, Terengganu, MALAYSIA

*mohdsabri@umt.edu.my

Abstract. Olive leaves extract (OLE) was evaluated as green corrosion inhibitor for stainless steel grade 316L (SS316L) in several media using scanning electron microscope (SEM), electrochemical impedance spectroscopy (EIS) and potentiodynamic polarization techniques. The Fourier Transform Infrared (FTIR) spectroscopy results reveals several active compound indicated by O-H stretch, C=O stretch, C-OH stretch and C-N stretch which can be attributed to oleuropein and hydroxtyrosol acting as the main inhibiting sources for corrosion. The results obtained also show the inhibition efficiency of OLE increase with the increase of OLE concentration. Through its inhibitive action elucidate from the electrochemical analysis, the extract was found to act as a mixed type inhibitor. Micrographs by SEM showed that the surface of steel which has been coated with 0% and 20% of OLE coating extract possess a lot of pin holes or pores while the steel with 10% of OLE coating extract shows the surface has multiple cracks. This study clearly shows the efficiency of OLE as anticorrosion coating for control of stainless steel in marine application.

1. Introduction

Corrosion is the degradation of metals resulting in an eventual failure of structures. As a ubiquitous problem face by most marine industry, complete elimination of the problem is difficult and thus a prevention methods to counter it are more practical [1]. The consequences of these inevitable threats have severely affected the operation cost causing a decrease in productivity. Reports from U.S. Department of Transportation for the period of 6 years between 1994 and 1999 suggest that approximately 25% of all reported accidents related to the pipeline system were due to corrosion [2]. Since industries depend heavily on the use of metals, practical corrosion protection of these metals has been widely studied in its usage whether in industrial application or products.

In the past, inorganic inhibitors such as phosphate and chromate are often used in marine coating due to their high efficiency [3]. However, these agents were known to cause hazard in environment due to their toxicity, non-degradable and/or bioaccumulation [4]. The first reported study which propels the beneficial property of plant extract as an inhibitor of corrosion was in 1930's when plant extract of *Chelidonium majus* (Celandine) and several others were tested for the corrosion inhibition in H_2SO_4 pickling baths [5]. Nowadays, interests in the use of an organic natural plant extract as a green inhibitor are steadily increased since these sources are much more acceptable, economical, readily available, practical, biodegradable and convenient. Various plant extract currently studied are found containing compound mixture of oxygen, sulfur and nitrogen which has been proven to be effective against corrosion [6-8]. Several studies suggest the inhibitor action by plant extract works through the adsorption on to the steel surface. Adsorption is a process of molecules attaching directly on to the surface without penetrating into the metals itself. Without an intermediate of paint, the compound will adsorb onto the surface through the displacement of water molecules and form a thin barrier on the surface [9]. Since the use of paint is applied in the study, the adsorption process is directly responsible by the binder component of the paint.

Olea sp. or olive is a popular medicinal plant across the globe. The leaf is considered as the waste from the farming of olive grove that gathers when the olive trees prune [10]. In the region of Mediterranean or Spain where the plant are extensively grow, the olive leaf is regarded as an agricultural wastes and usually be thrown away without further beneficial use. However, recent studies show the leaf is in fact possess the potential as an excellent corrosion inhibitor due to the abundant presence of phenolic compound [11].

The fresh pulp weight of olives usually produce 1-3% of phenolic compound which are considered as highly concentrate. Phenolic acid, phenolic alcohol, flavonoid and secoiridoids are the main classes of phenolic compound found in the extract of OLE. Generally, the most abundant of phenolic alcohol are hydroxytyrosol and tyrosol. The flavonoids of OLE are flavonol glycosides which are rutin, luteolin-7-glucoside and anthocyanins. Although phenolic acids, flavonoids and phenolic alcohol occur in many plants categorized to many botanical families, secoiridoids can only be exclusively identified in plants that belong to the Oleaceae family which is focused in the research [12-15].

Nowadays, application of coating in marine industry as well as others can be virtually seen everywhere. It can be seen on the wall of machines, bridges, pipelines, and superstructures. Coatings are not useful product by themselves, but they make other products better, more durable, more attractive, safer and more saleable [16]. In general, coating are made up of four constituents i.e. binder, pigment, additive and solvent [17]. Binder exists as to hold the pigment to the surface, pigments responsible in giving the coating its color and finish, additive serve to improve or modify the property of the coating and solvents are used to control the viscosity of the coatings for acceptable application [18]. The study focus on the efficiency of OLE integrated as an additive with the use of KTH 1K Alkyd paint in inhibit the corrosion progress on SS316L. Together with the inhibitive action of the paint, the extracts are expected to significantly increase the efficiency of the coating in reducing the corrosion rate. Although this type of paint are actually standardize and suitable for mild steel, but base on the study previously show this type of paint are suitable for the integration of plant extract for corrosion analysing process.

2. Materials and Methods

2.1 Chemicals

80% ethanol, acetone, distilled water and deionized water were supplied by Universiti Malaysia Terengganu. KTH 1K Alkyd and general thinner were purchased from a local Nippon Paint Holdings Co., Ltd.

2.2 Preparation of extract

Fresh olive leaves are plucked during early morning and cleansed using distilled water to remove impurities and dirt. The leaves were dried at ambient temperature for 2 to 3 days away from the exposure of sun light. Dry leaves are later grinded to obtain the powder for further extraction. The extraction methods to obtained OLE were done following the procedure conducted by Yateem et al. [19] with some modifications. Ten grams of olive leaves powder was weighed accurately using weighing scale and put into a conical flask. 100 ml of 80% ethanol was poured into it for maceration process. The immersed extract is left at ambient temperature for four hours. Next, the macerated solution is filtered using filter paper and evaporated using a rotary evaporator in vacuum at room temperature. The concentrated extract is stored in a refrigerator at 2-8°C before used.

2.3 FTIR Spectrum

Extract was characterized by FTIR spectroscopy for the determination of chemical functional group. Functional group of different types will absorb different characteristic of IR radiation frequencies. FTIR Spectroscopy was carried out using Thermo Nicolet Avatar 380. A small quantity of diluted extract was directly deposited onto the sample disk, creating a thin film. A spectrum was scanned in the frequency range of infrared light ranging from 4000 to 700 cm^{-1}.

2.4 Ultraviolet-Visible analysis

The absorbance of diluted extract was measured by using Lambda 25 UV-Vis spectroscopy. UV-Vis Spectroscopy is an instrument to analyze the absorption of the polymer and dyes. The 10 mm^3 transparent cube named cuvette was cleaned by using acetone solution and dropper was used to drop the extract solution into the cube. The cuvette filled with extract solution was put onto the sample compartment and the analysis begun. The process was repeated three times to get the average reading.

2.5 Preparation of Substrate

The substrate used is SS316L which was a marine grade stainless steel. The specimen was obtained from a supplier in Terengganu and was cut into small rectangular shape with a dimension of 25 mm x 25 mm x 1 mm. Then, polishing of the steel is done using polished machine with abrasive paper grade 100, 120 and 180. The necessity of polishing the SS316L was to ensure the layer of carbon oxide is removed so that it will not interfere with the later procedure. Simple preservation method for stainless steel was carried out regularly until it is coated using sponge, distilled water and soap respectively.

2.6 Coating Formulation

Coating formulated involved several materials such as KTH 1K Alkyd Finish, general thinner and OLE. Simple mixing method is done according to the formulation in table 1. The mixture was stirred in a beaker on stir plate for 1 hour until it is homogenous. Finish product was sealed and stored in desiccator until used. The volume of extract solution was calculated based on the Equation 1:

$$\frac{Volume\ of\ Solution\ (ml)}{Volume\ of\ Solvent\ (ml)}\ x\ 100 \qquad (1)$$

Table 1. Coating Formulation.

Type of Coating	Volume of Paint (ml)	Extract (%)	Thinner (ml)
1 (control)	350	0	35
2	350	10	35
3	350	20	35

2.7 Immersion Test

Several beakers containing a fix amount of distilled water, deionized water and seawater was prepared. Different concentrations of OLE (0%, 10% and 20%) coating are used to coat the SS316L and later the substrate are placed inside the beaker hanging. Experiment is conducted for a period of 30 days and in each subsequent 10, 20 and 30 days, substrates of different concentration were taken out for analysis. The immersion test was conducted in an enclosed environment to avoid any disturbance.

3. Results and Discussion

3.1 Functional Group of Olive Leaves

Fourier Transform Infrared (FTIR) spectroscopy results in figure 1 show several peaks indicating the presence of active compounds in OLE. The compound identified is an important aspect for further understanding on the inhibiting property of the extract. Through the study of compounds identified will pose strong clarification on the mechanism of OLE application as green inhibitor. Base on figure 1, OLE includes major components such as oleuropein and hydroxtyrosol. The primary components of compound can be identified from the structures of phenols O-H, carboxylic acid C=O and alkenes C=C stretching nodes which are observed at 3329 cm^{-1} and 1733 cm^{-1}. The band at 1112 cm^{-1} could be assigned to the C-OH vibrations of the protein of olive leaves or C-N stretching vibration of the amine. The characteristic peak at 2927 cm^{-1} may be attributed to the C-H stretching node while the peak of 1651 cm^{-1} corresponds to the presence of amide I. Finally, the peak observed at 1040 cm^{-1} could be the indicator of C-N stretching vibration of the amine group. Table 2 shows the major functional groups found in OLE.

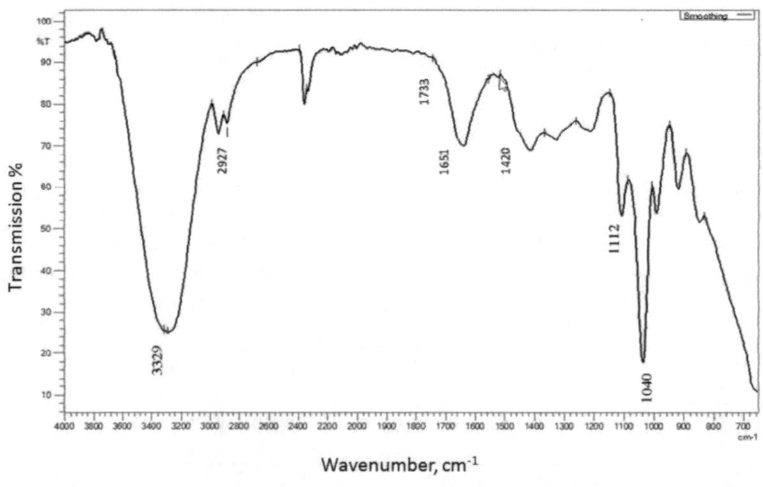

Figure 1. FTIR spectrum for olive leaf.

Table 2. Chemical Structures of the main Phenolic Compound in the Leaf of Olive [20].

Phenolic Compound	Chemical Formula	Chemical Class
Oleuropein	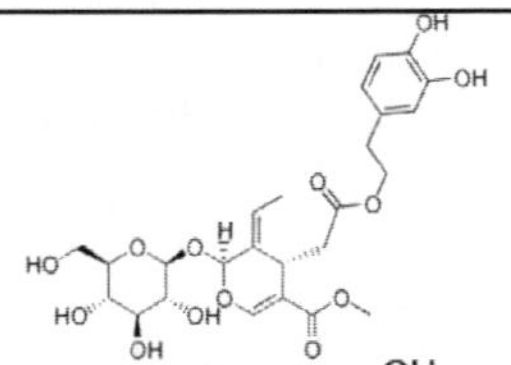	Secoiridoids
Hydroxytyrosol	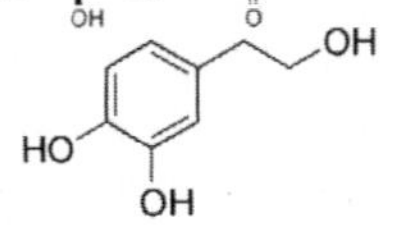	Phenolic alcohol
Verbascoside	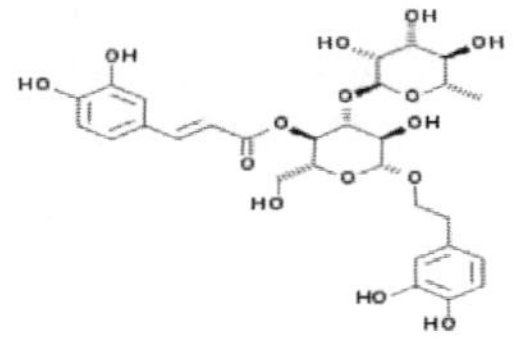	Flavonoids
Apigenin-7-glucoside	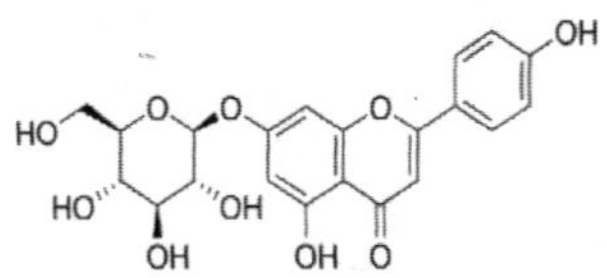	Flavonoids

3.2 Absorption Spectrum of Olive Leaves Extract

Absorbance is a measurable quantity that possesses a linear characteristic with the analyte concentration in comparison to the large range of concentration. By using UV-Vis spectroscopy, a measurement on the amount of light absorbed at a particular wavelength as the light passes through a sample or substances are recorded. Figure 2 shows the absorption spectra of diluted OLE. UV-spectrum was collected from 400 to 900 nm at 200 nm/min. The analysis was carried out using the lambda 25 software on the computer which connected to the equipment via cables. The absorption for the extract is 665 nm; respectively.

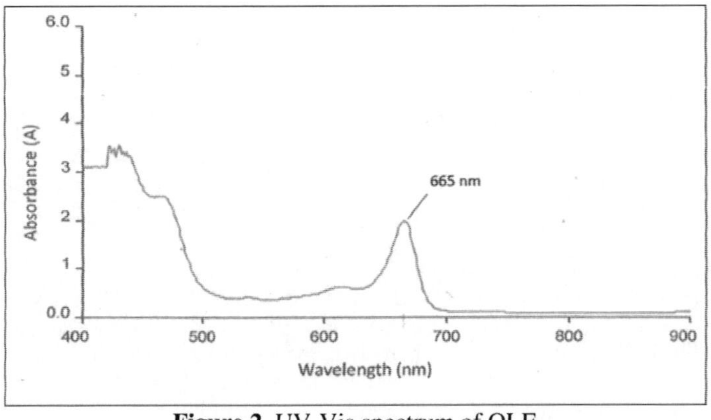

Figure 2. UV-Vis spectrum of OLE.

3.3 Electrochemical Impedance Spectroscopy (EIS)

Nyquist plots is used as the presentation of data collected for assessing the stability of the system. The real impedance (Z') was indicated at x-axis while the imaginary impedance (Z'') was indicated at y-axis. The parameter showed included the difference of OLE concentration use in the coating which was 0%, 10% and 20%. Base on the Nyquist plot in figure 3, capacitive semi-circle were shown by the curves obtained. Since there is no change in the shape of the curves with the difference of OLE concentration, it indicates that corrosion mechanism is maintained throughout the addition of inhibitor. The value the charge transfer resistance of the stainless steel as shown in table 3 significantly increased as the presence of OLE increased. However, the potential resistance (R_P) of 20% OLE at room temperature is lower than 10% OLE. The decrease of the R_P occurred due to the limitation in concentration. Increasing R_P leads to an increase of inhibitor efficiency. It can be conclude from the result that the ratio of OLE in the coating is found limited to 10% due to further increase of extract cause the potential resistance to reduce. Thus, increase in the concentration of extract beyond its optimum limit will results in inhomogeneity of coating sample [21].

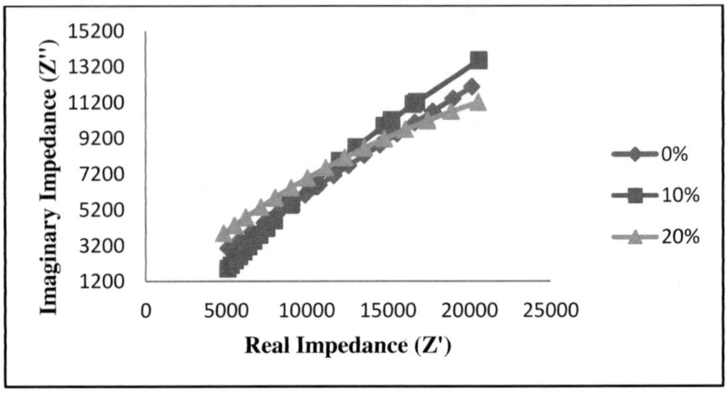

Figure 3. Nyquist plot for 30 Days in Seawater at room temperature.

Table 3. Impedance Parameter of SS316L in seawater for 30 days immersion.

Type of Coating	$R_P(\Omega)$
OLE 0%	75949
OLE 10%	105040
OLE 20%	75381

3.4 Potentiodynamic Polarization

Tafel plot is employed for the presentation of polarization measurement. From figure 4 and figure 5, the value of corrosion potential (E_{corr}) is seen to shift more towards the positive direction as the extract of olive leaf is introduced. The E_{corr} value of the samples in seawater and distilled water shown in table 4 and table 5 increases as there is presence of OLE which is -0.26573 V/SCE to -0.18793 V/SCE and -0.10672 to -0.06135 V/SCE respectively. Since the E_{corr} shift toward more positive direction, it elucidates that OLE act as a mixed-type inhibitor.

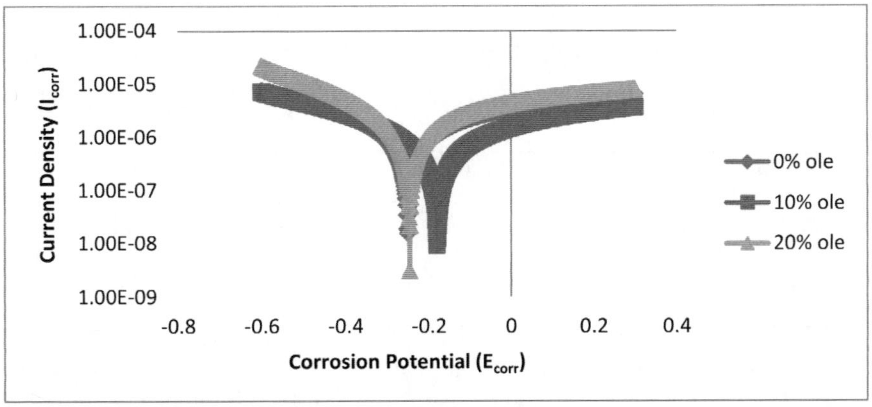

Figure 4. Potentiodynamic polarization of SS316L for 30 days in the medium of seawater.

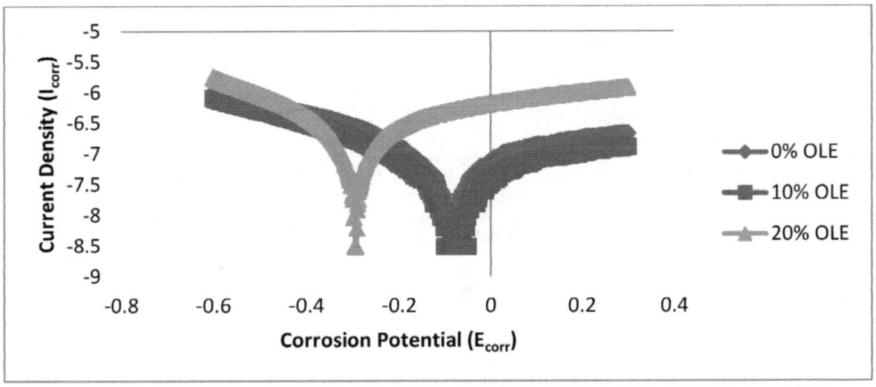

Figure 5. Potentiodynamic polarization of SS316L for 30 days in the medium of distilled water.

However, the value of E_{corr} drop to -0.25489 V/SCE and -0.29449 V/SCE for both samples in seawater and distilled water as the OLE concentration increases to 20%. As the value of corrosion potential increase the corrosion rate will decrease. Thus, increase in the concentration of OLE will results in higher corrosion potential. The low value of corrosion rate is due to the thin layer formed on steel substrate to protect the steel from corrosion [22]. The studied done previously on the inhibitor used was affected both in anodic and cathodic sites in Tafel plot with E_{corr} value towards more noble values and this phenomenon can be used to derived that inhibitor act as a mixed-type inhibitor [23].

Table 4. The electrochemical parameter of SS316L in seawater for 30 days immersion.

Type of Coating	E_{corr} (V/SCE)	I_{corr} (A.cm^2)	Corrosion rate (mm/year)
Olive Leaves Extract 0%	-0.26573	6.88E-06	0.079930
Olive Leaves Extract 10%	-0.18793	1.68E-06	0.019463
Olive Leaves Extract 20%	-0.25489	3.95E-06	0.045848

Table 5. The electrochemical parameter of SS316L in distilled water for 30 days immersion.

Type of Coating	E_{corr} (V/SCE)	I_{corr} (A.cm^2)	Corrosion rate (mm/year)
Olive Leaves Extract 0%	-0.10672	5.68E-07	0.006602
Olive Leaves Extract 10%	-0.06135	1.33E-07	0.001542
Olive Leaves Extract 20%	-0.29449	3.46E-07	0.004019

Table 4, Table 5 and Table 6 provide an elucidation that corrosion rate decrease as the value of current density (I_{corr}) becomes lower due to the presence of OLE. 10% of OLE has lower corrosion rate than 20% of OLE in both seawater and distilled water due to the ratio of the extract is proved to be limited at 10% in the study. According to another studies, the inhibition of steel increases with the increase of an organic extract concentration [24]. Moreover, through the samples immersed in seawater shows decrease in current density from 6.88 x 10-6 A.cm^2 to 1.68 x 10-6 A.cm^2. This behavior proved that OLE acts as a good corrosion inhibitor when integrated in paint for immersed stainless steel in marine medium.

Table 6. The electrochemical parameter of SS316L in deionized water for 30 days immersion.

Type of Coating	E_{corr} (V/SCE)	I_{corr} (A.cm^2)	Corrosion rate (mm/year)
Olive Leaves Extract 0%	-0.268934	4.13E-08	0.000480385
Olive Leaves Extract 10%	0.273840	3.53E-08	0.000410365
Olive Leaves Extract 20%	0.097405	4.21E-08	0.000489658

The potentiodynamic polarization curve obtained from the analysis of the samples immersed for 30 days in deionized water shows almost a similar rate of corrosion rate. Deionized water or demineralized water is ultrapure water usually used in microelectronics and instrument manufacture. Although most people hold the opinion that deionized water is a very corrosive medium, a study by National Aeronautics and Space Administration (NASA) shows the medium have no such effect on common metals such as aluminum and stainless steel [25].

3.5 Scanning Electron Microscope

The primary analysis was subjected to the coated surface of SS316L which has been immersed in seawater medium after 30 days; respectively. Both figure 6(a) and figure 6(b) show a similar result of the presence of pores or pin holes. Pore or pin holes are basically tiny spots indicating the absence of coated layer and varied in its dimension. There are several typical causes of this phenomenon such as the rupture of liquid coating caused by contamination, picking of coating by roll and non-uniform flow of coating when it is applied. Other related sources might be due incompatibility of the coating with the surface of the substrate. However, this theory is rejected since the selection of the components in the coating is selected based on the previous study conducted in several researches which used the same type of constituents and differs only in the type of extract use. Thus, it indicates that the incompatibility of the coating with the surface is not the factor that caused the surface to possess this state. Through the researcher observation, there are two possible causes of the phenomenon which are irregularities on metal and non-uniform flow of coating when it is applied. The evidence on the first suggestion are supported in figure 6(d) that showed the surface of a naked SS316L possessing several dark spots indicating the irregularities or roughness of the surface. The second causes occurred due to the unsuitable method of applying the coating onto the SS316L surface using a simple paint brush in several directions.

Generally, porosity is identified as an undesirable trait since it will expose the substrate to corrosive agents, deleteriously influence density, electrical property and diffusion characteristic as well as reduce mechanical properties, [26]. Various tiny pores are observed in figure 6(a) and figure 6(b) with dimension ranging from 10 μm to 40 μm. Figure 6(c) shows the SEM micrograph of the coated surface of SS316L with 10% OLE, it differ compare to previous two result in which the surface shows a muddy layer with cracks in several places. Cracks can be considered as another form of pores which extend usually in a direction parallel to the surface. Type of cracks includes transverse crack and enclosed crack. Probable causes of the phenomenon are due to the application of the coating was too thick, coating flash too quickly or at too high temperature. It may occur in an irregular pattern where the coating stresses are rather uniform. It may be linear, following brush marks occur in a curved pattern with the crack meeting and intersecting in a rather typical form. The problem can be overcome easily by using a proper resins and pigments with the addition of material specifically for reducing interior stresses within the coating. Through the observation, a few cracks can be seen on several places especially on the area where coating appeared thicker.

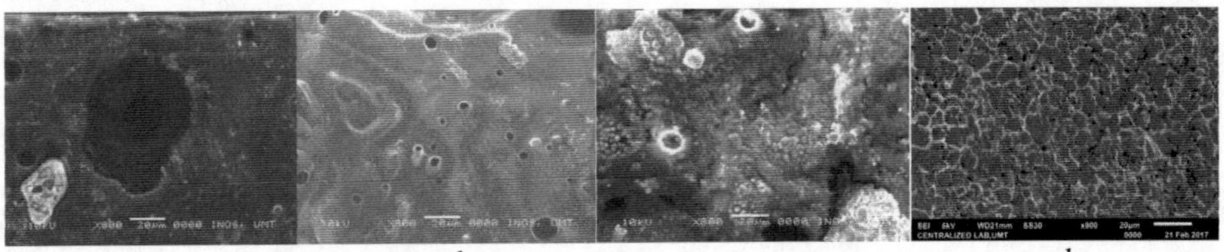

a b c d

Figure 6. SEM micrographs of SS316L surface (a) after 30 days of immersion with 0% OLE coating (b) after 30 days of immersion with 20% OLE coating (c) after 30 days of immersion with 10% OLE coating (d) naked.

4. Conclusion

In summary, inhibitor plays a significant role in endowing protection against corrosion. The olive leaves extraction was successfully prepared through a maceration method with an organic solvent. The absorption spectra by UV-Vis spectroscopy showed the absorption for the extract is 665 nm; respectively. Through the analysis of FTIR spectroscopy several functional groups which indicate the presence of antioxidant compound such as oleuropein and hydroxytyrosol was identified. The extract was applied as an additive providing the anti-corrosion property with different concentration (0%, 10% & 20%) for the means of investigating the efficiency of OLE as corrosion inhibitor. Immersion test for a month was carried out in the laboratory using three different media which were seawater, distilled water and deionized water. By the end of 30 days, sample was taken out and analyzed with EIS spectroscopy, PP and SEM for a detailed corrosion rate determination. EIS measurement indicates that the inhibitor of corrosion on SS316L is due to the protective film on metal surface and inhibition increases as the extract concentration increases. As the presence of OLE increases, the values of potential resistance (R_P) increase accordingly. However, ratio of OLE is found limited to only 10%. Further increase in the concentration of extract would cause inhomogeneity in coating paint sample. The pitting potential of SS316L shifted to more positive values with the addition of extract indicating resistance to pitting corrosion increase. The low value of corrosion potential

(E_{corr}) indicates higher corrosion rate. The value of corrosion current density (I_{corr}) increases with absence of OLE. Base on the results from all the samples tested, the substrate with 10% OLE coating showed the highest inhibition rate of corrosion followed by 20% and 0% OLE coating. In a conclusion, OLE possesses a good potential as an excellent inhibitor for stainless steel.

Acknowledgements

Financial support was received from the Look East Policy 2.0 (Vote No. 53168) and Niche Research Grant Scheme (NRGS/2015/53131/14) from Ministry of Higher Education Malaysia are gratefully acknowledged. All praise to Allah the Almighty who has given me the opportunity to pursue my passion in this field. Full appreciation to my parents and family who relentlessly supporting me until my undergraduate study done. Sincere gratitude to my main supervisor, Dr. Mohd Sabri Mohd Ghazali and my co-supervisors, Dr. Wan Rafizah Wan Abdullah and Prof. Dr. Wan Mohd Norsani Wan Nik for their willingness to motivate me contributed tremendously to my project. Special thanks to MSc students for their encouragement and help especially to Mr. Muhamad Syazlie Che Ibrahim and Ms. Hajar Hakimah Marzalan throughout this research. I also would like to thank the science officers and lab assistants, for all the help given in the lab. Finally, I also would like to thank to the staff at the Institute of Oceanography and Environment and School of Ocean Engineering for helping in analyzing the samples of my project.

References

[1] Rani, B.E.; Basu, B.B.J. Green Inhibitors for Corrosion Protection of Metals and Alloys: An Overview. Int. J. Corros. 2012, 2012:1-2012:15.
[2] Beavers, J.A.; Thompson, N.G. External Corrosion of Oil and Natural Gas Pipelines. Available online: http://www.asminternational.org (accessed on 5 September 2017).
[3] Chigondo, M.; Chigondo, F. Recent Natural Corrosion Inhibitors for Mild Steel: An Overview. J. Chem. 2016, 2016:1-2016:7.
[4] Ibrahim, T.H.; Chehade, Y.; Zour, M.A. Corrosion Inhibition of Mild Steel using Potato Peel Extracts in 2M HCl Solution. Int. J. Electrochem. Sci 2011, 6, 6542-6556.
[5] Sanyal, M. Organic Compounds as Corrosion Inhibitors in Different Environments—A Review. Prog. Org. Coat 1981, 9, 165-236.
[6] Louis, H.; Japari, J.; Sadia, A.; Philip, M.; Bamanga, A. Photochemical Screening and Corrosion Inhibition of Poupartia birrea Back Extracts as a Potential Green Inhibitor for Mild Steel in 0.5 M H2SO4 Medium. World News Nat. Sciences 2017, 10, 95-100.
[7] Popova, A.; Christov, M. Evaluation of Impedance Measurements on Mild Steel Corrosion in Acid Media in the Presence of Heterocyclic Compounds. Corros. Sci 2016, 48, 3208-3221.
[8] Reddy, C.M.; Sanketi, B.D.; Kumar, S.N. Corrosion Inhibition of Mild Steel by Capsicum annuum Fruit Paste. Perspectives in Science 2016, 8, 603-605.
[9] Arthur, D.E.; Jonathan, A.; Ameh, P.O.; Anya, C. A Review on the Assessment of Polymeric Materials used as Corrosion Inhibitor of Metals and Alloys. Int. J. Ind. Chem. 2013, 4:1-4:9.
[10] Nashwa, F.S.; Abdel-Aziz, M.E. Efficiency of Olive (Olea europaea L.) Leaf Extract as Antioxidant and Anticancer Agents. JAPT 2014, 20, 46-53.
[11] Somova, L.I.; Shode, F.O.; Ramnanan, P.; Nadar, A. Antihypertensive, Antiatherosclerotic and Antioxidant Activity of Triterpenoids Isolated from Olea europaea, Subspecies Africana Leaves. J Ethnopharmacol 2003, 84, 299-305.
[12] Robards, K.; Prenzler, P.D.; Tucker, G.; Swatsitang, P.; Glover, W. Phenolic Compounds and Their Role in Oxidative Processes in Fruits. Food Chem. 1999, 66, 401-436.
[13] Servili, M.; Montedoro, G. Contribution of Phenolic Compounds to Virgin Olive Oil Quality. Eur J Lipid Sci Technol 2002, 104, 602-613.

[14] Servili, M.; Selvaggini, R.; Esposto, S.; Taticchi, A.; Montedoro, G.; Morozzi, G. Health and Sensory Properties of Virgin Olive Oil Hydrophilic Phenols: Agronomic and Technological Aspects of Production that Affect Their Occurrence in the Oil. J. Chromatogr. A 2004, 1054, 113-127.

[15] De la Torre-Carbot, K.; Jauregui, O.; Gimeno, E.; Castellote, A.I.; Lamuela-Raventós, R.M.; López-Sabater, M.C. Characterization and Quantification of Phenolic Compounds in Olive Oils by Solid-Phase Extraction, HPLC-DAD, and HPLC-MS/MS. J. Agric. Food Chem. 2005, 53, 4331-4340.

[16] Aliofkhazraei, M. Developments in Corrosion Protection, InTech: Rijeka, Croatia, 2014; Chapter 12, pp. 241-270, ISBN 978-953-51-1223-5.

[17] Schiek, R. Paints and Pigments. J. Chem. Educ 1980, 57:1-57:19.

[18] Kilminster, A.E. Paints and Surface Coatings. AICCM Bulletin 1979, 5, 14-17.

[19] Yateem, H.; Afaneh, I.; Al-Rimawi, F. Optimum Conditions for Oleuropein Extraction from Olive Leaves. IJAST 2006, 4, 153-157.

[20] MacDonald-Wicks, L.K.; Wood, L.G.; Garg, M.L. Methodology for the Determination of Biological Antioxidant Capacity in Vitro: A Review. J. Sci. Food Agric. 2006, 86, 2046-2056.

[21] Abidin, Z.H.Z.; Nasir, K.M.; Jamari, S.K.M.; Saidon, N.; Lee, S.V.; Halim, N.A.; Yahya, R. The Characteristics of a Coating System Containing Lawsone Dye Colorant and PMMA-Acrylic Polyol Blended Resin. Pigm. Resin Technol. 2013, 42, 128-136.

[22] Hajar, H.M.; Zulkifli, F.; Suriani, M.J.; Sabri, M.M.; Nik, W.W. Lawsonialnermis Extract Enhances Performance of Corrosion Protection of Coated Mild Steel in Seawater. MATEC web conf. 2016, 78, 1-8.

[23] Umoren, S.A.; Solomon, M.M.; Eduok, U.M.; Obot, I.B.; Israel, A.U. Inhibition of Mild Steel Corrosion in H2SO4 Solution by Coconut Coir Dust Extract Obtained from Different Solvent Systems and Synergistic Effect of Iodide Ions: Ethanol and Acetone Extracts. JECE 2014, 2, 1048-1060.

[24] Ostovari, A.; Hoseinieh, S.M.; Peikari, M.; Shadizadeh, S.R.; Hashemi, S.J. Corrosion Inhibition of Mild Steel in 1M HCl Solution by Henna Extract: A Comparative Study of the Inhibition by Henna and its Constituents (Lawsone, Gallic acid, α-d-Glucose and Tannic acid). Corros. Sci. 2009, 51, 1935-1949.

[25] Johnson, B.A. Corrosion of Metals in Deionized Water at 38 °C (100 F). NASA 1969, 7:1-7:11.

[26] Nakahara, S. Microporosity in Thin Films. Thin Solid Films 1979, 64, 149-161.

Development and characterization of sugarcane bagasse fiber and nano-silica reinforced epoxy hybrid composites

A L Fong[1,2], N A N Khandoker[1] and S Debnath[1]*

[1] Mechanical Engineering Department, Curtin University, Miri, Malaysia
[2] Curtin Sarawak Collaborative Research (CSCR), Curtin University, Miri, Malaysia

*Corresponding author. Email: d.sujan@curtin.edu.my

Abstract. This paper presents an experimental study on the mechanical performance of sugarcane bagasse fiber reinforced epoxy composite. Tensile and flexural properties of the composites were investigated in this research. Different weightage of short fiber and fiber particulates were utilized to study their effects on the mechanical performance of the composites in terms of tensile and flexural properties. 1% of nano-silica was reinforced to investigate its effect on the mechanical performance of the composites. Hand lay-up composite molding process was used to fabricate the composite samples. During fabrication, ultrasonic mixing was carried out to study the effects on mechanical performance of the fiber particulate reinforced composites. In overall, ultrasonic mixing and addition of nano-silica particles has improved the mechanical performance of the fiber particulate composites. Morphology analysis on surface of composites has shown the removal of air bubbles and deagglomeration. 1wt% of short fiber reinforced composite exhibits the highest tensile and flexural properties among all the samples. Sugarcane bagasse particulates reinforced composites were shown to have better performance compared to short fiber reinforced composites when the wt% of the fiber increase.

Keywords: composite; sugarcane bagasse fiber; epoxy; nanosilica

1. Introduction

Nowadays, increase in sustainability and environmental concerns have evoked the interest in research and development of high performance and biodegradable composite. Efforts have been made to develop natural fibers and particulate reinforced composites as it is one of the most effective ways to satisfy the "eco-material" concept which serve as a basis to sustain the environment. One of the alternatives solution that attracts interest is utilization of abundantly natural fiber as reinforced materials in polymer matrix composite.

Composite is hybrid material which consists of two or more chemically distinct constituents. There is continuous phase which created by matrix and embedded by discontinuous phase, the reinforcement medium. The matrix phase is usually made up from fundamental material such as metals, ceramics and polymers. Meanwhile, the reinforcing phase is widely in the form of fibers, whiskers and particulates [1]. The reinforcing phase is usually stronger, lighter and stiffer than the matrix to enable the composites to possess better mechanical properties. Polymer matrix composites are utilized nowadays as they have good versatility, high performance and cost effective [2]. Epoxies are one of the most important matrices which are widely used for fiber-reinforced polymer due to its unique properties. Epoxy is relatively high strength, high modulus, low shrinkage, high chemical, heat and electrical resistance [3].

As the awareness has greatly increased the world responsiveness towards natural fibers, an investigation has been carried out to study natural fiber reinforcements. These natural fibers which have been utilized in reinforcement can be found in numerous applications in various fields such as automobile, furniture, packing and construction. Besides, compared to conventional reinforcing fibers, the advantages of natural fibers are found out to be low density, fully biodegradable, environmentally friendly, renewable, non-toxicity, low cost, high toughness, good insulation against heat and noise, good thermal properties, reduced tool wear, reduced dermal and respiratory irritation, ease of separation and lower abrasiveness [4].

Sugarcane is one of the major crops in tropical region which has a total plantation area of 34500 acres in Malaysia [4] while the amount of sugarcane produced is approximately 1.3 to 1.6 million tons annually [5]. Bagasse fibers gathered through the sugarcane milling process after the extraction of the sugar-bearing juice from sugarcane. Sugarcane bagasse fiber-reinforced composites are found to exhibit better specific mechanical properties, such as stiffness, flexibility, and modulus compared to those reinforced by glass fibers [6]. It can be perfect alternatives for reinforcing bio-composites as it is renewable and less expensive sources due to its abundance as well as low preparation cost of the fibers [7].

Table 1. Mechanical properties of commercially important lignocellulosic fibers (Khalil et al. 2010)

Fibers	Density (g/cm3)	Tensile strength (MPa)	Young's modulus (GPa)	Elongation at break (%)
OPEFB	0.7-1.55	248	3.2	2.5
Flax	1.4	800-1500	60-80	1.2-1.6
Hemp	1.48	550-900	70	1.6
Jute	1.46	400-800	10-30	1.8
Ramie	1.5	500	44	2
Coir	1.25	220	6	15-25
Sisal	1.33	600-700	38	2-3
Abaca	1.5	980	-	-
Cotton	1.51	400	12	3-10
Kenaf (bast)	1.2	295	-	2.7-6.9
Kenaf (core)	0.21	-	-	-
Bagasse	1.2	20-290	19.7-27.1	1.1
Henequen	1.4	430-580	-	3-4.7
Pineapple	1.5	170-1627	82	1-3
Banana	1.35	355	33.8	5.3

From the table above, sugarcane bagasse is shown to have a density of 1.2 g/cm3 and a maximum tensile strength of 290 MPa. While its Young's modulus achieved 27.1 GPa at its highest and the elongation of break is found to be 1.1%. The mechanical properties of sugarcane bagasse fibers are found to be the main focus of the researchers. Common mechanical properties attributed to the sugarcane bagasse fibers are the tensile strength, modulus of elasticity, density and elongation at break [8]. The age of the fibers, the source of the fibers and different techniques of surface treatment of the bagasse fibers are the factors which will affect the mechanical properties of the sugarcane fiber [9] .

IOP Conf. Series: Materials Science and Engineering **344** (2018) 012029 doi:10.1088/1757-899X/344/1/012029

Aspect ratio has a considerable effect on composite properties, hence it is important to conserve fiber length as much as possible during composite processing operations. Furthermore, the aspect ratio means an average length over diameter of the fibers is varied from plant species to species [8]. They [8] conducted a work regarding the effect of chemical compositions and particle size on reinforced polypropylene composite. They concluded that the improvement in mechanical properties achieved can be attributed to higher fiber length and aspect ratio. According to their finding, fiber aspect ratio must be in the range of 100–200 for optimum effectiveness. In a short fiber composite, tensile load is transferred into a fiber from the matrix through shear at the fiber/matrix interface. The tensile stress is zero at the ends of the fiber and increase along the fiber length. Thus, the length of the fiber should greater than its critical length for efficient reinforcement. The influence of fiber length on tensile strength and Young's modulus of sisal fiber reinforced polyester composite was investigated by [11]. They found that tensile strength and Young's modulus increase with an increase in the fiber length. Similar trend was reported by [12] in the case of short oil palm fiber with different lengths (2, 6, 10 and 14mm) into a natural rubber matrix. Tensile strength, elongation at break, and tensile modulus at 100% elongations were at a maximum when the length of the oil palm fiber was 6 mm. At higher fiber lengths, a decrease in the properties was found. This was due to the fiber entanglements prevalent at longer fiber length.

Apart from fibers, there are also other fillers such as particulates being used in reinforcement of composites. Nanoparticles have been utilized as reinforcing phase in composites. Several studies indicate that modulus, strength and toughness can be simultaneously increased with the addition of Nano-scale fillers [13]. They are generally organic polymer composites mostly filled with inorganic fillers, which combine the advantages of the inorganic filler material (i. e., rigidity, thermal stability) and of the organic polymer (i. e., flexibility, ductility, processability). There are different types of nanoparticles which have been developed and used in various areas due to their multifunctional properties, such as anti-bacteria, UV resistant, anti-wrinkle finishing and water repellent to fibers [14] [6].

Among different available nano-particles, silica nanoparticles are more useful for the fabrication of composites due to their amorphous structure, SiO2 content (more than 99 % purity in most products), and highly specific surface area (which leads to the super pozzolanic property) [19]. They are mostly reported in the literature and being extensively used in various applications including electronics, automotive and aerospace industries. Chen et al. (2008)[13] have investigated the thermal and mechanical properties as a function of the silica percentage with spherical silica particulates of size 12-nm. Characterization techniques utilized in order to investigate the morphology are transmission electron microscopy (TEM) and ultra-small-angle X-ray scattering (USAXS). It is found that nanoparticles can be dispersed with minimal aggregation when the amount used is up to 25 wt%. It is reported that there is an increase of 25% in tensile modulus and 30% in fracture toughness for samples less than 10 wt% of silica.

Yet there are some challenges faced by these natural fiber-polymer composites such as the poor interfacial adhesion between fibers and the binding matrix. Thus, the alkaline surface treatment is applied to improve the adhesion. Besides, use of longer fibers can further increase their tendency to agglomerate. The reinforcement efficiency could be reduced due to poor fiber dispersion as fiber may get tangled during mixing. As the fiber content is being increased in the matrix, the bonding between the fiber and the matrix start to deteriorate as the fiber become too close to each other. Besides, agglomeration of fibers within matrix creates poor stress transfer efficiencies due to poor interfacial bonding between them thus a decrease in the mechanical properties. The hand-mixing method also leads to agglomeration of fibers within matrix, which also contributes in decreasing the mechanical performance of the composite.

Therefore, this study will focus on minimizing the agglomeration problems of the particulates as well as enhancing interfacial properties between the fiber and the matrix, by investigating the mechanical properties (tensile and flexural properties) and morphology analysis of hybrid composite consisting of epoxy matrix reinforced with treated sugarcane bagasse (SCB) fibers and Nano-silica particulate. Besides, this research also includes the effect of ultrasonic mixing on mechanical performance of the fiber particulate reinforced composite.

2. Experimental Procedure

2.1. Materials

The bagasse was obtained from the local supplier. The fibers were washed and separated to obtain the inner parts, followed by drying in oven at 150 for 24 h. The fibers were extracted manually. Epoxy and hardener (Miracast 1517A/B) were procured from Miracon Sdn. Bhd. and were used as matrix resin while nano-silica particulate was procured from Sigma Aldrich Malaysia.

2.2. Preparation of composites

2.2.1. Alkaline Treatment of Bagasse Fiber

Bagasse fiber was cut into pieces and soaked in 2% NaOH for 24 h. The fibers were rinsed then soaked in distilled water for 1h. The alkaline-treated fibers were dried at 100˚C for 3h, and kept in air-tight container to prevent moisture.

2.2.2. Obtaining of bagasse in fibrous form and particulate form

Bagasse short fiber was first cut into 5-10mm. Bagasse fiber particulates were obtained through a disk mill machine. Endecott EFL 2000 test sieve shaker was used to obtained fiber particulates which were less than 0.6mm. All the bagasse was kept in containers separately until used to prevent moisture

2.2.3. Determination the weights of fillers and epoxy

The mix ratio of epoxy resin and hardener is 30:10. The different filler loading of sugarcane bagasse short fibers and particles are calculated by the weight percent (w/v) formula as shown as below:

Hand mixing method:

$$\textit{Mass of filler (g)} = \frac{\textbf{Resin and Hardener Required (g)}}{\textbf{100\%}} \times \textit{40g (mass of epoxy + hardener)} \quad (1)$$

Ultrasonic mixing:

$$\textit{Mass of filler (g)} = \frac{\textbf{Resin and Hardener Required (g)}}{\textbf{100\%}} \times \textit{160g (mass of epoxy + hardener)} \quad (2)$$

The weight percent (w/v) formula as shown as above was used to find the masses of filler loading of sugarcane bagasse fibers (g) required to mix with epoxy and hardener while using hand-mixing method and ultrasonic mixing. The required filler loading to be obtained are 1wt%, 2.5wt%, 3.5wt% and 5wt%. The only required filler loading for nano-silica particles is 1wt%, which is same as 1wt% of sugarcane bagasse fibers used in ultrasonic mixing.

2.3. Preparation of SCB short fiber and particulates composites by hand-mixing

Different weight percentage of 1%, 2.5%, 3.5%, and 5% of sugarcane bagasse fibers particles and short fibers were used in this experiment. Predetermined weight of epoxy resin and its hardener were poured into two separate disposable cups. Epoxy was shown to be in the bottom layer while hardener was shown in the top layer. The treated bagasse fibers particles were added according to their weightage into the epoxy and hardener mixture. The mixture was stirred for 2 minutes. The mold release agent was applied on the surface of the molds to serve as a parting agent between layers of the matrix and the mold surface, enable the hardened composite to be removed easily without causing damage to the mold. The composite mixture was then poured into the tensile and flexural molds by applying hand lay-up method. The specimens were put into the desiccator, and vacuum gas pump was switched on for 15 minutes to remove air bubbles in the composite. Later, the mold was left to air dry and cure for 24 hours then put into oven for post-curing for 3 hours.

2.4. Sonication of SCB fiber particulates/ nano-silica and epoxy mixture

In present work the SCB fiber particulates and nano-silica were dispersed in epoxy resin using a high intensity ultrasonic processor. The viscosity of the polymer increased while the weight percentage of SCB particulates increased. Although the weight percentage of nano-silica and fiber particulates is low, there is still high possibility to cause agglomeration. Thus, the sonication parameters stated below are carefully selected based on literature review to ensure better dispersion and deagglomeration:

- The sonicator probe is placed in the middle and the immersion depth is 2cm above from the bottom of the beaker. The sonication time was fixed to 20 min for epoxy, SCB fiber particulates and nano-silica.
- The mixture was placed inside the small diameter beaker (250ml), which immersed in ice-cooling bath to roughly the same level of the mixture to aid in the dissipation of heat.
- The sonication amplitude was set at 30% for 20 min to avoid damaging of SCB fiber particulates and nano-silica.

2.5. Mechanical Analysis

The tensile and flexural properties (strengths and moduli) of SCB short fiber composite, SCB fiber particulate composite, SCB fiber particulate composite using ultrasonic mixing and SCB fiber particulate composite with nano-silica are characterized experimentally. The mechanical tests are performed using LR Plus LLOYD Universal Testing Machine. The results obtained during the tensile and flexural tests were recorded. Morphology characterizations of sugarcane bagasse fibers were conducted by an optical light microscope. The resolution was adjusted to obtain the clearest micrographs for microstructure studies.

3. Results and Discussion

3.1. Effect of SCB short fiber and fiber particulates on mechanical performance of composites

Figure 1 shows the results of the tensile and flexural properties for various samples produced. SCB short fiber reinforced composites and SCB particulates reinforced composites are both shown to decrease gradually from the lowest fiber weightage (1%) to the highest fiber weightage (5%) in tensile strength. It was observed that the tensile strength and modulus is higher for 1wt% short fiber composite while compared to higher fiber content samples. 1wt% of short fiber composite has attained the best tensile strength and modulus, which is 7.3% and 14.6% better than 1wt% SCB fiber particulate composites. This might be happened as the fiber content increases, the tendency for the fiber/ matrix bonding strength to decrease is high. Both SCB short fiber and particulate reinforced composites shown to have a better result as at low fiber content, the bagasse fibers are wetted properly by the epoxy resin and the fibers are in contact with one another [12].

Excluding 1wt% of short fiber composite, the strength and modulus of 2.5wt%, 3.5wt% and 5wt% short fiber composite are shown to be lower compared with the particulate composite with the same fiber loadings. This is suspected that at higher short fiber content, the fibers are touching one another also reducing proper fiber wetting and bonding between the bagasse fiber and epoxy matrix. Besides, due to the light weight of the bagasse fiber, the mixture was overflowed when 5wt% of bagasse short fiber was used. The attachment between matrix and the reinforced fibers became lesser. It was observed that the epoxy resin was insufficient to cover all the surface of bagasse fiber in the composite [13]. The functional effectiveness of epoxy matrix to act as the binder material to transfer external loads to internal loads was affected. Therefore, the capabilities of stress transfer has been weaken at higher fiber loading. This was also proved by previous study that poor fiber-matrix adhesion was prominent at high fiber content [7].

In addition, minor entanglements and agglomeration has been induced among the fibers due to high evaporation of water in the drying process of SCB fiber. Therefore, the mechanical testing has been affected. Rough surface finish of the composite specimen due to the overflowed short fibers was also one of the factors that affect the mechanical properties of the composite.

However, in the case of particulate composites, the overflowing of the mixture is less likely to be happened and the epoxy resin was still sufficient to cover all the bagasse fiber in the composite due to the particle geometrical shape. The surface finish obtained is also shown to be better than short fiber composites thus the mechanical performance of particulate reinforced composite is higher than short fiber reinforced composite in 2.5wt%, 3.5wt% and 5wt%. Therefore, 2.5wt%, 3.5wt% and 5wt% shows better mechanical performance in tensile strength and modulus. The improvement is ranged from 8% to 23.8% as it overcome the shortage of short fiber composite during the increase of fiber weightage as mentioned above. The same case also applied in flexural strength, SCB fiber particulate composites were improved from 10% to 34%. However, the increase in fiber weightage still induce the problem of agglomeration therefore the mechanical performance of the particulates decreases gradually.

As for flexural properties, it was noticed that a higher flexural strength and modulus were obtained compared to tensile strength and modulus. This proved that the SCB short fiber composites were shown to give ideal performances in enduring forces than to resist opposing tensile stresses. Through this study, it was revealed that 1wt% of SCB fiber loading are more superior in giving relatively ideal mechanical properties while compared to specimens with other fiber loading. SCB short fiber has shown improvement in flexural strength and modulus, which is 9.7-18.5% higher while compared to SCB fiber particulates at different weightage of fiber. Short fibers were more capable in preventing the deformation under the presence of bending loads and fatigue stresses. It is more effective for the fibers to lock into the epoxy matrix thus improved the stress transfer from the matrix to the fibrous fillers.

IOP Conf. Series: Materials Science and Engineering **344** (2018) 012029 doi:10.1088/1757-899X/344/1/012029

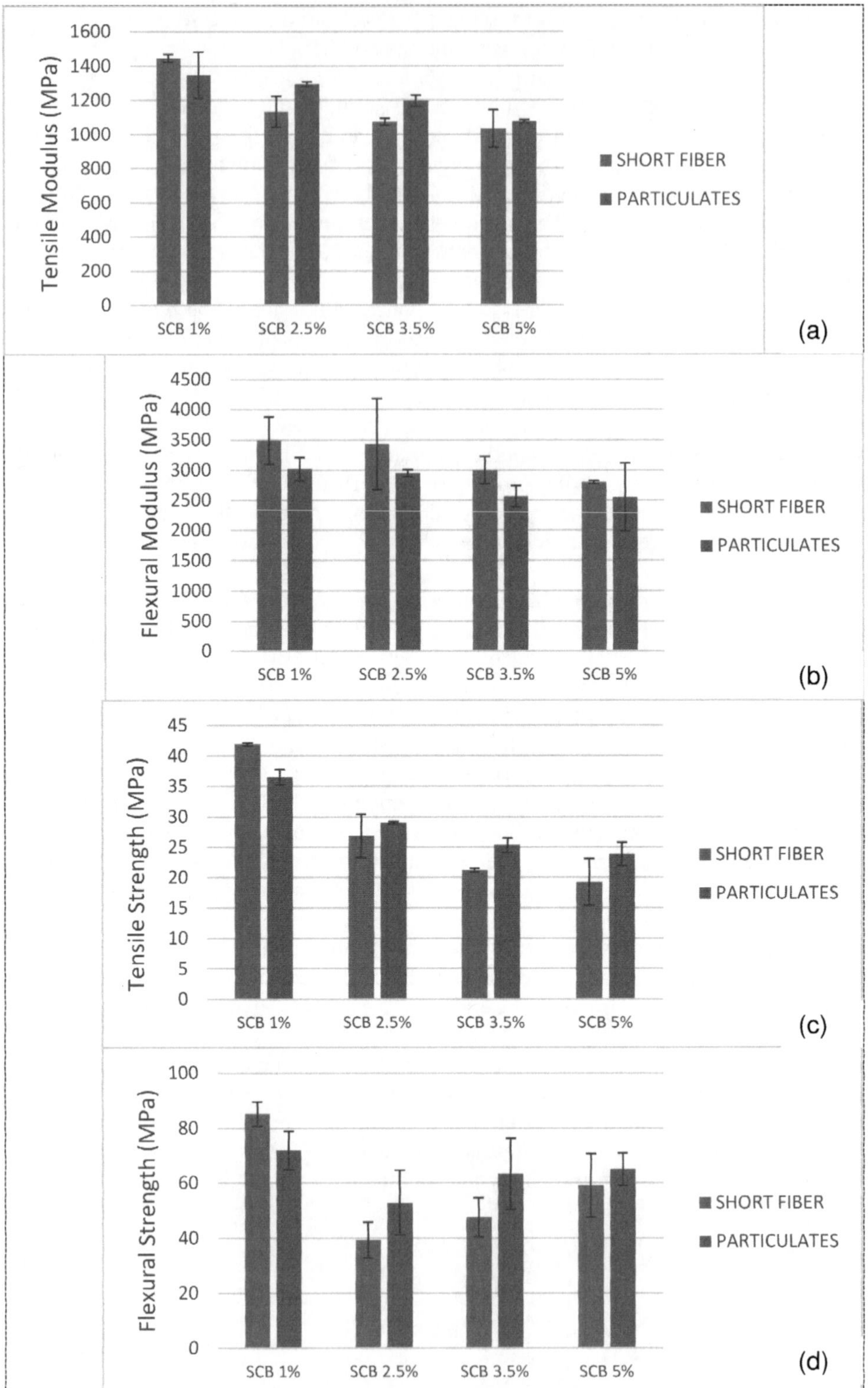

Figure 1. Tensile and flexural properties of SCB short fiber and particulate composite

SCB fiber particulates were shown to have less function in improving flexural properties compared to tensile properties. According to Aruniit et al 2011 [2], the flexural strength could only remain steady or make decrease due to their unfavourable geometrical features. However, samples which reinforced with fiber particulates has shown significant improvement in flexural strength, which is 34.2% and 33.2% higher for 2.5wt% and 3.5wt% respectively. This could be said that SCB particulates has overcome the problems of short fiber being too compact to be dispersed well within the matrix while increasing the fiber weightage. As the weightage of the short fiber increased, the capabilities of stress transfer weakened.

Deploying the hand lay-up method which require low cost tooling and minimal power usage will reduce the pollution to the environment. It is also economical as it requires low cost manufacturing tools compared to other conventional method. However, a random orientation alignment of the fibers in the composite is obtained and overflowing of fibers are unpreventable. The maximum reinforcement efficiency can hardly be achieved. Therefore, it is hard to obtain maximum strength due to the impossibility to achieve 100% uniformity in fiber orientation and distribution. According to Luz S.M et al (2007)[9], the compression and injection moulding processes were performed for better evaluation of sugarcane bagasse. Through injection moulding under vacuum process, a homogenous distribution of fibers without blister could be obtained. According to Torres et al. (2017) [12], composite panels were manufactured by vacuum assisted resin transfer moulding (VARTM) on a rectangular tool plate. The fibers distribution in the composite are controlled in the manner that the overflowing of fibers in the composite were prevented. This overcame the issue that is caused by the hand lay-up method.

In overall, the short fiber reinforced composite only show better mechanical performance at lowest fiber weightage (1wt%) due to minor entanglement, agglomeration, insufficient coverage of epoxy resin with the fiber at higher fiber weightage. Specimen manufacturing and technical issues are also one of the factors. Maximum reinforcement efficiency can hardly be achieved due to the restriction of hand lay-up method. The mechanical performance of the composites at higher fiber weightage could be improved using fiber in particulate form.

3.2. Effect of ultrasonic mixing and addition of nano-silica on mechanical performance of composites

In this study, tensile and flexural properties of SCB fiber particulates composites were compared using two different fabrication methods and one with addition of nano-silica. Refer to Figure 2, tensile strength and modulus of the composites using ultrasonic mixing were improved, which is ranged from 3.5-15.9% and 3-9% respectively. This could prove that ultrasonic processor had aid in deagglomeration as well as in enhancing dispersion of fiber particles. From the results, 5% weightage of fiber particles was shown to have 9% and 15.9% of improvement in tensile strength and tensile modulus respectively, which were the highest among all. In the case of flexural strength and modulus, the mechanical performance of SCB fiber particulate reinforced composite has also improved, which ranged from 3.8-20.8%. In overall, it can be concluded that the mechanical performance of the composites was elevated as the agglomeration was minimized while ultrasonic mixing was applied.

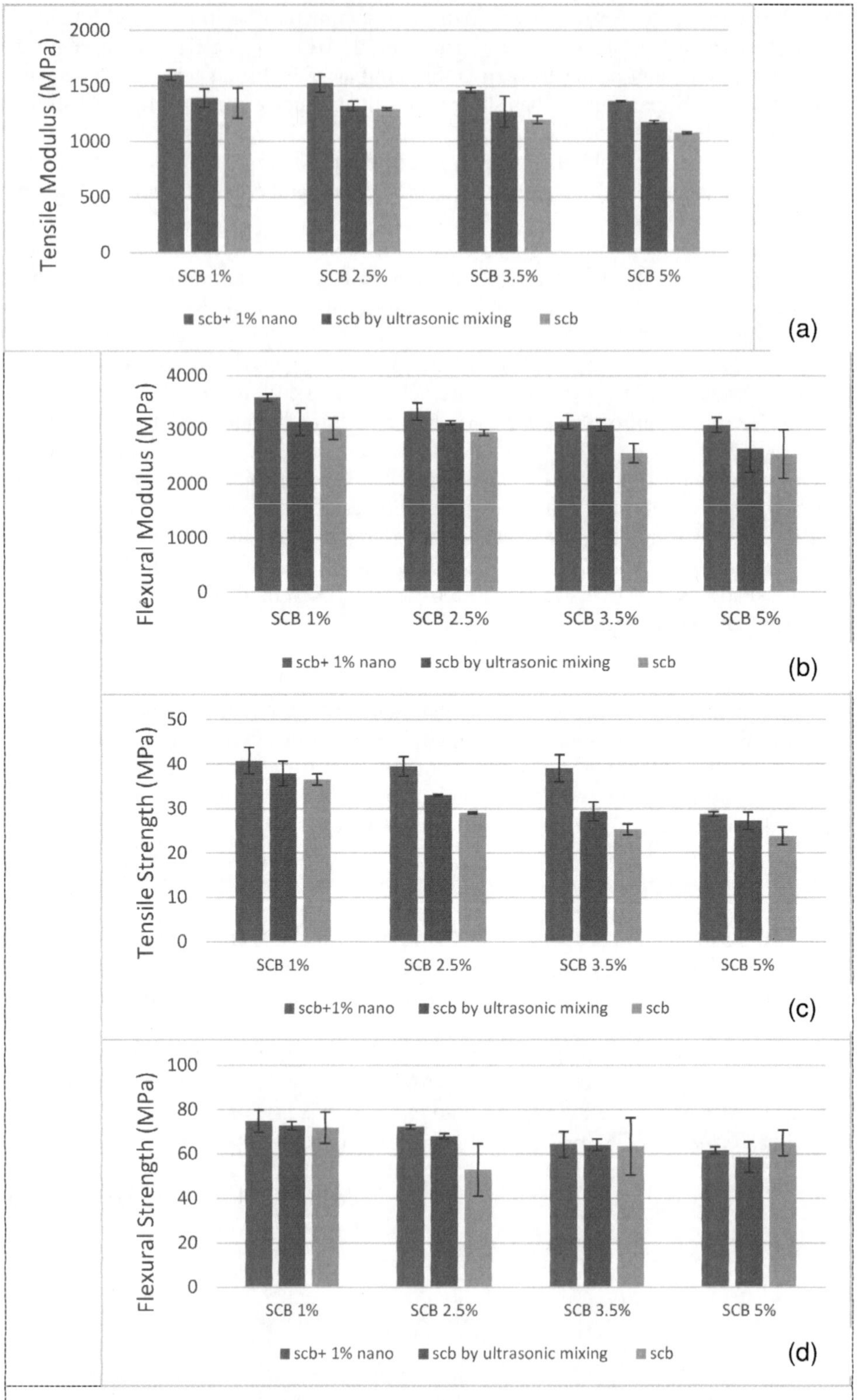

Figure 2. Tensile and Flexural Properties of SCB composites using hand -mixing, ultrasonic mixing and with addition of nano-silica

IOP Conf. Series: Materials Science and Engineering **344** (2018) 012029 doi:10.1088/1757-899X/344/1/012029

With the addition of nano-silica, it can be observed that the tensile strength and tensile modulus of the SCB fiber particulate composite have been successfully improved, which ranged from 5.4% to 33.17% and 14.7% to 16.2% respectively. As for results of flexural testing, it was noticed that the trends were in line with the tensile properties testing results. The improvement ranged from 0.6% to 5.2% for flexural strength and 2% to 14.4% for flexural modulus. From the results, it can be concluded that the nano-silica particles have elevated mechanical performance especially tensile properties of the composite due to its extremely large surface area and its smooth nonporous surface. The results obtained was in line with the research done by Dittenber et al. 2012 [13], which indicated that the addition of nano-silica particles is able to increase stiffness around 10% as the particles fill in the voids in the composites and minimized the agglomeration. Although the improvement was insignificant for flexural strength, it was concluded that the addition of the nano-silica particles has improved the mechanical performance of the composite. The agglomeration was minimized and the presence of the voids in the composite reduced.

4. Morphology Analysis

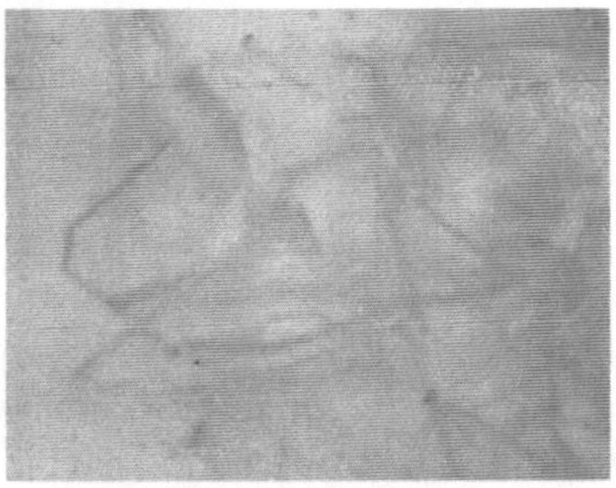

a) 1wt% SCB short fiber reinforced composite

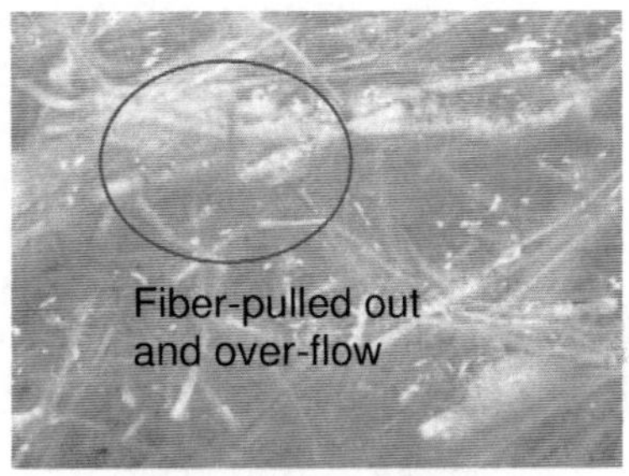

b) 5wt% SCB short fiber reinforced composite

c) 1wt% SCB fiber particulate reinforced composite

d) 1wt% SCB fiber particulate reinforced composite using ultrasonic mixing

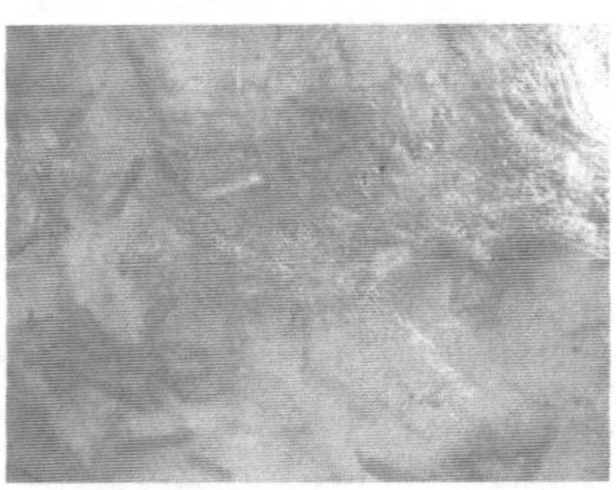

e) 1wt% SCB fiber particulate reinforced composite

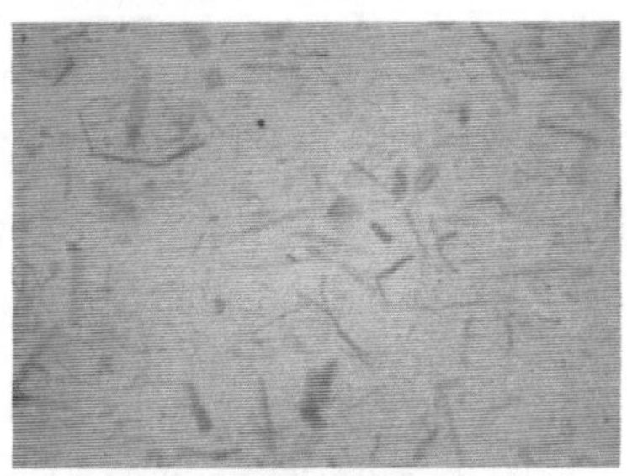

f) 1wt% SCB fiber particulate reinforced composite using ultrasonic mixing

Figure 3. Optical microscopy pictures taken of the composites

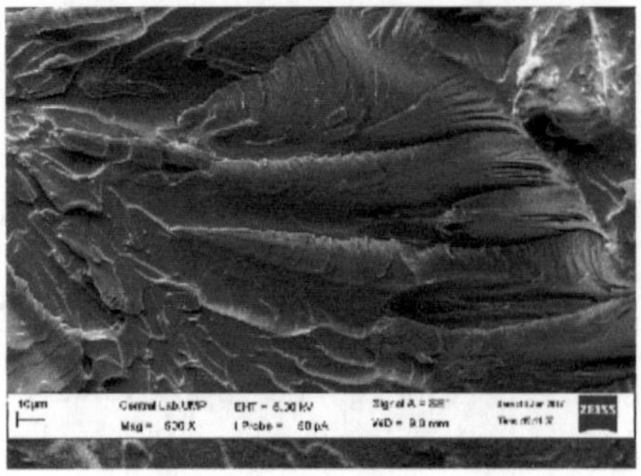

a) 3.5wt% SCB fiber particulate reinforced composite using ultrasonic mixing

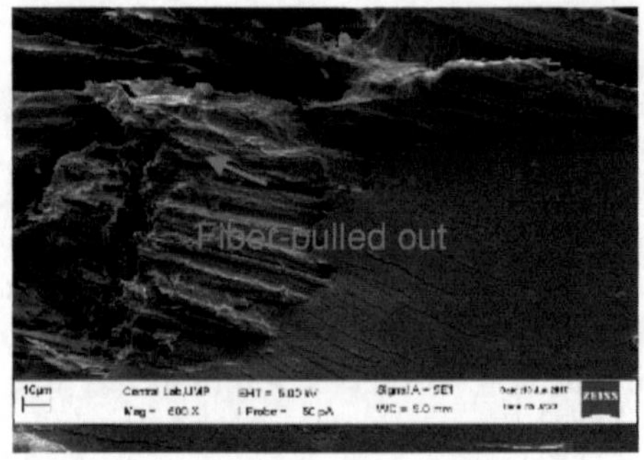

b) 3.5wt% SCB short fiber reinforced composite

Figure 4. SEM micrograph of the fracture surface of composites

It can be observed from Figure 3 that 5wt% SCB short fiber composite showed overflowing of the fiber and results in a rough surface finish as well as some fiber pulled out while compared to 1wt% SCB short fiber composite. The attachment between matrix and the reinforced fibers is lesser in 5wt% SCB short fiber composite. The epoxy matrix in 1wt% SCB short fiber composite was shown to be more sufficient to cover all the surface of bagasse fiber which results in higher capabilities of stress transfer. The bagasse fibers are wetted properly by the epoxy resin. Secondly, 1wt% and 5wt% SCB fiber particulate composite was significantly shown to have more air bubbles while compared to the SCB fiber composites using ultrasonic mixing. The use of ultrasonic mixing has proved to reduce air bubbles successfully. Moreover, ultrasonic mixing was also attributed to better fiber dispersion and reducing agglomeration in 1wt% SCB fiber particulate composite. Figure 4 presents the fractured region after tensile tests were performed to the composites. The differences between particulate SCB and fibrous SCB composites were observed. It was once again verified that fibers distribution in the epoxy matrix and pull out of fibers, characterized the mechanism of fragile fracture. It would appear that the bonding between matrix and particulate SCB is superior to that of the fibrous SCB composite due to the dispersion of particulates was better compare to fibers form when the weightage of SCB is higher.

5. Conclusion

The summarization of the findings from the study on the mechanical performance (tensile and flexural properties) of the sugarcane short fiber reinforced epoxy composites, sugarcane fiber particulate reinforced epoxy composites, sugarcane fiber particulate with nano-silica reinforced epoxy composites and the effect of using ultrasonic mixing are stated as below:

- All the sugarcane short fiber reinforced epoxy composites obtained higher flexural modulus compared to sugarcane fiber particulates reinforced epoxy composites.
- Among all the sugarcane short fiber composites, 1wt% of sugarcane short fiber composite exhibits the highest tensile and flexural properties compared to 2.5wt%, 3.5wt% and 5wt%.
- Sugarcane fiber particulates reinforced composites were shown to have better performance when the wt% of the fiber increase while compared to short fiber reinforced composites.
- The sugarcane fiber particulates composites fabricated using ultrasonic processor showed improvement in both tensile and flexural properties due to the deagglomeration of the fiber particulates.

- The addition of nano-silica particles has further elevated the mechanical performance of the alkali treated sugarcane fiber particulate composites due to the reduction of micro-voids in the epoxy resin and agglomeration is reduced.
- The morphology study has revealed that ultrasonic mixing has successfully reduced the air bubbles and agglomeration. The epoxy matrix in 1wt% SCB short fiber composite was shown to be more sufficient to cover all the surface of bagasse fiber compared to 5wt% of SCB short fiber composite.

Acknowledgment

The study was supported by Curtin Sarawak Collaborative Research grant (CSCR).

References

[1] Callister, William D. (2007). *Materials Science and Engineering: An Introduction.* https://abmpk.files.wordpress.com/2014/02/book_maretial-science-callister.pdf

[2] Aruniit A., Kers J., Tall K. (2011). "Influence of filler proportion on mechanical and physical properties of particulate composite." *Agronomy Research Biosystem Engineering Special.*

[3] Alamri, H., and I. M. Low. (2013). "Effect of Water Absorption on the Mechanical Properties of Nanoclay Filled Recycled Cellulose Fibre Reinforced Epoxy Hybrid Nanocomposites." *Composites Part A: Applied Science and Manufacturing* 44: 23-31.

[4] Salit, M. S. (2014). Tropical natural fibres and their properties In *Tropical natural fibre composites* (pp. 15-38). Springer Singapore.

[5] Kadir, A. and N. Maasom. (2013). "Recycling sugarcane bagasse waste into fired clay brick." *International Journal of Zero Waste Generation*, 1(1), 21-26.

[6] Mulinari, Daniella R., Herman J. C. Voorwald, Maria Odila H. Cioffi, Maria Lúcia C. P. da Silva, and Sandra M. Luz. (2009). "Preparation and properties of HDPE/sugarcane bagasse cellulose composites obtained for thermokinetic mixer." *Carbohydrate Polymers* 75 (2):317-321. doi: 10.1016/j.carbpol.2008.07.028.

[7] Cao, Y., S. Shibata, and I. Fukumoto. (2006). "Mechanical properties of biodegradable composites reinforced with bagasse fibre before and after alkali treatments." *Composites Part A: Applied Science and Manufacturing* 37 (3):423-429. doi: 10.1016/j.compositesa.2005.05.045.

[8] Jawaid, M., and H. P. S. Abdul Khalil. 2011. "Cellulosic/Synthetic Fibre Reinforced Polymer Hybrid Composites: A Review." *Carbohydrate Polymers* 86 (1): 1-18.

[9] Luz, Sandra Maria da, Adilson Roberto Goncalves and A.P. Del'Arco. (2007). "Mechanical behaviour and microstructural analysis of sugarcane bagasse fibers reinforced polypropylene composites." *Composites Part A Applied Science and Manufacturing* 38(6):1455-1461. doi: 10.1016/j.compositesa.2007.01.014

[10] Oladele, I.O. (2013). "Effect of Bagasse Fibre Reinforcement on the Mechanical Properties of Polyester Composites", *The Journal of the Association of Professional Engineers of Trinidad and Tobago*, Vol.42, pp.12-15.

[11] Shibata, Shinichi, Rahman M. Bozlur, Isao Fukumoto and Yasuyuki Kanda. (2010). "Effects of Injection Temperature on Mechanical Properties of Bagasse/ Polypropylene Injection Molding Composites" *BioResources* 5(4).

[12] Torres, J., Vandi, L.-J., Veidt, M., & Heitzmann, M. (2017). The mechanical properties of natural fibre composite laminates: A statistical study. *Composites Part A: Applied Science and Manufacturing*, 98, 99-104.

[13] Dittenber, David B and Gangarao HVS. (2012). "Critical review of recent publications on use of natural composites in infrastructure." *Composites Part A Applied Science and Manufacturing* 43(8):1419–1429. doi: 10.1016/j.compositesa.2011.11.019

Institute of Physics
Dirac House, Temple Back
Bristol BS1 6BE UK

ISSN: 1757-8981
ISBN 978-1-5108-6160-2

9 781510 861602

CURRAN ASSOCIATES INC.
proceedings.com

2nd International Conference on Advanced Materials (ICAM-2017)

IOP Conference Series: Materials Science and Engineering
Volume 305

Irbid, Jordan
10 – 13 July 2017

IOP
Institute of Physics